Lecture Notes in Computer Science   13984

The series Lecture Notes in Computer Science (LNCS), including its subseries Lecture Notes in Artificial Intelligence (LNAI) and Lecture Notes in Bioinformatics (LNBI), has established itself as a medium for the publication of new developments in computer science and information technology research, teaching, and education.

LNCS enjoys close cooperation with the computer science R & D community, the series counts many renowned academics among its volume editors and paper authors, and collaborates with prestigious societies. Its mission is to serve this international community by providing an invaluable service, mainly focused on the publication of conference and workshop proceedings and postproceedings. LNCS commenced publication in 1973.

Erik Blasch · Frederica Darema · Alex Aved
Editors

Dynamic Data Driven Applications Systems

4th International Conference, DDDAS 2022
Cambridge, MA, USA, October 6–10, 2022
Proceedings

 Springer

Editors
Erik Blasch
MOVEJ Analytics
Fairborn, OH, USA

Frederica Darema ⓘ
InfoSymbiotic Systems Society
Bethesda, MD, USA

Alex Aved ⓘ
Air Force Research Laboratories
Canastota, NY, USA

ISSN 0302-9743   ISSN 1611-3349 (electronic)
Lecture Notes in Computer Science
ISBN 978-3-031-52669-5   ISBN 978-3-031-52670-1 (eBook)
https://doi.org/10.1007/978-3-031-52670-1

This Springer imprint is published by the registered company Springer Nature Switzerland AG
The registered company address is: Gewerbestrasse 11, 6330 Cham, Switzerland

Paper in this product is recyclable.

Preface

The Dynamic Data Driven Applications Systems (DDDAS)/InfoSymbiotics 2022 (or "DDDAS 2022") conference showcased research advances and science and technology capabilities stemming from the Dynamic Data Driven Applications Systems (DDDAS) paradigm, whereby instrumentation data are dynamically integrated into an executing application model and in reverse the executing model controls the instrumentation.

DDDAS 2022 continued and expanded the path set by prior DDDAS forums on research advances and Science and Technology capabilities integrating data with modeling. DDDAS/InfoSymbiotics[1] plays a key role in enabling advances creating new capabilities in many application areas and is also driving advances in foundational methods, through system-level (as well as subsystems-level) approaches that include comprehensive, principle- and physics-based models and instrumentation, uncertainty quantification, estimation, observation, optimization, and sampling. DDDAS focuses on cognizant methods for dealing with big-data, for awareness, control, (machine) learning, planning, and decision support. DDDAS encompasses "Systems" analysis, prediction of behaviors, and operational control, all of which entail multidisciplinary collaborative research and advances. The DDDAS paradigm has shown the ability to engender new capabilities in (but not limited to) aerospace, bio-, cyber-, geo-, space-, and medical sciences, as well as critical infrastructure security, resiliency, and performance; the scope of application areas ranges "from the nano-scale to the extra-terra-scale". This year's expanded conference scope showcased additional topics, including wildfires and other natural hazards, and climate grand challenges, and the role of DDDAS in enabling and supporting the optimized design and operation of 5G and Beyond5G communications and networking infrastructures.

The DDDAS community has made significant progress in closing the loop among data, information, and knowledge, through DDDAS-based improved modeling processes, system-cognizant models learning with the aid of instrumentation measurements, and, in return, controlling the instrumentation to turn the big-data deluge into information for improved understanding and mitigation of model errors.

Participants from academia, industry, government, and international counterparts reported original work where DDDAS research is advancing scientific frontiers, engendering new engineering capabilities, and adaptively optimizing operational processes. The conference and these proceedings span a broad set of topics and interests as delineated above.

October 2022

Erik Blasch
Frederica Darema
Alex Aved

---

[1] InfoSymbiotic Systems or InfoSymbiotics are terms introduced to denote DDDAS.

Organization

General Chairs

Erik Blasch   MOVEJ Analytics, USA
Sai Ravela   Massachusetts Institute of Technology, USA

Program Committee Co-chairs

Frederica Darema   InfoSymbiotic Systems Society, USA
Alex Aved   Air Force Research Laboratories, USA
Nurcin Celik   University of Miami, USA
Carlos Varela   Rensselaer Polytechnic Institute, USA

Program Committee

Ankit Goel   University of Maryland, Baltimore County, USA
Salim Hariri   University of Arizona, USA
Thomas Henderson   University of Utah, USA
Fotis Kopsaftopoulos   Rensselaer Polytechnic Institute, USA
Artem Korobenko   University of Calgary, Canada
Zhiling Lang   Illinois Institute of Technology, USA
Richard Linares   Massachusetts Institute of Technology, USA
Dimitri Metaxas   Rutgers University, USA
Asok Ray   Penn State University, USA
Sonia Sacks   Booz-Allen-Hamilton, USA
Themistoklis Sapsis   Massachusetts Institute of Technology, USA
Ludmilla Werbos   University of Memphis, USA

Contents

Main-Track Plenary Presentations - Security

Main-Track Plenary Presentations - Distributed Systems

Main-Track: Keynotes

Main-Track: Wildfires Panel

Workshop on Climate, Life, Earth, Planets

Keynotes-Appendix

Wildfires-Appendix

Introduction to the DDDAS2022 Conference

Introduction to the DDDAS2022 Conference Infosymbiotics/Dynamic Data Driven Applications Systems

Erik Blasch[1(✉)] and Frederica Darema[2]

[1] MOVEJ Analytics, Fairborn, OH, USA
erik.blasch@gmail.com
[2] InfoSymbiotic Systems Society, Bethesda, MD, USA

Abstract. The 4th International DDDAS 2022 Conference, convened on October 6–10, featured presentations on Dynamic Data Driven Applications Systems (DDDAS)-based approaches and capabilities, in a wide set of areas, with an overarching theme of "*InfoSymbiotics/DDDAS for human, environmental and engineering sustainment*". The topics included aerospace mechanics and space systems, networked communications and autonomy, biomedical and environmental systems, and featured recent techniques in generative Artificial Intelligence, theoretical Machine Learning, and dynamic Digital Twins. Capturing the tenets of the DDDAS paradigm across these areas, solutions were presented to address challenges in systems-of-systems' approaches providing analysis, assessments and enhanced capabilities in the presence of complex and big data. The conference comprised of the main track that featured 31 plenary presentations of peer reviewed papers, five keynotes, an invited talk, and a panel on wildfires monitoring. In conjunction with the main track of the DDDAS conference, a Workshop on Climate and Life, Earth, Planets (CLEPs) was conducted, which featured 20 presentations on environmental challenges and a panel on Seismic and Nuclear Explosion monitoring. In addition to the papers of the plenary presentations in the main track of the conference, the DDDAS2022 Proceedings feature an overview of the conference, a synopsis of the main-track papers, and summaries of the keynotes and the wildfires panels followed by corresponding contributed papers by the the speakers in these sessions. Additional information and archival materials, including the presentations' slides and recordings, are available in the DDDAS website: www.1dddas.org.

Keywords: Dynamic Data Driven Applications Systems · DDDAS · InfoSymbiotic Systems · InfoSymbiotics · DDDAS-based Dynamic Digital Twin

1 DDAS Developments

DDDAS has an over two-decades history of engaging the broader community, and has been featured in numerous forums (conferences, workshops, panels), including the DDDAS conference series started in 2016 (with the most recent in the series, the

E. Blasch et al. (Eds.): DDDAS 2022, LNCS 13984, pp. 3–13, 2024.
https://doi.org/10.1007/978-3-031-52670-1_1

DDDAS2020 Conference [1]) and in a series of Springer DDDAS Handbooks publications (Volume 1 – 1st and 2nd Editions [2–4], with Volumes 2 and 3 expected to be published in 2023 and 2024 respectively). The concepts that have been addressed under the DDDAS rubric cover a wide set of areas and domains, facilitated from the advances in computer hardware and data processing broadening the use of simulation methods and machine analytics and enhancing the opportunities to exploit the DDDAS paradigm.

The DDDAS2022 Conference Proceedings consist of five Parts. This Introduction is followed by Part 1, which contains the 31 papers of the main track of the conference. Overviews of the keynotes and invited talk, and the Wildfires panel sessions, with corresponding abstracts and papers are presented in Parts 2 and 3; respectively. Part 4, includes a paper contributed by the organizer of a Workshop on Climate, Life, Earth, and Planet (CLEPs), run in-tandem with the main track of the conference. The conference Agenda is included in Part 5.

The remaining Sections of this Introduction begins with a review of the DDDAS paradigm in Sect. 2. The outline of the proceedings is presented in Sect. 3. Section 4 provides an overview of the 30 plenary presentation papers. Section 5 of this Introduction presents summary remarks on the conference content.

2 DDDAS Concept Review

Dynamic Data Driven Applications Systems (DDDAS) is a framework for natural, engineered, or societal systems analysis and design that dynamically integrates comprehensive, first-principles, and high-dimensional models of systems with the corresponding data instrumentation, simulation, and augmentation. The application of the DDDAS framework has facilitated more accurate modeling methods providing (1) efficient orchestration of big data and sensor systems, (2) effective situation and state awareness for prediction of behaviors, (3) adaptive management of systems' operational conditions, (4) robust response of systems to external conditions, and (5) scalable measurement analysis of multimodal information in a wide set of application areas such as biomedical, surveillance, and internet of things.

As shown in Fig. 1, DDDAS encompasses adaptive state estimation that uses an *instrumentation reconfiguration loop* (IRL).The IRL loop seeks to reconfigure the sensors (instrumentation) in order to enhance the information content of the measurements. The sensor reconfiguration is guided adaptively by the simulation (modeling) of the physical process. Consequently, the sensor reconfiguration is *dynamic*, and the overall process is (dynamic) *data driven*.

The two key aspects of DDDAS are (1) the *dynamic data augmentation loop* (DAL), which integrates instrumentation data to drive the physical system simulation so that the trajectory of the simulation follows moreclosely the trajectory of the physical system. The dynamic data assimilation (which combines theory with observations), the DAL loop uses input data if input sensors are available. The second key aspect is the additional innovative feature of DDDAS, which goes beyond the traditional "data assimilation" methods; that is the additional *instrumentation reconfiguration loop* (IRL), shown in Fig. 1, controls the instrumentation system (such as the set of sensors, actuators, sensor executing models and sensors signal analysis), to collect additional data in targeted

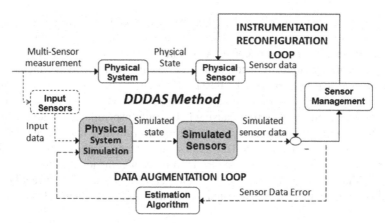

Fig. 1. A Depiction of Dynamic Data Driven Applications Systems (DDDAS) feedback loop.

ways, where data are needed to improve or speedup the model, or apply coordinated control of sensors or actuators. For example if the instrumentation consists of a set of physical sensors, in DDDAS the executing model (or simulation) adaptively guides the physical sensors, in a coordinated way, in order to enhancethe information content of the collected data. The *dynamic data augmentation* and the *instrumentation* reconfiguration feedback loops are **computational** and **physical** feedback loops. The simulation guides the instrumentation reconfiguration and what datato be collected, and in turn, uses this additional, selected data, to improve the accuracy of the physical system simulation model. This "meta" (positive) feedback loop is central in DDDAS.

Key aspects of DDDAS include the algorithmic and statistical methods that can incorporate dynamically the measurement data with that of the high-fidelity modeling and simulation, and as needed, invoke models of higher or lower levels of fidelity (or resolution) depending on the dynamic data inputs. DDDAS-based approaches have been applied, presently for more than twenty years, in many applications areas and have shown to enable more accurate and faster modeling and analysis of the characteristics and behaviors of a system.

Figure 2 highlights advances in DDDAS. For advances in data processing, much of the DDDAS developments have been expanded along the lines from measurements to context and simulations to generative approaches. Typically, the DDDAS concept utilizes measurements to augment first-principle physics or other, system-cognizant models. Hence, more recent work and techniques in DDDAS expanded the scope of measurements to include sensor-networks data as well as the environment and *situational context*, utilizing collected data in cognizant, adaptive and coordinated ways. An example is that of tracking an object from which estimation techniques include information and signal processing (and machine learning (ML)) over the sensor measurements; however, the knowledge of the terrain and weather are both contextual knowledge to augment the following of an object. The second advancement comes from deep learning in the generative approaches. With a large corpus of static data, known and variational (unknown) data can be generated to train a system. The *generative approaches* are different than simulations in that the simulation is based on first-principle modeling; whereas the generative approaches utilize the known data (without the mathematical rigor) to discover features from an encoder and decoder. As pointed-out in [4], the DDDAS paradigm, where-by data are dynamically incorporated into the executing model, which can also be considered a "learning" process by the (DDDAS-based) model; however unlike ML and static NN (neural networks) methods which have limited cognizant selectivity of the data used for their learning thus causing issues of transparency and interpretability, in the DDDAS-based learning is safeguarded from incorporating data incongruent with its knowledge of the model. In fact, DDDAS-based methods can be employed to detect errors or failure of sensors and take actions to mitigate the impact of failure (for example, [5] on failure of the pitot tubes in Air France crash incident). The current DDDAS approaches combine measured, simulated, generative, and contextualized data.

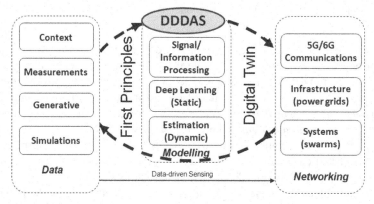

Fig. 2. Advances in DDDAS from first principles to (dynamic) digital twins.

Not only are the advances in DDDAS about the data, but also the *emerging applications*. Developments in the community include devices, internet of things, and swarms of platforms. The devices include emerging developments in quantum and 5G/6G communications. As the data is fundamental to DDDAS techniques, access to the data is facilitated from the channel of delivery, such as going from 4G to 5G/6G communications protocols. Another area of prominent research in DDDAS is in the Internet of Things (IoT) infrastructure grid such as for power monitoring of the grid. Many DDDAS researchers are exploring the power grid from the DDDAS-based Digital Twin (Dynamic Digital Twin [4]) models, the many sensors, and the processing techniques. The need for such systems is paramount within the current complex engineered infrastructures and systems interacting and affected by their environments affected by environmental disasters – such as wildfires and other climate change adverse situations. DDDAS-based advances in infrastructures support response to environmental adversities. For example, when a power grid is destroyed by wildfires, the DDDAS model can effectively determine how and when to alter the grid access to minimize damage to the entire systems or delivery of service to critical points in the infrastructure. The third area is that of swarms of ground (e.g., cars), air (e.g., unmanned aerial vehicles), and space (e.g. satellite) platforms. The DDDAS concept leverages the models and data for effective control of these platforms.

3   DDDAS2022 Conference – Proceedings Outline

The Conference consisted of the DDDAS *main-track* which comprised of plenary presentations of 31 peer-reviewed papers, five Keynotes, an invited talk, and a Panel on Wildfires. Inconjunction with the main-track, a Workshop on Climate, Life, Earth, and Planet (CLEPs) was conducted which featured 20 presentations and a Panel on Seismic and Nuclear Explosion Monitoring.

The conference proceedings are organized as follows. This Introduction paper presents an overview of the main-track plenary presentations, and the subsequent Part 1 contains the 31 papers organized along the 10 main-track Sessions: Aerospace Systems – I and II, Space Systems, Networked Systems, Systems Support Methods, Deep Learning I and II, Tracking, Security, and Distributed Systems. In Part 2, labeled "Keynotes", an overview is provided of the main track keynotes and invited talk, highlighting the salient points in the presentations by Yuri Bazilevs, Sertac Karaman, Manolis Kellis, Nathaniel Bastian, Sreeja Nag, and Andreas Savakis, which covered DDDAS-based advances in the areas of aerospace and mechanical systems modeling, autonomic operations, genomics, adversarial artificial networks, geosciences observations, and multimodal domain adaptation [6–8]. The Keynotes overview is followed by abstracts and papers contributed by the speakers. Part 3, on the "Wildfires Panel", includes an overview of the panel scope and summary of the discussions by panelists Ilkay Altintas, Fatemeh Afgah, Milton Halem, Milt Halem, Thomas Huang, Mrinal Kumar, Jan Coen, and Kamran Mohseni; the overview is followed by abstracts and papers contributed by the panelists.

The CLEPs Workshop featured presentations on the use of DDDAS for big data environmental challenges. The CLEPs discussions stimulated additional discussions in utilizing the DDDAS methods to support analysis utilizing physics modeling, data collections, and coordination of instrumentation systems. Part 4 presents a paper contributed by Sai Ravela, the CLEPs Workshop organizer.

Figure 3 depicts the inter-relation of the various Sessions of the DDDAS2022 Conference discussions and the inter-relations of the work presented across Sessions. The book highlights use-cases of DDDAS across multiple domains (space, air, ground, network, and infrastructure) applications from which data is collected and combined with the modeling, at model execution time and in reverse the executing model guides cognizant collection of data and use. New methods include advances in estimation theory combined with deep learning. These results then inform and provide enhanced solutions to applications challenges, such as real-time decision support for monitoring and assessments in adverse events conditions as for example seismic and wildfire monitoring, which were the subject of the panels.

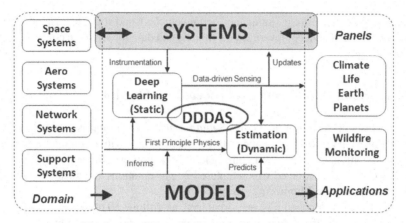

Fig. 3. Pictorial Outlineof the DDDAS Proceedings

4   DDDAS2022 – Main Track Plenary Presentations Overview

From the papers submitted for peer review, 31 papers selected for presentations at the DDDAS2022 Conference are included in the present Proceedings of the conference; the papers represent a wide range of important application areas and science and technology methods, and in their scope demonstrate (1) use of the DDDAS modeling-instrumentation feedback-control approach, (2) theoretical advances, and (3) applications that encompass big data challenges.

Based on the accepted papers, the submissions presented in the ten Plenary Presentations Sessions span several systems domains of interest (e.g., space, aero, networks and support) as well as those concerning the emerging areas of deep learning and estimation techniques. Given the multidisciplinary nature of DDDAS-based applications and their environments, many of the papers in their scope and impact straddle more than one area. Thus, the overview following here captures such relations across-sessions presentations as well as their nexus to keynotes presentations. The overview begins with presented work relating to space environments, and progressively goes to aerospace and mechanical systems, to distributed networked systems and supporting computational and communications infrastructures, and then to foundational methods and domain applications and areas.

4.1   Space Systems

For **space systems**, the proliferation of the space environment offers many opportunities for DDDAS to contribute, especially when considering networked systems [9]. *Utkarsh Mishra* et al., in their paper "Probabilistic Admissible Region Based Track Initialization", focused on the challenges for conjunctive resident space object analysis for space situational awareness (SSA) with a focus on track initiation.The tracking of a space object includes initiation, maintenance/custody, and termination. Hence, being able to use DDDAS modeling to know where to look to start a resident space object (RSO) track is a key DDDAS demonstration. Typically, the space observations come from telescopes from optical imagery. Optical imagery is limited when the climate atmosphere changes, so other sensing modalities are need for robust assessment; *Justin Henry* et al., as discussed in their "Radar cross-section modeling of space debris" paper, designed a theoretical approach to utilize radar to determine the debris radar cross section. The ability to operate in the space environment will include a constellation of satellites working together. Xiaohua Li, et al. provide a way to adaptively coordinate the collection of the satellites and keep control as a swarm to increase the coverage from high resolution imagery. For a new development in monitoring cislunar, *David Schwab* et al. presented their work on "Reachability Analysis to Track Non-cooperative Satellite in Cislunar Regime." The space-research topic is important as further climate challenges will require Earth observation and the movement of the satellite would require a coordination since the many RSOs would operate at low earth orbit (LEO). These papers (presented in Sessions 3 and 4 of the main track) highlight some of the areas where DDDAS is spurting RSO estimation, debris avoidance, cislunar, and swarm-based space-to-ground observations; all that requiring first-principle models and real-time data control.

4.2 Aerospace Systems

DDDAS for **aerospace systems** has engendered many new capabilitiesfor structural health analysis and coordination of aerialplatforms, well tested with the many UAVs available that each include different sensor packages. The papers in Sessions 1 and 2, provided a range of such advances. *Anirban Chaudhuri,* et al. extend a common DDDAS approach on reduced order modeling (ROM) using data assimilation methods. The approach is extended for "Generalized multi-fidelity active learning for Gaussian-process-based reliability analysis," which is critical to determine the deployability, safer, and insurable of current AI/ML systems. *Peiyuan Zhou,* et al. complements with "Towards the formal verification of data-driven flight awareness: Leveraging the Cramér-Rao lower bound of stochastic functional time series models." Another growing area is hypersonics research which require high-fidelity physics models to determine the effects of fast moving aircraft. *Zachary Mulhollan,* et al. provide a general analysis in their paper "Essential Properties of a Multimodal Hypersonic Object Detection and Tracking System". The coordination of air platforms is developed by *David Sacharny,* et al. in "Dynamic Airspace Control via Spatial Network Morphing". An additional approach is presented by *Chase St. Laurent* et al. for "Coupled Sensor Configuration and Path-Planning in a Multimodal Threat Field." In summary, the DDDAS method was extended for air platform control using learning techniques for certifiable systems, the development of control strategies for swarms of UAVs, and a general overview of single platforms travelling at high speeds.

4.3 Networked Systems and Security

DDDAS applications include large systems of sensors from areas of embedded sensing to the internet of things (IoT) [10], and distributed sensing, computation and communication infrastructures that enable smart surveillance at the edge [11]. For **network systems**, including cloud and edge architectures [12], there are three areas of interest, including power (micro) grids systems analysis adaptive operational management, use of digital twin for edge devices, and security systems. Papers covering these areas were presented in Sessions 4, 5, 9, and 10, in the main track.

For power grid resilience, there is a need to fill measurement gaps with modeled data to account for disruptions as presented by *Maureen Golan*, et al., in their paper "Power Grid Resilience: Data Gaps for Data-Driven Disruption Analysis". *Nurcin Celik* et al., in "DDDAS for Optimized Design and Management of Wireless Cellular Networks", described the use of DDDAS for optimized design and operational management (including resilience) of wireless cellular networks. Leveraging DDDAS-based learning *Temitope Runsewe*et al. highlight the ability towards new methods for edge computing with 5G and beyond 5G networked systems.

For digital twin analysis of edge devices, DDDAS provides a good development and testing strategy [13]. *Masud Rana*, et al., in "Attack-resilient Cyber-physical System State Estimation for Smart Grid Digital Twin Design", discussed their work on attack-resilient cyber-physical systems (CPS) combining state estimation and smart grid digital twin design. As with the edge-based designs afforded by DDDAS design, *Robert Canady*, et al., in "Applying DDDAS Principles for Realizing Optimized and Robust Deep Learning Models at the Edge", highlighted principles for realizing optimized and deep learning modeling at the edge. As a use case, in "Passive Radio Frequency-based 3D Indoor Positioning System via Ensemble Learning", *Liangqi Yuan*, et al. presented a DDDAS-based ensemble learning using passive radio frequency-based measurements for 3D indoor positioning.

Advances in security include the use of DDDAS-based methods in blockchain [14] and assessment of system challenges discerning between sensor failures and physical attacks. *Georgios Diamatopoulos*, et al. presented their work where they used DDDAS-based approaches to address the so-called trilemma tradeoff between decentralization, scalability, and security, in their paper on "Dynamic Data-Driven Digital Twins for Blockchain Systems". *Tahir Ekin*, et al. followed up with their work on "Adversarial Forecasting through Adversarial Risk Analysis within a DDDAS Framework", creating capabilities to thwart adversaries' ability of manipulation and corruption of digital data streams with ensuing degradation in performance of forecasting algorithms and adverse impact in decision quality. Additionally, as a use case, *Maria Pantopoulou*, et al., in their paper on "Monitoring and Secure Communications for Small Modular Reactors", showcased the ability for on-line continuous surveillance and ensured secure transmission of sensor readings for detecting onset of anomalies or failure potential in reactor operations.

4.4 Deep Learning Systems

Advances in deep learning have permeated many research communities such as sensor data fusion [15, 16]. The DDDAS papers focus on the areas of (1) computational efficient sensing analysis and (2) machine interaction with software, hardware, and humans. The papers below, were presented in Sessions 5, 6 and 7, of the main track. Each paper has their independent analysis based on the sensing and paradigm focus. *Jie Wei* et al. opened with many deep learning approaches and focused on contrastive learning for "Deep Learning Approach for Data and Computing Efficient Situational

Assessment and Awareness in Human Assistance and Disaster Response and Damage Assessment Applications." Another detailed effort analyzed the challenges of learning without labeled data such as *Joel B. Harley,* et al. with "Unsupervised Wave Physics-Informed Representation Learning for Guided Wavefield Reconstruction". Another special applications for DDDAS-enhanced deep learning (DDDAS-DL) methods for data-specific areas include *Aneesh Rangnekar,* et al. with "SpecAL: Towards Active Learning for Semantic Segmentation of Hyperspectral Imagery" and *Peng Cheng* et al. with "Multimodal IR (infrared) and RF (radio frequency) based sensor system for real-time human target detection, identification, and geolocation." Thus, DDDAS traditional methods are using the deep learning for data exploitation to update the models of feature analysis for sensor exploitation.

The second area of DDDAS-DL is that assessing the use of multimodal measurements using *DL within a system* [17]. At the basic level, the theory and mathematics need to be assessed to determine the provable bounds of performance. *Song Wen,* et al. focused on "Learning Interacting Dynamic Systems with Neural Ordinary Differential Equations". The updates of the processing with ODEs was informative on the usability of DL methods. In support of the mathematical analysis, the types of data and features being processed is important. Hence, the discussion of the features was highlighted as critical in determine the alignment of features from *Nandini Ramanan,* et al. in "Relational Active Feature Elicitation for DDDAS." Finally, since DDDAS is a systems concept, the human user is important in the analysis. To discuss the systems questions such as human-machine teaming and machine-based digital twin, *Nan Zhang* et al., in their paper "Explainable Human-in-the-loop Dynamic Data-Driven Digital Twins", discussed how exploiting the DDDAS-based bidirectional symbiotic sensing feedback loops and its continuous updates enhance explainability of the rationale, and utility gained in implementing the decision to allow humans-in-the-loop to evaluate the decisions in interactive decision making.

Overall, concepts of DL instantiation within a DDDAS system will continue to evolve with the variations in DL methods, and these papers begin the discussion.

4.5 Estimation and Tracking

Advances in objet tracking technology rely on signal processing and estimation analysis that has long been developed and enabled with DDDAS for future capability for coordinating measurements [18].The estimation papers cover a range of domains focused on the theoretical analysis of processing data either from measurements, simulated models, or trained from deep learning methods [19]. Papers covering estimation and tracking were presented in main-track Sessions 4, 5, 8 and 9. The paper by *Yinsong Wang,* et al. titled "Tracking Dynamic Gaussian Density with a Theoretically Optimal Sliding Window Approach", deals with the density estimation of the data to be processed. Using the density analysis to determine the need to transmit all the data or a representation of the information as a distribution is key for efficient communications and sensor processing. To accommodate nonlinear estimation methods *Ruixin Niu* presented a paper on "Transmission Censoring and Information Fusion for Communication-Efficient Distributed Nonlinear Filtering".

Building on the estimation analysis, a series of long-time DDDAS participates show-cased methods in support specific domains. For the space domain, *Damien Gueho and Puneet Singla* discussed estimation methods in their paper "Towards a data-driven bilinear Koopman operator for controlled nonlinear systems and sensitivity analysis". Likewise, collecting information from an air platform supports object recognition, so *SevgiGurbuz* presented research on "Physics-Aware Machine Learning for Dynamic, Data-Driven Radar Target Recognition". To support the ground domain that overlaps with the keynote presentation from *Yuri Bazilevs, Celso Do Cabo, Mark Todisco and Zhu Mao* continued the DDDAS research with "Data Augmentation of High-Rate Dynamic Testing via a Physics-Informed GAN Approach," for mechanical systems assessing the dynamic performance of loads. Finally for the ocean domain and supporting the need to monitor the environment, *Cong Wei and Derek A. Paley* concludes the book with "Distributed Estimation of the Pelagic Scattering Layer using a Buoyancy Controlled Robotic System".

5   Summary Remarks

The DDDAS2022 Conference, with an overarching theme of *"InfoSymbiotics/DDDAS for human, environmental and engineering sustainment",* highlighted recent themes that enhanced the DDDAS paradigm, especially with deep learning, domain processing, and climate analysis. For the systems approach from DDDAS, many common architectures can augment the simulated data with that of generative DL approaches that are refined when deployed with measurement systems, especially for cases from multimodal sensing. The systems domain approaches include applications for space, air, ground, sea, and cyber from which theoretical alignments between the instrumented data and data collections are present to develop technology for responsive nonlinear systems. Many of the papers follow the theme that DDDAS enables awareness of the system [1, 20]. Finally, the conference focused on using DDDAS technology to deal with the many challenges facing the Earth from climate change. Together, the conference was well received with new advances from which to utilize the wealth of knowledge from first-principle physics, instrumented sensor data collections, and generative DL simulations.

References

1. In:Darema, F., Blasch, E., Ravela, S., Aved, A. (eds.). Dynamic Data Driven Applications Systems: Third International Conf., DDDAS 2020 Boston, 2–4 October 2020 (2020)
2. Blasch, E., Ravela, S., Aved, A. (eds.): Handbook of Dynamic Data Driven Applications Systems. Springer, Cham (2018). https://doi.org/10.1007/978-3-319-95504-9
3. In:Blasch, E.P., Darema, F., Ravela, S., Aved, A.J.: (eds.) Handbook of Dynamic Data Driven Applications Systems, vol. 1, 2nd ed. Springer, Cham (2022). https://doi.org/10.1007/978-3-030-74568-4
4. In:Darema, F., Blasch, E.P., Ravela, S., Aved, A.J. (eds.), Introduction (Chapter-1), Handbook of Dynamic Data Driven Applications Systems, vol. 2, Springer, Cham (2023). https://doi.org/10.1007/978-3-031-27986-7
5. Imai, S., Blasch, E., Galli, A., Zhu, W., Lee, F., Varela, C.A.: Airplane flight safety using error-tolerant data stream processing. IEEE Aerosp. Electron. Syst. Mag. **32**(4), 4–17 (2017)

6. Nagananda, N., Taufique, A.M.N., Madappa, R., Jahan, C.S., Minnehan, B.: Benchmarking domain adaptation methods on aerial datasets. Sensors **21**(23), 8070 (2021)

7. Chen, H.-M., Savakis, A., Diehl, A., Blasch, E., Wei, S., Chen, G.: Targeted adversarial discriminative domain adaptation. J. Appl. Remote. Sens. **15**(3), 038504 (2021)

8. Jahan, C.S., Savakis, A.: Cross-modal knowledge distillation in deep networks for SAR image classification. In: Proceedings SPIE, vol. 12099, pp. 20–27 (2022)

9. Xu, R., Chen, Y., et al.: Exploration of blockchain-enabled decentralized capability-based access control strategy for space situation awareness. Opt. Eng. **58**(4), 041609 (2019)

10. Kamhoua, C.A., Njilla, L.L., Kott, A.: Modeling and Design of Secure Internet of Things, Wiley (2020)

11. Chen, N., Chen, Y., et al.: Enabling smart urban surveillance at the edge. In: SmartCloud (2017)

12. Munir, A., Kwon, J., Lee, J.H., Kong, J., et al.: FogSurv: a fog-assisted architecture for urban surveillance using artificial intelligence and data fusion. IEEE Access **9**, 111938–111959 (2021)

13. Kapteyn, M.G., Pretorius, V.V.R., Willcox, K.E.: A probabilistic graphical model foundation for enabling predictive digital twins at scale. Nat. Comput. Sci. **1**, 337–347 (2021)

14. Qu, Q., Xu, R., Chen, Y., Blasch, E., Aved, A.: Enable fair Proof-of-Work (PoW) consensus for blockchains in IoT by Miner Twins (MinT). Future Internet **13**(11), 291 (2021)

15. Liu, S., Gao, M., John, V., Liu, Z., et al.: Deep learning thermal image translation for night vision perception. ACM Trans Intell. Syst. Technol. **12**(1), 1–18 (2020)

16. Vakil, A., Liu, J., Zulch, P., et al.: A Survey of multimodal sensor fusion for passive RF and EO information integration. IEEE Aerosp. Electron. Syst. Mag. **36**(7), 44–61 (2021)

17. Munir, A., Blasch, E., Kwon, J., Kong, J., Aved, A.: Artificial intelligence and data fusion at the edge. IEEE Aerosp. Electron. Syst. Mag. **36**(7), 62–78 (2021)

18. Hammoud, R.I., Sahin, C.S., et al.: Automatic association of chats and video tracks for activity learning and recognition in aerial video surveillance. Sensors **14**, 19843–19860 (2014)

19. Blasch, E.: Machine learning/artificial intelligence for sensor data fusion–opportunities and challenges. IEEE Aerosp. Electron. Syst. Magazine **36**(7), 80–93 (2021)

20. Munir, A., Aved, A., Blasch, E.: Situational awareness: techniques, challenges, and prospects. AI, **3**, 55–77 (2022)

Main-Track Plenary Presentations - Aerospace

Generalized Multifidelity Active Learning for Gaussian-process-based Reliability Analysis

Anirban Chaudhuri[✉] and Karen Willcox

Oden Institute for Computational Engineering and Sciences,
The University of Texas at Austin, Austin, TX 78712, USA
{anirbanc,kwillcox}@oden.utexas.edu

Abstract. Efficient methods for achieving active learning in complex physical systems are essential for achieving the two-way interaction between data and models that underlies DDDAS. This work presents a two-stage multifidelity active learning method for Gaussian-process-based reliability analysis. In the first stage, the method allows for the flexibility of using any single-fidelity acquisition function for failure boundary identification when selecting the next sample location. We demonstrate the generalized multifidelity method using the existing acquisition functions of expected feasibility, U-learning, targeted integrated mean square error acquisition functions, or their *a priori* Monte Carlo sampled variants. The second stage uses a weighted information-gain-based criterion for the fidelity model selection. The multifidelity method leads to significant computational savings over the single-fidelity versions for real-time reliability analysis involving expensive physical system simulations.

Keywords: contour location · failure boundary · optimal experimental design · AK-MCS · EGRA · Kriging · adaptive sampling · multiple information source · risk

1 Introduction

The contour location problem is encountered in many engineering design applications in the form of stability boundary identification or failure boundary identification for reliability analysis. Such methods are critical for achieving real-time risk assessment of dynamic data-driven application systems (DDDAS) [5,8]. The computational expense associated with accurate contour location can quickly become prohibitive because of the requirement of many expensive high-fidelity model evaluations. This is a particular challenge for the complex physical systems

This work has been supported in part by Department of Energy award number DE-SC0021239, ARPA-E Differentiate award number DE-AR0001208, and AFOSR DDIP award FA9550-22-1-0419.

envisioned to benefit from the DDDAS paradigm. In such applications, there are often cheaper lower-fidelity models available that can be used to estimate the targeted contour at a lower cost, but these low-fidelity models lack the accuracy of the high-fidelity models. A multifidelity method combines these low-cost low-fidelity models together with a small number of evaluations of the expensive high-fidelity model. In this way, multifidelity methods leverage the availability of multiple models to achieve a given task, just as sensor fusion methods leverage the availability of multiple data sources [4]. In this paper, we propose a multifidelity method for contour location through extensions of active learning methods using Gaussian process (GP) surrogates that can fuse information from multiple models of different fidelity to estimate target contours with comparable accuracy to that of a high-fidelity model, but at significantly reduced computational cost.

This work deals with reliability analysis of a system with the N_z random variables $Z \in \Omega \subseteq \mathbb{R}^{N_z}$ with the probability density function π as the inputs, where Ω denotes the random sample space. The vector of a realization of the random variables Z is denoted by z. The probability of failure of the system is $p_F = \mathbb{P}(g(Z) > 0)$, where $g : \Omega \mapsto \mathbb{R}$ is the limit state function. Reliability analysis typically requires Monte Carlo sampling for strongly nonlinear systems for estimating the probability of failure estimate $\hat{p}_F$ as $\hat{p}_F = \frac{1}{m} \sum_{i=1}^{m} \mathbb{I}_\mathcal{G}(z_i)$, where $z_i, i = 1, \ldots, m$ are m samples from probability density π, $\mathcal{G} = \{z \mid z \in \Omega, g(z) > 0\}$ is the failure set, and $\mathbb{I}_\mathcal{G}(z) = \begin{cases} 1, z \in \mathcal{G} \\ 0, \text{else} \end{cases}$ is the indicator function.

The Monte Carlo estimation of p_F can be computationally prohibitive due to the numerous expensive high-fidelity model evaluations. One way of tackling the issue of computational cost is to improve Monte Carlo convergence rates through variance reduction techniques, such as, importance sampling [20], cross-entropy method [15], subset simulation [1], etc. The second class of methods uses approximations for the failure boundary to reduce computational cost through the first- and second-order reliability methods, which can be efficient for mildly nonlinear problems and cannot handle systems with multiple failure regions [14, 26]. The focus of this paper is on the third class of methods that use active learning for creating adaptively refined surrogates around the failure boundary to replace the high-fidelity model simulations.

In this work, without loss of generality, the failure of the system is defined by $g(z) > 0$. The failure boundary is defined as the zero contour of the limit state function, $g(z) = 0$, and any other failure boundary, $g(z) = c$, can be reformulated as a zero contour (i.e., $g(z) - c = 0$). Reliability estimation requires accurate classification of samples to fail or not, which needs surrogates to accurately predict the failure boundary. This boils down to a contour location problem. We concentrate on the Gaussian-process (GP)-based active learning methods (a.k.a. kriging-based methods) that use the GP prediction mean and prediction variance to decide the next sample location for a target contour location. Various acquisition functions have been defined in the literature with target contour as the failure boundary. Efficient Global Reliability Analysis (EGRA) uses the the expected feasibility function (EFF) as the acquisition function [3]. Picheny et al. [24] proposed a targeted integrated mean square error (TIMSE) criterion for refining the GP surrogate around the target contour. Adaptive Kriging with Monte Carlo simulation (AK-MCS) uses the U-learning acquisition function and restricts the search space to a predefined Monte Carlo (MC) sample set. A population-based adaptive sampling technique was developed by Dubourg et al. [10]. A review of some surrogate-based methods for reliability analysis can be found in Ref. [21]. However, the above methods can only use a single fidelity model. Recently a few multifidelity active learning methods for contour location have been developed that can further accelerate the reliability estimation. Another approach is the hierarchical bi-fidelity adaptive support vector machine developed by Dribusch et al. [9] and used for locating the aeroelastic flutter boundary. A one-step lookahead multifidelity contour location acquisition function was proposed by Marques et al. [18,19]. In our previous work, we developed multifidelity EGRA (mfEGRA) [6,7], a multifidelity extension of the EGRA method. This work presents a generalized multifidelity method that can extend most single-fidelity GP-based active learning contour location method to utilize multiple information sources.

In a multifidelity method, we must determine: (a) how to combine different information sources, (b) where to sample next, and (c) what information source to use for evaluating the next sample. The mfEGRA method is a two-stage method that first selects the sample location using EFF and then selects the optimum information source for that location through a weighted information gain acquisition function [7]. The key attribute of the mfEGRA method exploited here is the ability of the two-stage method to combine existing user-preferred acquisition functions for location selection with an information source selection strategy. We show the effectiveness of the generalized multifidelity method using three popular acquisition functions — EFF, AK-MCS, and TIMSE — for location selection in the first stage. The second stage uses a weighted one-step lookahead information gain acquisition function for information source selection developed in our previous works [6,7]. The key features of the proposed multifidelity active learning method are as follows:

- flexibility to extend any existing single-fidelity acquisition function to multifidelity;
- natural model selection in the multifidelity setup based on the correlations and the relative cost between high- and low-fidelity models;
- non-hierarchical multifidelity surrogate except knowledge about the highest fidelity model [25]; and
- significant computational savings for accurate reliability estimation.

2   Multifidelity Active Learning for Reliability Analysis

In this section, we present the multifidelity active learning method using different acquisition functions for contour location, specifically, failure boundary identification. The multifidelity method shows the ease of extending the tools developed for mfEGRA [7] to other single-fidelity contour location acquisition functions.

2.1   Multifidelity Active Learning Method Overview

The multifidelity method consists of three parts: (1) multifidelity surrogate for combining different fidelity models, (2) selecting the next sampling location as described in Sect. 2.2, and (3) selecting the information source used for evaluating the next sample as described in Sect. 2.3. Figure 1 shows an overview of the proposed multifidelity method. This work builds on our previous work on mfEGRA [7] to create a generalized multifidelity method that can extend any existing or new single-fidelity acquisition function to use multiple models with different fidelities and costs instead of only using the high-fidelity model. Let $g_l : \Omega \mapsto \mathbb{R}, l \in \{0, \ldots, k\}$ be a collection of $k + 1$ models for g with associated cost $c_l(z)$ at location z, where the subscript l denotes the information source. We define the model g_0 with $l = 0$ to be the high-fidelity model for the limit state function. The k low-fidelity models of g are denoted by $l = 1, \ldots, k$.

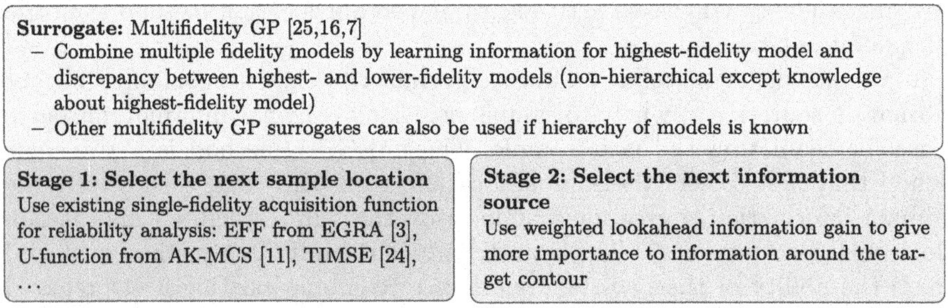

Surrogate: Multifidelity GP [25,16,7]
- Combine multiple fidelity models by learning information for highest-fidelity model and discrepancy between highest- and lower-fidelity models (non-hierarchical except knowledge about highest-fidelity model)
- Other multifidelity GP surrogates can also be used if hierarchy of models is known

Stage 1: Select the next sample location	**Stage 2: Select the next information source**
Use existing single-fidelity acquisition function for reliability analysis: EFF from EGRA [3], U-function from AK-MCS [11], TIMSE [24], ...	Use weighted lookahead information gain to give more importance to information around the target contour

Fig. 1. Generalized multifidelity active learning method overview.

Similar to mfEGRA [7], we use the multifidelity GP surrogate from Poloczek et al. [25], which was built on earlier work by Lam et al. [16], to combine information from the $k + 1$ information sources into a single GP surrogate, $\widehat{g}(l, z)$. A more detailed description about the assumptions and the implementation of the multifidelity GP surrogate can be found in Refs. [7,25]. We chose this specific multifidelity GP surrogate since it can handle fidelity models without a known hierarchy except knowledge about the highest fidelity model. However, any other multifidelity GP surrogate (e.g., Refs. [17,23]) can be used, if hierarchy of models is fully known. Consider that n samples $\{[l^i, z^i]\}_{i=1}^n$ have been evaluated and these samples are used to fit the multifidelity GP surrogate. Then the surrogate is refined around the target contour by sequentially adding samples. The next sample z^{n+1} and the next information source l^{n+1} used to refine the surrogate are found using the two-stage multifidelity method described below. In each iteration, the stage 1 decision feeds into the stage 2 decision that in turn affects the stage 1 decision of the next iteration. The iterations continue till some user-defined stopping criterion is met. The final adaptively refined multifidelity surrogate predictions for high-fidelity model $\hat{g}(0, z)$ are used for the probability of failure estimation. The MC estimate of the probability of failure is then estimated using the refined multifidelity surrogate as given by $\hat{p}_F^{MF} = \frac{1}{m}\sum_{i=1}^{m}\mathbb{I}_{\tilde{\mathcal{G}}}(z_i)$, where $z_i, i = 1,\ldots,m$ are m samples from probability density π and $\tilde{\mathcal{G}} = \{z \mid z \in \Omega, \hat{g}(0, z) > 0\}$.

2.2 Sample Location: EFF/AK-MCS/TIMSE

The first stage of multifidelity GP-based contour location involves selecting the next sample location z^{n+1} using some acquisition function $J(z)$ to iteratively refine the surrogate. We use existing single-fidelity contour location methods from the literature: EFF [3], AK-MCS [11], and TIMSE [24]. *Note that any other acquisition functions can also be used.* As mentioned before, we describe the method considering the zero contour as the failure boundary for convenience. The multifidelity GP surrogate prediction for the high-fidelity model at any z is defined by the normal distribution $\mathcal{Y}_z \sim \mathcal{N}(\mu(0, z), \sigma^2(0, z))$, where $\mu(0, z)$ is the posterior mean and $\sigma(0, z)$ is the posterior variance.

In this work, the acquisition functions used for selecting the next sample location are:

1. **EFF:** The expected feasibility within an ϵ-band around the contour of interest is given by [3]

$$
\begin{aligned}
J(z) = \mathbb{E}[F(z)] = \mu(0, z) &\left[2\Phi\left(\frac{-\mu(0, z)}{\sigma(0, z)}\right) - \Phi\left(\frac{-\epsilon(z) - \mu(0, z)}{\sigma(0, z)}\right) - \Phi\left(\frac{\epsilon(z) - \mu(0, z)}{\sigma(0, z)}\right) \right] \\
&- \sigma(0, z)\left[2\phi\left(\frac{-\mu(0, z)}{\sigma(0, z)}\right) - \phi\left(\frac{-\epsilon(z) - \mu(0, z)}{\sigma(0, z)}\right) - \phi\left(\frac{\epsilon(z) - \mu(0, z)}{\sigma(0, z)}\right) \right] \\
&+ \epsilon(z)\left[\Phi\left(\frac{\epsilon(z) - \mu(0, z)}{\sigma(0, z)}\right) - \Phi\left(\frac{-\epsilon(z) - \mu(0, z)}{\sigma(0, z)}\right) \right],
\end{aligned}
\tag{1}
$$

where $F(z) = \epsilon(z) - \min(|y|, \epsilon(z))$ is the feasibility function with y as a realization of $\mathcal{Y}_z$, and Φ is the cumulative distribution function and ϕ is the probability density function of the standard normal distribution. Similar to [3], we define $\epsilon(z) = 2\sigma(0, z)$ to trade-off between exploration and exploitation. Note that multifidelity EGRA has already been shown to be an efficient method in our previous work [7].

2. **AK-MCS:** This is an *a priori* MC sampled variant for the EFF (Eq. (1)) and the U-learning function (Eq. (2)). The search space for z is limited to a discrete set of m MC samples $\mathcal{Z}$ drawn from a given distribution of the input random variables [11]. The U-learning function was proposed as an alternative for EFF and is given by [11]

$$
J(z) = -U(z) = -\frac{|\mu(0, z)|}{\sigma(0, z)}.
\tag{2}
$$

Limiting the search space to the *a priori* MC samples helps refine the surrogate in regions of higher probability. In this work, we denote AK-MCS with U-learning function by AK-MCS-U and with EFF function as EGRA-MCS.

3. **TIMSE:** The TIMSE approach is a one-step lookahead strategy that seeks to improve the accuracy of the surrogate around the targeted contour. TIMSE

minimizes the weighted overall posterior variance based on adding one more sample at z as given by

$$J(z) = -\text{IMSE}_T(\{[l^i, z^i]\}_{i=1}^n, [0, z]) = \int_\Omega \sigma_F^2(0, z' \mid \{[l^i, z^i]\}_{i=1}^n, [0, z])W(z')dz', \tag{3}$$

where $W(.)$ is the weight function that gives more importance to the targeted contour and $\sigma_F^2(0, z' \mid \{[l^i, z^i]\}_{i=1}^n, [0, z])$ is the posterior variance of a hypothetical future GP surrogate augmented with sampling location at z with a high-fidelity evaluation ($l = 0$) [24]. Note that the posterior variance of the GP surrogate only depends on the location of the samples and does not need any new evaluations. The weights depend on the current GP surrogate predictions and is given by

$$W(z') = \frac{1}{\sqrt{2\pi(s_\epsilon^2 + \sigma^2(0, z'))}} \exp\left(-\frac{1}{2}\frac{\mu^2(0, z)}{s_\epsilon^2 + \sigma^2(0, z')}\right), \tag{4}$$

where the parameter s_ϵ denotes the size of the domain of interest around the target contour (here, zero contour). As suggested by [24], $s_\epsilon = 10^{-3}$ is set to a small number. We use the variant of TIMSE using *a priori* MC samples that is also presented in Ref. [24].

When the contour location is done over a given space uniformly, the next sample location is selected by maximizing the acquisition function $J(z)$ by

$$z^{n+1} = \arg\max_{z \in \Omega} J(z). \tag{5}$$

For the cases when search space is limited to *a priori* MC samples, the next sample location is selected from the discrete set $\mathcal{Z}$ by

$$z^{n+1} = \arg\max_{z \in \mathcal{Z}} J(z). \tag{6}$$

2.3   Information Source: Weighted Lookahead Information Gain

Given the next sample location at z^{n+1} obtained using Eqs. (5) or (6), the second stage of the multifidelity method selects the information source l^{n+1} to be used for simulating the next sample by maximizing the information gain. The idea of using a weighted one-step lookahead information gain in the context of information selection for active learning GP-based contour location was first used in the mfEGRA method using EFF for location selection [6,7]. The same criterion can be readily combined with any existing acquisition function for contour location since it is detached from the location selection process. This gives us an easily generalized multifidelity extension for most existing active learning GP-based contour location methods. In this section, we recap the weighted one-step lookahed information gain criterion. *Here, we use EFF as the weights but any other weighting scheme that can target the specific contour of interest are also applicable, such as the TIMSE weights given by Eq. (4).* A detailed derivation can be found in Ref. [7].

We measure the Kullback-Leibler (KL) divergence between the current surrogate predicted GP, $G_C(z) \sim \mathcal{N}(\mu_C(0, z), \sigma_C^2(0, z))$, and a hypothetical future surrogate predicted GP, $G_F(z|z^{n+1}, l_F, y_F)$, when a particular future information source is used to simulate the sample at z^{n+1} to quantify the information gain. The hypothetical future simulated data $y_F \sim \mathcal{N}(\mu_C(l_F, z^{n+1}), \sigma_C^2(l_F, z^{n+1}))$ is obtained from the current GP surrogate prediction at the location z^{n+1} using a possible future information source $l_F \in \{0, \ldots, k\}$. The subscript C indicates the current GP surrogate and the subscript F indicates the hypothetical future GP surrogate.

The total lookahead information gain is obtained by integrating over all possible values of y_F as described below. Since both G_C and G_F are Gaussian distributions, we can get the KL divergence $D_{KL}(G_C(z) \parallel G_F(z|z^{n+1}, l_F, y_F))$ between them explicitly. The total lookahead information gain for any z is given by

$$D_{IG}(z^{n+1}, l_F) = \mathbb{E}_{y_F} \left[\int_\Omega D_{KL}(G_C(z) \parallel G_F(z|z^{n+1}, l_F, y_F)) dz \right] = \int_\Omega D(z \mid z^{n+1}, l_F) dz, \quad (7)$$

where

$$D(z \mid z^{n+1}, l_F) = \log\left(\frac{\sigma_F(0, z|z^{n+1}, l_F)}{\sigma_C(0, z)} \right) + \frac{\sigma_C^2(0, z) + \bar{\sigma}^2(z|z^{n+1}, l_F)}{2\sigma_F^2(0, z|z^{n+1}, l_F)} - \frac{1}{2},$$

$\bar{\sigma}^2(z|z^{n+1}, l_F) = (\Sigma_C((0, z), (l_F, z^{n+1})))^2 / \Sigma_C((l_F, z^{n+1}), (l_F, z^{n+1}))$, and $\sigma_C^2(0, z)$ is the posterior prediction variance of the current GP surrogate for the high-fidelity model built using the available training data of n samples. The posterior variance of the hypothetical future GP surrogate $\sigma_F^2(0, z|z^{n+1}, l_F, y_F)$ depends only on the location z^{n+1} and the source l_F, and can be replaced with $\sigma_F^2(0, z|z^{n+1}, l_F)$. Note that we don't need any new evaluations of the information source for constructing the future GP. In practice, we choose a discrete set $\mathcal{Z} \subset \Omega$ via Latin hypercube sampling to numerically integrate Eq. (7) as given by $D_{IG}(z^{n+1}, l_F) = \int_\Omega D(z \mid z^{n+1}, l_F) dz \approx \sum_{z \in \mathcal{Z}} D(z \mid z^{n+1}, l_F)$. Note that in the case of location selection acquisition functions implemented using *a priori* MC samples, $\mathcal{Z}$ is naturally the same *a priori* MC samples since those are the samples of interest.

We are interested in gaining more information for our target contour, which is the failure boundary. We use a weighted version of the lookahead information gain normalized by the cost of the information source to give greater importance to gaining information around the target contour. In this work, we use weights defined by by the EFF, $w(z) = \mathbb{E}[F(z)]$ similar to Ref. [7]. As mentioned before, we can also use the TIMSE weights given by Eq. (4) as $w(z) = W(z)$ instead of EFF.

The next information source l^{n+1} is selected by maximizing the weighted lookahead information gain normalized by the cost of the information source as given by

$$l^{n+1} = \arg \max_{l \in \{0, \ldots, k\}} \sum_{z \in \Omega} \frac{1}{c_l(z)} w(z) D(z|z^{n+1}, l_F = l). \quad (8)$$

The optimization problem in Eq. (8) is a one-dimensional discrete variable problem that needs $k + 1$ (number of available models) evaluations of the objective function to solve the optimization problem exactly and typically k is a small number.

3   Numerical Experiment: Reliability Analysis of Acoustic Horn

We use the two-dimensional acoustic horn problem [13,22] to demonstrate the effectiveness of the proposed two-stage multifidelity active learning method method. The multifidelity method is denoted by attaching 'mf' in front of the respective single-fidelity method name. The inputs to the system are the three random variables wave number $k \sim \mathcal{U}[1.3, 1.5]$, upper horn wall impedance $Z_u \sim \mathcal{N}(50, 3)$, and lower horn wall impedance $Z_l \sim \mathcal{N}(50, 3)$. The output of the model is the reflection coefficient s, which is a measure of the horn's efficiency. We define the failure of the system to be $s(z) > c$, where $c = 0.1$. The limit state function is defined as $g(z) = s(z) - c$ so that the failure boundary is $g(z) = 0$. The high-fidelity model is a finite element model with 35895 nodal grid points. The low-fidelity model is a reduced basis model with $N = 100$ basis vectors [13,22]. The low-fidelity model evaluations are 40 times faster than the high-fidelity model evaluations on an average. A more detailed description of the acoustic horn model used in this work can be found in Ref. [22]. All the results are repeated for 10 different initial design of experiments (DOE) with an initial sample size of 10 generated using Latin hypercube sampling.

In this case, using 10^5 MC samples with the high-fidelity model evaluations leads to a probability of failure of $p_F = 0.3812$. Figure 2 compares the convergence of the relative error in probability of failure estimation for the multifidelity active learning method with the respective single-fidelity versions. In all the cases, the multifidelity extensions lead to significant computational savings

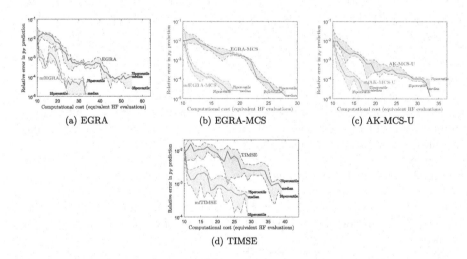

(a) EGRA   (b) EGRA-MCS   (c) AK-MCS-U

(d) TIMSE

Fig. 2. Comparing relative error in the estimate of probability of failure using the multifidelity extensions and the single-fidelity methods for the acoustic horn application with 10 different initial DOEs[1](The results for EGRA and EGRA-MCS first appear in our previous publication on mfEGRA [7] and are repeated here for completeness.) .

as compared to the single-fidelity methods. To achieve a median relative error below $10^{-3}(= 0.1\%)$, mfEGRA, mfEGRA-MCS, mfAK-MCS-U, and mfTIMSE lead to 24%, 45%, 33%, and 50% reduction in computational effort, respectively.

4   Conclusions and Future Directions

We presented a two-stage multifidelity active learning method for GP-based reliability analysis to leverage cheaper lower-fidelity information sources. The method offers the flexibility to extend any single-fidelity acquisition function for contour location to multifidelity leading to substantial computational savings. The method uses the existing acquisition functions for location selection in the first stage followed by a weighted information gain criterion for fidelity model selection in the second stage. For the acoustic horn reliability analysis problem, the multifidelity versions led to 24–50% reduction in computation costs compared to the different single-fidelity versions.

There are a few possible future directions. First, one can explore other ways to weight the information gain metric, such as the TIMSE weights. Second, the proposed multifidelity method can be easily extended to use MC variance reductions techniques, such as importance sampling, cross-entropy. Such extensions have been shown to be effective for single-fidelity versions [2,12]. Third, in the context of reliability-based design optimization under uncertainty, the proposed multifidelity method can be combined with the information reuse method proposed in Ref. [6].

References

1. Au, S.K., Beck, J.L.: Estimation of small failure probabilities in high dimensions by subset simulation. Probab. Eng. Mech. **16**(4), 263–277 (2001)
2. Balesdent, M., Morio, J., Marzat, J.: Kriging-based adaptive importance sampling algorithms for rare event estimation. Struct. Saf. **44**, 1–10 (2013)
3. Bichon, B.J., Eldred, M.S., Swiler, L.P., Mahadevan, S., McFarland, J.M.: Efficient global reliability analysis for nonlinear implicit performance functions. AIAA J. **46**(10), 2459–2468 (2008)
4. Blasch, E., Lambert, D.A.: High-level information fusion management and systems design. Artech House (2012)
5. Blasch, E., Ravela, S., Aved, A. (eds.): Handbook of Dynamic Data Driven Applications Systems. Springer, Cham (2018). https://doi.org/10.1007/978-3-319-95504-9
6. Chaudhuri, A., Marques, A.N., Lam, R., Willcox, K.E.: Reusing information for multifidelity active learning in reliability-based design optimization. In: AIAA Scitech 2019 Forum, p. 1222 (2019)
7. Chaudhuri, A., Marques, A.N., Willcox, K.: mfEGRA: multifidelity efficient global reliability analysis through active learning for failure boundary location. Struct. Multidiscip. Optim. **64**(2), 797–811 (2021)
8. Darema, F.: Dynamic Data Driven Applications Systems: A New Paradigm for Application Simulations and Measurements. In: Bubak, M., van Albada, G.D., Sloot, P.M.A., Dongarra, J. (eds.) ICCS 2004. LNCS, vol. 3038, pp. 662–669. Springer, Heidelberg (2004). https://doi.org/10.1007/978-3-540-24688-6_86

9. Dribusch, C., Missoum, S., Beran, P.: A multifidelity approach for the construction of explicit decision boundaries: application to aeroelasticity. Struct. Multidiscip. Optim. **42**(5), 693–705 (2010)

10. Dubourg, V., Sudret, B., Bourinet, J.M.: Reliability-based design optimization using kriging surrogates and subset simulation. Struct. Multidiscip. Optim. **44**(5), 673–690 (2011)

11. Echard, B., Gayton, N., Lemaire, M.: AK-MCS: an active learning reliability method combining kriging and monte Carlo simulation. Struct. Saf. **33**(2), 145–154 (2011)

12. Echard, B., Gayton, N., Lemaire, M., Relun, N.: A combined importance sampling and kriging reliability method for small failure probabilities with time-demanding numerical models. Reliab. Eng. Syst. Saf. **111**, 232–240 (2013)

13. Eftang, J.L., Huynh, D., Knezevic, D.J., Patera, A.T.: A two-step certified reduced basis method. J. Sci. Comput. **51**(1), 28–58 (2012)

14. Hohenbichler, M., Gollwitzer, S., Kruse, W., Rackwitz, R.: New light on first-and second-order reliability methods. Struct. Saf. **4**(4), 267–284 (1987)

15. Kroese, D.P., Rubinstein, R.Y., Glynn, P.W.: The cross-entropy method for estimation. In: Handbook of Statistics, vol. 31, pp. 19–34. Elsevier (2013)

16. Lam, R., Allaire, D., Willcox, K.: Multifidelity optimization using statistical surrogate modeling for non-hierarchical information sources. In: 56th AIAA/ASCE/AHS/ASC Structures, Structural Dynamics, and Materials Conference (2015)

17. Le Gratiet, L.: Multi-fidelity Gaussian process regression for computer experiments. Ph.D. thesis, Université Paris-Diderot-Paris VII (2013)

18. Marques, A., Lam, R., Willcox, K.: Contour location via entropy reduction leveraging multiple information sources. In: Advances in Neural Information Processing Systems, pp. 5217–5227 (2018)

19. Marques, A.N., Opgenoord, M.M., Lam, R.R., Chaudhuri, A., Willcox, K.E.: Multifidelity method for locating aeroelastic flutter boundaries. AIAA J., 1–13 (2020)

20. Melchers, R.: Importance sampling in structural systems. Struct. Saf. **6**(1), 3–10 (1989)

21. Moustapha, M., Sudret, B.: Surrogate-assisted reliability-based design optimization: a survey and a unified modular framework. Struct. Multi. Optim. **60**(5), 2157–2176 (2019). https://doi.org/10.1007/s00158-019-02290-y

22. Ng, L.W., Willcox, K.E.: Multifidelity approaches for optimization under uncertainty. Int. J. Numer. Meth. Eng. **100**(10), 746–772 (2014)

23. Perdikaris, P., Raissi, M., Damianou, A., Lawrence, N.D., Karniadakis, G.E.: Nonlinear information fusion algorithms for data-efficient multi-fidelity modelling. Proc. R. Soc. A **473**(2198), 20160751 (2017)

24. Picheny, V., Ginsbourger, D., Roustant, O., Haftka, R.T., Kim, N.H.: Adaptive designs of experiments for accurate approximation of a target region. J. Mech. Des. **132**(7), 071008 (2010)

25. Poloczek, M., Wang, J., Frazier, P.: Multi-information source optimization. In: Advances in Neural Information Processing Systems, pp. 4291–4301 (2017)

26. Rackwitz, R.: Reliability analysis-a review and some perspectives. Struct. Saf. **23**(4), 365–395 (2001)

Essential Properties of a Multimodal Hypersonic Object Detection and Tracking System

Zachary Mulhollan[1,2], Marco Gamarra[2], Anthony Vodacek[1], and Matthew Hoffman[1(✉)]

[1] Rochester Institute of Technology, Rochester, NY 14623, USA
mjhsma@rit.edu
[2] Air Force Research Laboratory, Rome, NY 13441, USA

Abstract. Hypersonic object detection and tracking is a necessity for the future of commercial aircraft, space exploration, and air defense sectors. However, hypersonic object detection and tracking in practice is a complex task that is limited by physical, geometrical, and sensor constraints. Atmospheric absorption and scattering, line of sight obstructions, and plasma sheaths around hypersonic objects are just a few factors as to why an adaptive, multiplatform, multimodal system are required for hypersonic object detection and tracking. We review recent papers on detection and communication of hypersonic objects which model hypersonic objects with various solid body geometries, surface materials, and flight patterns to examine electromagnetic radiation interactions of the hypersonic object in the atmospheric medium, as a function of velocity, altitude, and heading. The key findings from these research papers are combined with simple gas and thermal dynamics classical physics models to establish baselines for hypersonic object detection. In this paper, we make a case for the necessity of an adaptive multimodal low-earth orbit network consisting of a constellation of satellites communicating with each other in real time for hypersonic detection and tracking.

Keywords: Hypersonic · Object Detection · Tracking · Multimodal · Adaptive · System · Remote Sensing

1 Introduction

An object is travelling at hypersonic speed if it is travelling equal to or greater than five times the speed of sound (Mach 5 or greater). A hypersonic object may be naturally occuring such as a meteor, or it may be a manned or unmanned aircraft. Detection and tracking of hypersonic objects requires a multidisciplinary effort from communications, sensors, aerodynamics, thermodynamics, and material science tech divisions constructing a system that welcomes innovation, and

© The Author(s), under exclusive license to Springer Nature Switzerland AG 2024
E. Blasch et al. (Eds.): DDDAS 2022, LNCS 13984, pp. 27–34, 2024.
https://doi.org/10.1007/978-3-031-52670-1_3

challenges current operational conventions in each field. There is no "silver bullet" for hypersonic object detection and tracking. Rather, we demonstrate that a dynamic data driven application system (DDDAS) that has 1) contextually aware sensing capability, 2) physics-based modeling, 3) multimodal sensors real time communication, and 4) efficient data processing and compression is a necessity for robust detection and tracking.

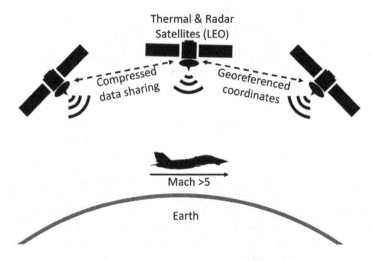

Fig. 1. A constellation of satellites sharing real time observational and self referential information with each other to maintain consistent coverage for hypersonic object detection and tracking.

Hypersonic object detection and tracking requires a multidisciplinary effort from communications, sensors, aerodynamics, thermodynamics, and material science tech divisions constructing a system that welcomes innovation, and challenges current operational conventions in each field. There is no "silver bullet" for hypersonic object detection and tracking. Rather, we demonstrate that a dynamic data driven application system (DDDAS) that has 1) contextually aware sensing capability, 2) physics-based modeling, 3) multimodal sensors real time communication, and 4) efficient data processing and compression is a necessity for robust detection and tracking.

2   Sensing Modalities for Hypersonic Detection

To detect a hypersonic object we must first characterize how it could look from a remote observation point. To do this, we must understand basic principles of electromagnetic radiation (EM) interactions with the hypersonic object's surface, and model how the atmospheric medium can transmit, absorb, or scatter the EM on its propagation path to and from the detection sensor. The sensors may be airborne or terrestrial, but in this paper we will focus on airborne sensors because modern hypersonic objects are capable of flying at low altitudes, which significantly reduces the operational range of terrestrial sensors due to line of sight obscurations from the horizon. In defense applications where early detection is crucial for interception of hypersonic weapons, terrestrial sensors alone without coordination from airborne sensors will have difficulty with providing enough lead time to detect and respond to a hypersonic threat. For our analysis we assume our target is boost-glide vehicles since they have increased maneuverability, can operate with minimal exhaust, and can fly at low altitude.

What EM frequencies are transmitted through the atmosphere and therefore detectable by a sensor on a remote satellite platform? Visible light radiation, infrared radiation, and radio-frequency radiation are the only viable choices for detecting hypersonic objects in the atmosphere [8]. We propose that utilizing radio-frequency and thermal radiation is preferred over visible light for hypersonic object detection because visible light requires an external light source which makes it ineffective at nighttime, and visible light could not detect hypersonic objects when occlusions are present such as clouds [11]. Although clouds can absorb and obscure thermal radiation before it reaches the sensor, thermal energy is a great resource to use for detection because hypersonic objects emit a significant amount of heat flux on their surface due to atmospheric friction [6]. Radio-frequency radiation compared to thermal and visible light is more invariant to inclement weather and clouds, and is used in practice as an active sensor such as conventional radar. Therefore, we propose that an effective hypersonic detection system should utilize both thermal and radio-frequency radiation sensing to provide redundancy and situational coverage in adverse atmospheric conditions.

2.1   Radar

A concern for radio-frequency detection is that hypersonic objects during flight undergo an energy interaction with atmospheric molecules that ionizes the air in the immediate vicinity, creating a plasma sheath that blankets the object [4,9,10]. In the case of water vapor, the hypersonic object will disassociate the molecule into $H_2O \longrightarrow H_2O^+ + e^-$, which creates an electrical conductive cloud of positive ions and excited electrons. This plasma sheath attenuates the radio-frequency signature of the object itself which adds a natural stealth property to the hypersonic object [1,7,12,13]. Plasma absorbs any incident electromagnetic energy that is below the plasma frequency ω_p, which is approximated in Eq. 1.

$$\omega_p \equiv \sqrt{9000 * \rho_p[e^-/cm^3]} \qquad \text{(Hz)} \qquad (1)$$

The electron density ρ_p of a hypersonic plasma sheath depends on many factors such as surface geometry, atmospheric pressure, wall temperature, and velocity [15]. In a hypersonic plasma sheath the electron density will be highest at the nose of the object and bow edge extremities of the surface geometry (i.e. fins or wings) [2]. Not only does the plasma sheath absorb any frequency $< \rho_p$, the time varying structure of the plasma sheath will cause inter-pulse and intra-pulse phase modulation to the radar pulse, leading to degradation of hypersonic target detection performance [3,14]. Target detection performance of radar is most corrupted when the carrier frequency is approximate to the plasma frequency.

2.2 Thermal

Although plasma sheath attenuates detection signals for radar detection, it can improve the thermal signature of a hypersonic object. When an object travels at hypersonic speed in the atmosphere it generates thermal energy through friction that strips away electrons from air molecules, creating a hot plasma. Depending on surface geomtery, the plasma is often at its most dense 2 to 7 cm away from the hypersonic object surface, which in turn contributes to a heat flux transfer from the kinetically excited hot plasma to the wall of the hypersonic object [6]. The simple formula for heat flux is

$$\phi_q = -k\frac{dT(x)}{dx}, \qquad (\mathrm{W/m^2}) \qquad (2)$$

where ϕ_q is the heat flux, k is thermal conductivity, and T is temperature. Thermodynamics of hypersonic flow is difficult to define as simple equations because there are a myriad of dependencies and parameters that will affect total heat flux and the spatial distribution of impinged heat flux. Such dependencies and parameters include 1) surface geometry, 2) atmospheric pressure, 3) chemical makeup of atmosphere, 4) turbulent or laminar flow, 5) distance between bow shock and object surface, and 6) pitch of object [2,16]. Therefore, most hypersonic thermodynamic research relies on numerical modeling and lab experiments to gather thermal heating estimates of specific hypersonic object geometries and flight patterns. Since it is beyond the scope of this paper, we will focus on an example hypersonic thermal scenario instead of an investigation of thermodynamic numerical simulations.

We see in Fig. 2 the thermal flux of a hypersonic object travelling at Mach 7 velocity with an angle of attack $\alpha = 2$ degrees above the horizon [5]. Since the object is increasing in altitude, the windward side is ventral and the leeward side is dorsal. In both laminar and turbulent flow cases the windward side has greater thermal flux and therefore greater thermal signature. This fact is unfortunate for early detection via satellite thermal sensors because after launch while object is gaining altitude, low earth orbit sensors will only visibly have line-of-sight to the colder leeward (top) side. However, except for near perfect smooth flying conditions, airflow around the hypersonic object will almost always be turbulent flow and not laminar flow. Conversely, where $\alpha = 2$ degrees, the low earth orbit sensors will see the windward side of the object with greater heat flux.

The final consideration we will note for thermal detection of a hypersonic object is the surface material's thermal emissivity ε_λ, where λ is the wavelength or frequency of electromagnetic radiation of interest. Thermal emissivity is a material's effectiveness of emitting its kinetic energy as thermal radiation. A simplistic estimation of the emitted thermal energy of a hypersonic that can be detected by a remote satellite sensor is

$$M_{e,\lambda} = \phi_{q,\lambda} * \varepsilon_\lambda * t_{atmos,\lambda}(x) \qquad (\mathrm{W/m^2}) \qquad (3)$$

Φ (W/cm^2)	Windward min	Windward max	Leeward min	Leeward max
Laminar	3	25	1.5	8
Turbulent	22	27	11	20

Fig. 2. A probable range of thermal heat flux impinged on a hypersonic object's surface by hot ionized atmospheric plasma sheathed around the object. Note the vast discrepancy in heat flux between the windward and leeward sides. Values in table collected from [5]

where $M_{e,\lambda}$ is the radiant emittance (the radiant flux emitted by a surface per unit area), and $t_{atmos,\lambda}(x)$ is the spectral atmospheric transmission (between 0 and 1) along path length x. This approach should only be used for simple benchmarking of sensors for feasibility, and we encourage numerical simulations paired with physics modeling of atmosphere for a rigorous performance characterization of thermal sensors for hypersonic detection.

3   Tracking

Effective flight path tracking of hypersonic boost-glide vehicles requires a constellation of low earth orbit (LEO) satellites with real time inter-satellite communications (see Fig. 1). Low earth orbit is necessary due to the potential low altitude flying capabilities of boost-glide hypersonic vehicles (altitude less than 30km); a higher orbit would suffer from low unresolvable spatial resolution and higher atmospheric attenuation of detectable signal. For hypersonic tracking applications and average orbit altitude of about 1200 km above Earth is considered ideal. A LEO orbit for a satellite also means that each satellite has less instantaneous ground coverage than a satellite observing Earth from a distance, which means that a full coverage hypersonic object tracking system needs multiple satellites relaying information to each other as the object travels outside one satellites field of view and into the next satellites zone.

At least two satellites need to have simultaneous vision on the boost-glide vehicle to pinpoint its three dimensional location in the airspace. This information is crucial for commercial airspace to prevent collisions, and for military defense to intercept hypersonic threats. Each satellite will be equipped with a precise internal Global Navigation Satellite System and inertial measurement unit (GNNS-IMU) system and receive streams of external GNSS positional data to measure its own position in space, and to convert its thermal and radar sensor data into a georeferenced world coordinate system that each satellite in the constellation can understand. Not only will georeferenced data pinpoint the hypersonic objects position, but it also provides an efficient communication language between satellites. Small data packets containing coordinates, object heading, and time of acquisition can keep each satellite linked to each other with fast wireless transmissions. This is what we define as internal situational awareness, where each satellite of the system is aware of what itself and the other satellites are doing and observing in an iterative updated loop.

External situational awareness is also key for a hypersonic object tracking system. We define external as any information that is collected and utilized for tracking that is not sourced from the satellite constellation. Examples of external situational awareness for this system could include current weather and atmospheric conditions, as well as updated information regarding commercial aircraft traffic control systems. Weather and atmospheric information is key for hypersonic tracking since thermal and radar modalities are weather dependent. For example, in a cloudy overcast atmospheric setting, radar will most likely have a higher weighted contribution than thermal for detecting and tracking hypersonic objects. Additionally, access to updated air traffic control data can assist in reducing false positives for hypersonic tracking.

4   Iterative Sensor Operation in DDDAS System

We define a conceptual adaptive multimodal iterative sensor operation loop as part of a dynamic data driven application system (DDDAS) for hypersonic object detection and tracking. A constellation of low earth orbit satellites utilize active sensing radar and passive sensing thermal provides the capability to detect even the most elusive hypersonic targets, using boost-glide vehicles as the example object that is the most difficult to detect and track. We concisely summarize our envisioned DDDAS hypersonic object detection and tracking system in Fig. 3.

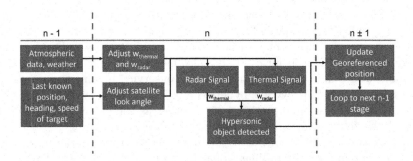

Fig. 3. Flow chart of a multimodal iterative sensor operation loop as part of a dynamic data driven application system (DDDAS) from the perspective of a single satellite among the constellation. In an iteratively looping operation, a satellite among the constellation receives initializing information (n-1 time step). Then in the (n) execution time step, we execute changes in the satellites attention by adjusting the satellites look angle towards a probable target, and adjust weights to its thermal and radar sensors based on contextual awareness.

We consider an example execution of the detection and tracking system from the perspective of a single satellite. In an iteratively looping operation, a satellite among the constellation receives initializing information such as 1) atmospheric and weather data, 2) air traffic data, and 3) last known position, heading, and speed of hypersonic target (if applicable). All that is considered initializing information is organized in the (n-1) time step in Fig. 3. Then in the (n) execution time step, the initializing information informs changes in the satellites attention by adjusting the satellites look angle towards a target of interest, and adjust weights to its thermal and radar sensors based on contextual awareness. A weighted decision from the fusion of both the thermal and radar data is used to decide if there is evidence to conclude that a hypersonic object is detected.

Fast communication between satellites leads to executive decisions based on incoming streams of processed data from the radar and thermal sensors on each satellite, that is then converted to a georeferenced position of a target hypersonic object. When a hypersonic object is detected by satellite(s), an alert signal is then broadcasted to other satellites in the network so that they can adjust their attitude and attention to the predicted location of the moving target. This process keeps repeating as necessary to maintain a track of a hypersonic target from launch (or atmospheric reentry) to landing.

5   Conclusion

We discussed the main physics, sensor, geometric, and communication limitations for a hypersonic object detection and tracking system. We used the flight maneuvering capabilities of a boost-glide hypersonic vehicle as our reference for the most difficult object to detect and track. These limitations lead to our decision that a contextually aware, multimodal (thermal and radar) low earth orbit constellation of satellites is the most probable system hardware framework for continuous hypersonic object detection and tracking. Iteratively updating the system with a stream of internal and external data in a DDDAS loop provides versatility to adapt, react, and execute hypersonic detection and tracking operations in various environmental conditions, which helps maintain a robust performance amongst the intricate network of satellites. Hypersonic object detection and tracking is an open problem that is large in scale and multidisciplinary, which makes DDDAS a great paradigm to tackle this problem. We are hopeful that some of the attributes of a hypersonic detection and tracking system provided in this paper will be implemented in the near future, enabling advancements in the commercial aircraft, space exploration, and defense sectors.

6   Disclaimer

This report was prepared as an account of work sponsored by an agency of the United States Government. Neither the United States Government nor any agency thereof, nor any of their employees, makes any warranty, express or implied, or assumes any legal liability or responsibility for the accuracy, completeness, or usefulness of any information, apparatus,product, or process disclosed, or represents that its use would not infringe privately owned rights. Reference herein to any specific commercial product, process, or service by trade name, trademark, manufacturer, or otherwise does not necessarily constitute or imply its endorsement, recommendation, or favoring by the United States Government or any agency thereof. The views and opinions of authors expressed herein do not necessarily state or reflect those of the United States Government or any agency thereof.

References

1. Ding, Y., et al.: An analysis of radar detection on a plasma sheath covered reentry target. IEEE Trans. Aerosp. Electron. Syst. **57**(6), 4255–4268 (2021). https://doi.org/10.1109/TAES.2021.3090910
2. Jiangting, L., et al.: Influence of hypersonic turbulence in plasma sheath on synthetic aperture radar imaging. IET Microwaves Antennas Propag. **11**, 2223–2227 (2017). https://doi.org/10.1049/iet-map.2017.0303
3. Korotkevich, A.O., et al.: Communication through plasma sheaths. J. Appl. Phys. **102** (2007). https://doi.org/10.1063/1.2794856
4. Li, J.Q., Hao, L., Li, J.: Theoretical modeling and numerical simulations of plasmas generated by shock waves. Sci. Chin. Technol. Sci. **62**(12), 2204–2212 (2019). https://doi.org/10.1007/s11431-018-9402-3
5. Mifsud, M., et al.: A case study on the aerodynamic heating of a hypersonic vehicle. Aeronaut. J. **116**, 873–893 (2012). https://doi.org/10.1017/S0001924000007338
6. Moran, J.H., et al.: Joule heated hot wall models in hypersonic wind tunnel facilities for energy deposition studies. Aerosp. Sci. Technol. **126**, 107627 (2022). https://doi.org/10.1016/j.ast.2022.107627
7. Of, I., Waves, T., Turbulent, W.A.: Interaction of transverse waves with a turbulent plasma 22 (1966)
8. Primer, A.T.: Remote radiation detection by electromagnetic air breakdown **4957**, 828–834 (2017)
9. Sha, Y.X., et al.: Analyses of electromagnetic properties of a hypersonic object with plasma sheath. IEEE Trans. Antennas Propag. **67**, 2470–2481 (2019). https://doi.org/10.1109/TAP.2019.2891462
10. Shang, J.J., Yan, H.: High-enthalpy hypersonic flows. In: Advances in Aerodynamics 2 (2020). https://doi.org/10.1186/s42774-020-00041-y
11. Shaw, J.A., Nugent, P.W.: Physics principles in radiometric infrared imaging of clouds in the atmosphere. Eur. J. Phys. **34** (2013). https://doi.org/10.1088/0143-0807/34/6/S111
12. Shi, L., et al.: Unconventionally designed tracking loop adaptable to plasma sheath channel for hypersonic vehicles. Sensors (Switzerland) **21**, 1–15 (2021). https://doi.org/10.3390/s21010021
13. Smith, F.G.: Atmospheric Propagation of Radiation Approved for Public Release Atmospheric Propagation of Radiation The Infrared and Electro-Optical, vol. 2
14. Song, L., et al.: Effect of time-varying plasma sheath on hypersonic vehicle-borne radar target detection. IEEE Sens. J. **21**, 16880–16893 (2021). https://doi.org/10.1109/JSEN.2021.3077727
15. Vayner, B.V., Ferguson, D.C.: Electromagnetic environment radiation around in the plasma the shuttle
16. Zank, G.P., et al.: The interaction of turbulence with shock waves. AIP Conf. Proc. **1216**, 563–567 (2010). https://doi.org/10.1063/1.3395927

Dynamic Airspace Control via Spatial Network Morphing

David Sacharny$^{(\boxtimes)}$ ⓘ, Thomas Henderson ⓘ, and Nicola Wernecke

The University of Utah, Salt Lake City, UT 84112, USA
{sacharny,tch}@cs.utah.edu, u1369785@utah.edu
http://www.cs.utah.edu/ tch

Abstract. In the coming years, a plethora of new and autonomous air-craft will fill the airspace, approaching a density similar to the ground traffic below. At the same time, human pilots that use the prevailing navigational tools and decision processes will continue to fly along flexible trajectories, now contending with inflexible non-human agents. As the density of the airspace increases, the number of potential conflicts also rises, leading to a possibly disastrous cascade effect that can fail even the most advanced tactical see-and-avoid algorithms. Any engineered solution that maintains safety in the airspace must satisfy both the computational requirements for effective airspace management as well as the political issue that human-pilots should maintain priority in the airspace. To this end, the research presented here expands on a concept of air traffic management called the *Lane-Based Approach* and describes a method for morphing the underlying spatial network to effectively deal with multiple potential conflicts. The spatial-network, which represents a model of the airspace occupied by autonomous aircraft, is mutated with respect to extrapolated human-piloted trajectories, leading to a real-world execution that modifies the trajectories of multiple vehicles at once. This reduces the number of pairwise deconfliction operations that must occur to maintain safe separation and reduces the possibility of a cascade effect. An experiment using real Automatic Dependent Surveillance-Broadcast (ADS-B) data, representing human-piloted aircraft trajectories, and simulated autonomous aircraft will demonstrate the proposed method.

Keywords: UAS Traffic Management · Tactical Deconfliction · DDDAS · Lane-Based Approach

1 Introduction

The problem of tactically deconflicting multiple flights and automating air traffic management fundamentally requires a dynamic data driven application system (DDDAS). Human air traffic controllers must fuse multiple data sources into a running model of the airspace before making critical decisions. What the Federal Aviation Administration's air traffic controller manual often refers to as "common sense" is actually a complex feedback loop involving multiple cyber-physical systems and sensors. This paper presents the case for this in terms

© The Author(s), under exclusive license to Springer Nature Switzerland AG 2024
E. Blasch et al. (Eds.): DDDAS 2022, LNCS 13984, pp. 35–43, 2024.
https://doi.org/10.1007/978-3-031-52670-1_4

of the fundamental problem of motion planning for multiple agents, which can quickly become intractable for both humans and machines. The experimental section will demonstrate how the proposed DDDAS system can cope with at least hundreds of conflicts, whereas the prevailing commercial tactical collision avoidance system has sparse evidence that it can deal with more than seven.

2   Background

The airspace represents an environment where multiple agents execute trajectories according to their mission goals and constraints. It is a dynamic and uncertain environment: with respect to any individual agent, other agents are essentially moving obstacles, and the perceived state-space is subject to a host of possible errors. Planning conflict-free trajectories for agents, represented as point objects (i.e., in configuration space) with constraints on its velocity, is provably NP-hard even when the moving obstacles are convex polygons with constant linear velocity [4,8]. By including uncertainty in controls and perception, the motion-planning problem becomes non-deterministic exponential time hard [4]. In practical terms, the problem of generating a conflict-free trajectory has a worst-case time complexity that is at least exponential in the degrees-of-freedom (the union of all agent's degrees-of-freedom), and can easily become intractable for both humans and machines. For tactical deconfliction, generating new trajectories in response to unplanned conflicts during the execution of a mission, dynamic constraints add to an already non-holonomic system and increase the base of the exponential in the worst-case time complexity. The prevailing method for dealing with this complexity is by limiting the number of aircraft that have the potential for conflicts in a given area, either by requiring minimum separation of aircraft or maximum sector capacities [1]. Both methods functionally reduce the density of vehicles in the airspace and therefore reduce conflict probabilities. As the density of the airspace increases, the number of potential conflicts also increases and could lead to a cascading effect where the conflict resolution procedure produces further conflicts [6]. Other methods for reducing this complexity involve decomposing the problem into sub-problems that can then be distributed among agents, for example sequencing aircraft into a single corridor for landing and launching. This reduces the degrees of freedom that both air traffic controllers and pilots must consider individually, however without a complete structuring of, and sequencing throughout the airspace, cascading conflicts are still possible.

   The National Aeronautics and Space Administration (NASA), the Federal Aviation Administration (FAA), and the International Civil Aviation Organization (ICAO) have all advocated for the concept of *Strategic Deconfliction* for the next generation of air traffic control. The new paradigm assumes the density of the airspace will continue to increase, and with the introduction of Advanced Air Mobility (AAM) and autonomous vehicles of all sizes, the density could eventually rival that of ground-traffic. Among the strategies encompassed by strategic deconfliction, two methods include structuring the airspace, and/or requiring that all planned trajectories are deconflicted prior to executing a mission. The latter strategy removes or reduces the dynamic constraints associated with tactical deconfliction, however it does not guarantee that cascading conflicts are impossible because contending operations must still resolve conflicts in a virtual airspace. In previous works, the authors have described a *Lane-Based Approach* for an unmanned aircraft systems (UAS) traffic management system (UTM), whereby the airspace is structured as a spatial network, coupled with a simple and efficient deconfliction algorithm [9] (see [10] for a comparison of this approach with that currently proposed by the FAA and NASA). However, there will likely always be aircraft (human-piloted or otherwise) not subject to this kind of standardization, and therefore, tactical scenarios must still be considered.

The prevailing tactical deconfliction system required on all large passenger and cargo aircraft worldwide is the Traffic Alert and Collision Avoidance System (TCAS), and its forthcoming successor is the Airborne Collision Avoidance System X (ACAS X). TCAS receives telemetry from nearby aircraft once-per-second, tracking their slant range, altitude, and bearing to calculate the closest point of approach (CPA) and issue alerts to the pilot should they become a threat. Resolution advisories (RA) from TCAS recommend altitude adjustments to the flight trajectory, increasing the vertical separation between aircraft while minimizing the effect on the current trajectory [2]. Data describing scenarios where more than seven aircraft are in conflict at a time is sparse for (TCAS) [3], stemming from the fact that the airspace density has remained relatively low. Additionally, TCAS was designed for human-piloted aircraft that are assumed to be capable of achieving climb and descend rates of 1500ft/min and 2500ft/min respectively [7]. While TCAS has been considered for UAS given it's long track record (in particular, for the Global Hawk UAS [2]), the inflexibility of it's algorithms to adapt to aircraft with more limited climb and descend rates, or provide heading and airspeed guidance, limit it's applicability. ACAS X, on the other hand, is designed to be more flexible with support for global positioning system (GPS) data, small aircraft, and new sensor modalities [7]. ACAS X applies dynamic programming to a model of the airspace and collision problem formulated as a partially observable Markov decision process (POMDP) [7]. However, this formulation may still suffer from combinatorial explosions of possible trajectory states if multiple aircraft are involved. Many methods for planning multiple conflict-free trajectories exist, for example the two-phase decoupled approach [12] which first computes a path for each robot individually, then applies operations to the resulting path set to avoid collisions. The advantage of this approach is that the "search space explored by the decoupled planner has lower dimensionality than the joint configuration space explored by the centralized planner" [12]. However, the cascade effect is still prevalent since the local planners do not consider the global configuration of vehicles. Structuring the airspace can help mitigate this potential by reducing the dimensionality of the configuration space. In other words, agents can make assumptions about the probable trajectories of all aircraft.

3   Structured Airspace and Aircraft Trajectories

The Lane-Based approach reduces the dimensionality of the motion planning problem for individual agents by predefining all the possible trajectories through the airspace. It is then the agent's responsibility to select the "lanes" it must traverse to accomplish its mission, and reserve a time/space slot in those lanes. The lane vertexes represent waypoints, or control points, for a vehicle's actual trajectory. The vertical and lateral separation constraints between aircraft are managed by the lane network, while the longitudinal separation between aircraft (termed *headway*) is the responsibility of the individual agents. The cascade potential is reduced because a tactical resolution procedure need only consider the predefined paths of the lane system (effectively encapsulating multiple agents as a single "lane system") and perhaps a few aircraft that are unable or unwilling to follow the structure. The encapsulation is intended to enable system designers and regulators to control which aircraft are allowed to fly where, given their instrumentation and performance capabilities. To model the interaction between aircraft both inside and outside the structured airspace, we consider a generalized trajectory model formed from the interpolation of waypoints and time-of-arrivals (TOAs). The trajectories within the structured airspace are encapsulated by the lane-system, and can be treated as a single object with a reduced number of degrees-of-freedom. As conflicts arise between aircraft inside and outside of the lane-network, the conflict resolution procedure can effectively control multiple aircraft without the combinatorial explosion or cascading conflicts that would result in the same system without any airspace structure.

The generalized trajectory model considered here is a shape-preserving piecewise cubic interpolation of waypoints and time-of-arrivals. This type of interpolation uses a polynomial $P(x)$ with the following properties (from [13]):

- Each subinterval, $x_k \leq x \leq x_{k+1}$, is a cubic Hermite interpolating polynomial for the given waypoints and specified derivatives at the interpolation points.
- $P(x)$ interpolates the waypoints y such that $P(x_j) = y_j$, and the first derivative $\frac{dP}{dx}$ is continuous. The derivative may not be continuous at the control points.
- The slopes at the x_j are chosen in such a way that $P(x)$ preserves the shape of the data and respects monotonicity.

Figure 1 shows an example of this type of interpolation for a single aircraft traversing a lane system. This trajectory model was selected because there are readily available software implementations (e.g., the *waypointTrajectory* object in MATLAB) and its behavior under a range of constraints is appealing. Additionally, the homotopy equivalence class containing these trajectories enables many different constructions, for example, using wavefront and optimal control algorithms (see a survey of mathematical models for aircraft trajectories in [5]).

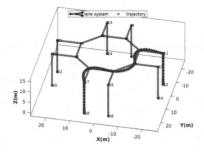

Fig. 1. Example trajectory interpolation performed over lane system vertexes.

4   A DDDAS Approach to Multiple Aircraft Deconfliction

A challenge for the automated air traffic management system proposed here is to efficiently address tactical conflicts (those that occur during the execution of missions) between aircraft using the structured lane-based airspace and those that are not. Tactical conflict resolution between aircraft within the lane system is described in [11]. To handle tactical conflicts, the proposed system vertically displaces the lane vertexes, analogous to the human-pilots vertical displacements initiated by TCAS advisories. To determine the required vertical displacement and climb/descend rates, sensor data (e.g., telemetry or radar) that describes the state of the airspace is fed into a simulated model of the airspace. The simulation has the capability of testing multiple scenarios before deciding on a control for the vertical displacement of the structured airspace. A diagram for the overarching dynamic data-driven application system (DDDAS) is shown in Fig. 2.

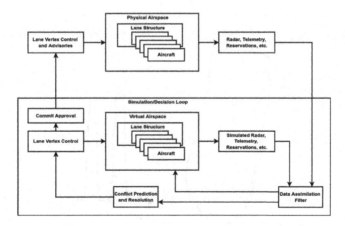

Fig. 2. Dynamic data driven application system with a feedback loop for lane vertex control.

5   Experiments

The foundation of the proposed DDDAS system is computational; many publications exploring the automation of air traffic management neglect the computational complexity of the motion planning problem and often begin with modeling vehicles or software architectures. This research instead begins with the problem of intractability and builds upon it an architecture that supports efficient deconfliction. However, the physical limitations of the agents are important to consider and therefore the experiment described here is designed to explore how those constraints can be considered within the framework proposed. To this end, a dataset was collected using an Automatic Dependent Surveillance-Broadcast (ADS-B) receiver placed atop one of the authors roofs in Salt Lake City, Utah, USA.

The particular implementation chosen here assumes a discrete proportional controller for the lane vertexes with a single parameter K and a sample rate of one second. A low-flying trajectory from the ADSB data was chosen as an example non-structured airspace flight for generating the conflict, shown in Fig. 3 and in relation to the experimental lane system in Fig. 4. Within the lane system 300 aircraft were scheduled using lane-based strategic deconfliction. Once the vehicles have begun executing their trajectories, they must traverse a spatial network that is morphing to direct traffic away from the threat posed by the intruder aircraft. The ADSB telemetry is used to create a shape-preserving piecewise cubic interpolation, and an estimate of the closest point of approach is used to inform a vertical displacement control input. Figure 5 shows a plot of the calculated closest-point-of-approach between the initial lane system (only a single node is chosen in this case) and the ADSB trajectory interpolation. The vehicles within the lane system recalculate their trajectories at each time-step when any of their trajectory waypoints is updated by the morphing process.

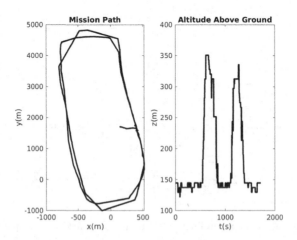

Fig. 3. Selected ADSB trajectory, translated to experiment position.

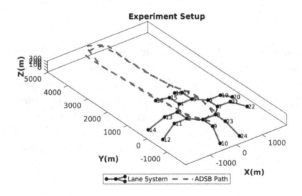

Fig. 4. Experiment setup showing conflict between lane system and ADSB path.

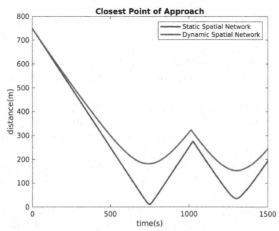

Fig. 5. Closest point of approach between lane system and ADSB trajectory (K=0.01).

These tactical updates result in changes to the originally planned climb/descent rates, which is a critical parameter to consider before applying the control to the physical system. Table 1 shows the maximum climb and minimum descent rate for each simulation trial.

Table 1. Climb rates for tested proportional control constants.

K	Max Climb Rate (ft/min)	Min Descend Rate (ft/min)
no control	374.253	−374.914
0.0001	374.253	−99.1532
0.001	374.253	−157.1
0.004	374.253	−216.575
0.01	374.253	−114.992
0.1	374.253	−88.4352

6 Conclusion

The computational complexity of coordinating multiple vehicles is one of the major reasons why air traffic control has yet to be automated. The proposed DDDAS system presented here provides a standardized way to resolve multiple conflicts by morphing the underlying spatial network of a lane-based system, and by simulating controls prior to execution in a physical system. Future research should continue to build on the structured airspace approach, adding more complexity to the physical models of aircraft and validating decisions before applying them.

References

1. Bertsimas, D., Patterson, S.S.: The air traffic flow management problem with enroute capacities. Oper. Res. **46**(3), 406–422 (1998). https://doi.org/10.1287/opre.46.3.406
2. Billingsley, T.B.: Safety Analysis of TCAS on Global Hawk using Airspace Encounter Models. Massachusetts Institute of Technology, Cambridge, MA, USA (2006)
3. Billingsley, T.B., Espindle, L.P., Griffith, J.D.: TCAS multiple threat encounter analysis. Tech. Rep. ATC-359, MIT Lincoln Laboratory (2009)
4. Canny, J.F.: The Complexity of Robot Motion Planning. The MIT Press, Cambridge, MA, USA (1988)
5. Delahaye, D., Puechmorel, S., Tsiotras, P., Feron, E.: Mathematical models for aircraft trajectory design: a survey. In: ENRI International Workshop ATM/CNS, Tokyo, Japan (2013)
6. Jardin, M.R.: Analytical relationships between conflict counts and air-traffic density. J. Guid. Control. Dyn. **28**(6), 1150–1156 (2005). https://doi.org/10.2514/1.12758
7. Kochenderfer, M.J., et al.: Decision Making Under Uncertainty: Theory and Application, 1st edn. The MIT Press, Cambridge (2015)
8. Reif, J., Sharir, M.: Motion planning in the presence of moving obstacles. J. ACM **41**(4), 764–790 (1994)
9. Sacharny, D., Henderson, T.: A lane-based approach for large-scale strategic conflict management for UAS service suppliers. In: International Conference on Unmanned Aircraft Systems, Atlanta, GA, 11–14 Jun 2019, pp. 937–945 (2019). https://doi.org/10.1109/icuas.2019.8798157
10. Sacharny, D., Henderson, T., Cline, M., Russon, B., Guo, E.: FAA-NASA vs. lane-based strategic deconfliction. In: IEEE International Conference on Multisensor Fusion and Integration for Intelligent Systems, Karlsruhe, Germany, 14–16 Sep 2020, pp. 13–18 (2020)
11. Sacharny, D., Henderson, T.C., Guo, E.: A DDDAS protocol for real-time large-scale UAS flight coordination. In: Darema, F., Blasch, E., Ravela, S., Aved, A. (eds.) DDDAS 2020. LNCS, vol. 12312, pp. 49–56. Springer, Cham (2020). https://doi.org/10.1007/978-3-030-61725-7_8

12. Saha, M., Isto, P.: Multi-robot motion planning by incremental coordination. In: IEEE International Conference on Intelligent Robots and Systems, Beijing, China, Oct 9–15 2006, pp. 5960–5963 (2006). https://doi.org/10.1109/IROS.2006.282536
13. The Mathworks, Inc.: Piecewise cubic hermite interpolating polynomial (PCHIP) (2022). https://www.mathworks.com/help/matlab/ref/pchip.html. Accessed 21 Jul 2022

On Formal Verification of Data-Driven Flight Awareness: Leveraging the Cramér-Rao Lower Bound of Stochastic Functional Time Series Models

Peiyuan Zhou[✉], Saswata Paul, Airin Dutta, Carlos Varela, and Fotis Kopsaftopoulos

Rensselaer Polytechnic Institute, Troy, NY 12180, USA
{zhoup2,pauls4,duttaa5,varelc,kopsaf}@rpi.edu

Abstract. This work investigates the application of the Cramér-Rao Lower Bound (CRLB) theorem, within the framework of Dynamic Data Driven Applications Systems (DDDAS), in view of the formal verificationof state estimates via stochastic Vector-dependent Functionally Pooled Auto-Regressive (VFP-AR) models. The VFP-AR model is identified via data obtained from wind tunnel experiments on a "fly-by-feel" wing structure under multiple flight states (i.e., angle of attack, velocity). The VFP-based CRLB of the state estimates is derived for each true flight state reflecting the state estimation capability of the model considering the data, model, and estimation assumptions. Apart from the CRLB obtained from pristine data and models, CRLBs are estimated using either artificially corrupted testing data and/or sub-optimal models. Comparisons are made between CRLB and state estimations from corrupted and pristine conditions. The verification of the obtained state estimates is mechanically verified the formal proof of the CRLB Theorem using Athena, which provides irrefutable guarantee of soundness as long as specified assumptions are followed. The results of the study indicate the potential of using a CRLB-based formal verification framework for state estimation via stochastic FP time series models.

Keywords: stochastic models · firmal verification · Cramér-Rao lower bound · state awareness

1 Introduction

Future intelligent aerial vehicles will be able to "feel," "think," and "react" in real time based on high-resolution ubiquitous sensing leading to autonomous operation based on unprecedented self-awareness and self-diagnostic capabilities. This concept falls within the core of *Dynamic Data-Driven Application Systems (DDDAS)* concept as they have to dynamically incorporate real-time

© The Author(s), under exclusive license to Springer Nature Switzerland AG 2024
E. Blasch et al. (Eds.): DDDAS 2022, LNCS 13984, pp. 44–52, 2024.
https://doi.org/10.1007/978-3-031-52670-1_5

data into the modeling, learning, and decision making application phases, and in reverse, steer the data measurement process based on the system's dynamic data integration and interpretation. [3,4,7].

In this study, state awareness is tackled within a data-driven framework based on functionally pooled stochastic time series models [8]. The main contribution of this work involves the postulation of a formal verification scheme for stochastic state awareness leveraging the derivation and mathematical proof of the Cramér-Rao lower bound (CRLB) for the proposed state estimators. The proposed method is experimentally assessed on a prototype self-sensing wing structure subjected to a series of wind tunnel experiments under multiple flight states –defined by a pair of angle of attack (AoA) and airspeed [8]. Initially, parametric vector-dependent functionally pooled auto-regressive (VFP-AR) models are used to represent the dynamics of the wing as it undergoes different flight states. Model parameter estimation is based on a weighted least squares (WLS) approach. In addition, the CRLB for both the WLS-based model parameter estimator as well as the inverse state estimation is derived and introduced as means to reflect on the quality of the data and models. The CRLB theorem is formally proved (verified) within the Athena language. For prior work by the authors the reader is referred to [4,5,13]. The main idea outlined and assessed in this work is that we can have a prior knowledge of the VFP-AR-based state estimation effectiveness and quantify the uncertainty via the use of the CRLB proof accounting for the data quality, as well as the model identification and parameter estimation characteristics. It is also demonstrated how the state estimation correctness can be monitored via a CRLB-based runtime scheme that is checked experimentally via wind tunnel data and artificially introduced data and model abnormalities.

2 Functionally Pooled Time Series Models

In this study, response-only AutoRegressive (AR) models are employed to represent the wing dynamics under varying flight states via vibration data. Each flight state is defined by a pair of airspeed and angle of attack values. The VFP representation uses data pooling techniques to identify the structural dynamics captured under multiple states by AR models to be treated as one entity in model identification. Furthermore, the stochasticity in the data set is characterized via a residual covariance matrix and related to a flight state vector via estimated functional dependencies [6,9]. The general form of VFP-AR$(na)_p$ model is given by [9]:

$$y_k[t] = \sum_{i=1}^{na} a_i(\boldsymbol{k}) \cdot y_k[t-i] + e_k[t] \qquad a_i(\boldsymbol{k}) = \sum_{j=1}^{p} a_{i,j} G_j(\boldsymbol{k}) \qquad (1)$$

with t designating the normalized discrete time ($t = 1, 2, 3, \ldots$ with absolute time being $(t-1)T_s$, where T_s stands for the sampling period), na designating the AR order, p the number of function basis. $y_k[t]$ designates the data under various states specified by state vector $\boldsymbol{k} = [k_1, \ k_2, \ldots, k_n]$. $e_k[t]$ is the residual (one-step-ahead prediction error) sequence of the model, which is assumed white (serially uncorrelated) zero mean with variance $\sigma_e^2(\boldsymbol{k})$. $G_j(\boldsymbol{k})$ is the function basis (e.g., Chebyshev, Legendre, Jacobi and etc.), where the model parameters $a_i(\boldsymbol{k})$ are modeled as explicit functions of the state vector $\boldsymbol{k}$. Model parameter estimation is achieved within a linear regression framework and weighted least squares (WLS) estimators [8,9].

2.1   CRLB Formulation for VFP-AR Models

The VFP-AR model is identified via the process introduced in [9]. The regression form of equations (1) is obtained by expressing Eq. (1) in terms of the parameter vector ($\boldsymbol{\theta} = [\ a_{1,1}\ a_{1,2}\ \dots\ a_{i,j}\ \vdots\ \sigma_e^2(\boldsymbol{k})\ ]^T$) to be estimated via available data via WLS [8,9]. The flight state is parametrized via the operating state vector *boldsymbolk* to be estimated via the use of signals corresponding to an unknown state and re-parametrization of the identified VFP-AR model. Next, the CRLB of the estimated operating state vector $\widehat{\boldsymbol{k}}$ is obtained. The state estimator $\widehat{\boldsymbol{k}}$ can be expressed as:

$$\widehat{\boldsymbol{k}} = \arg \min_{\boldsymbol{k} \in \mathbf{R}^m} \sum_{i=1}^{N} e_u^T[t, \boldsymbol{k}] e_u[t, \boldsymbol{k}]\ , \quad \sigma_u^2(\widehat{\boldsymbol{k}}) = \frac{1}{N} \sum_{t=1}^{N} e_u[t, \widehat{\boldsymbol{k}}] e_u^T[t, \widehat{\boldsymbol{k}}] \qquad (2)$$

where e_u is the model residual sequence and $\sigma_u^2(\widehat{\boldsymbol{k}})$ is the corresponding model residual variance for the estimated state. $\widehat{\boldsymbol{k}}$ can be shown to be asymptotically ($N \to \infty$) Gaussian distributed with $\widehat{\boldsymbol{k}} \sim \mathcal{N}(\boldsymbol{k}, \boldsymbol{\Sigma}_k)$ [6]. The log-likelihood function of $\widehat{\boldsymbol{k}}$ based on N samples of an unknown signal and is given by:

$$\ln \mathcal{L}(\widehat{\boldsymbol{k}}, \sigma_u^2(\widehat{\boldsymbol{k}})) = -\frac{N}{2} ln(2\pi) - \frac{N}{2} ln(\sigma_u^2) - \frac{1}{2} \sum_{t=1}^{N} \frac{e_u^T(\widehat{\boldsymbol{k}}), t) e_u(\widehat{\boldsymbol{k}}), t)}{\sigma_u^2(\widehat{\boldsymbol{k}})} \qquad (3)$$

Then, the Cramér Rao Lower bound for $\boldsymbol{\Sigma}_k$ is given by:

$$\boldsymbol{\Sigma}_{CRLB} = \left[\mathbf{E} \Big[\big(\frac{\delta \ln \mathcal{L}(\boldsymbol{k}, \sigma_u^2)}{\delta \boldsymbol{k}} \big) \big(\frac{\delta \ln \mathcal{L}(\boldsymbol{k}, \sigma_u^2)}{\delta \boldsymbol{k}} \big)^T \Big] \right]^{-1}. \qquad (4)$$

3   Machine-Checked Proof of the CRLB Theorem

Data corruption caused by imprecision in avionics engineering or hostile cyber activities may affect the decision chain of DDDAS systems and can lead to catastrophic errors in safety-critical aerospace systems. For this reason, it is necessary to rigorously verify all aspects of such systems to ensure that they will behave correctly within some acceptable bounds. Informal proofs of mathematical theorems can have errors [10], but formal methods allow the use of precise logical and mathematical techniques for the rigorous verification of such proofs. This is important in avionics engineering where even subtle imprecision can be

```
define CRLB-Theorem :=
(forall theta X .
  (not
    (=
      (Random.expected (Random.pow
                              (Derivative.parDifLog (Logarithm.ln (Vector.
      jointDen X theta)) theta) 2)) 0.0 ))
  ==>
  (((RealExt.pow
      (Derivative.parDif
        (Random.expected
          (Estimator.estOut (Estimator.consEst (Model.consMod theta X)))) theta)
      2)
      /
      (Random.expected (Random.pow
                              (Derivative.parDifLog (Logarithm.ln (Vector.
      jointDen X theta)) theta) 2)))
    <=
    (Random.var (Estimator.estOut (Estimator.consEst (Model.consMod theta X)))))))
conclude CRLB-Theorem
pick-any theta
pick-any X
assume (not (= (Random.expected (Random.pow (Derivative.parDifLog (Logarithm.ln (
    Vector.jointDen X theta)) theta) 2)) 0.0 ))
let{
  W := (Estimator.estOut (Estimator.consEst (Model.consMod theta X)));
  Y := (Derivative.parDifLog (Logarithm.ln (Vector.jointDen X theta)) theta);
  V_Y_not_zero := (not (= (Random.expected (Random.pow (Derivative.parDifLog (
    Logarithm.ln (Vector.jointDen X theta)) theta) 2) ) 0.0 ) );
  conn-2-Covariance-Inequality-THEOREM := (!uspec (!uspec
    Covariance-Inequality-THEOREM theta) X);
  cov-inequality-expression := (!mp conn-2-Covariance-Inequality-THEOREM
    V_Y_not_zero);
  r1 := (RealExt.pow  (Derivative.parDif (Random.expected (Estimator.estOut (
    Estimator.consEst (Model.consMod theta X)))) theta) 2);
  r2 := (Random.var (Estimator.estOut (Estimator.consEst (Model.consMod theta X))));
  r3 := (Random.expected (Random.pow (Derivative.parDifLog (Logarithm.ln (Vector.
    jointDen X theta)) theta) 2) );
  conn-2-swap-sides-axiom := (!uspec (!uspec (!uspec RealExt.swap-sides-axiom r1) r2
    ) r3);
  cov<=VXVY-implies-cov_by_VY=<=VX := (!mp conn-2-swap-sides-axiom V_Y_not_zero)
}
(!mp cov<=VXVY-implies-cov_by_VY=<=VX cov-inequality-expression)
```

Fig. 1. Specification and proof of the CRLB Theorem in Athena.

catastrophic. A mechanically verified formal proof can provide an irrefutable guarantee that the property in question will be true as long as the assumptions made during proof development hold.

Athena [2] is an *interactive proof assistant* that can be used to develop and mechanically verify formal proofs of properties. It is based on *many-sorted first order logic* [11] and uses *natural deduction* [1] style proofs, which is an intuitive way of reasoning. Athena also provides a *soundness guarantee* that any proven theorem will be a logical consequence of sentences in Athena's *assumption base*, which is a set of sentences that have either been proven or asserted to be true. We have developed a mechanically verified proof (Fig. 1) of the Cramér-Rao Lower Bound Theorem (`CRLB-THEOREM`) using Athena. Mathematically, `CRLB-THEOREM` states that:

$$\left(\mathbb{E}\left[\left(\frac{\partial}{\partial \theta} \log f(X;\theta) \right)^2 \right] \neq 0.0 \right) \implies \left(\mathbb{V}[T(X)] \geq \frac{\left(\frac{\partial}{\partial \theta} \mathbb{E}[T(X)] \right)^2}{\mathbb{E}\left[\left(\frac{\partial}{\partial \theta} \log f(X;\theta) \right)^2 \right]} \right)$$

where $X = (X_1, X_2, ... X_N) \in \mathbb{R}^N$ is a random vector with joint density $f(X;\theta)$, $\theta \in \Theta \subseteq \mathbb{R}$, and $T(X)$ is a biased estimator of θ.

Formally specifying and proving the correctness of complex mathematical statements, such as `CRLB-THEOREM`, in interactive proof assistants like Athena requires access to formal constructs that are sufficiently expressive to correctly specify such statements. Furthermore, for formal reasoning about such high-level statistical properties, it is also necessary to have access to formal definitions from across mathematics, such as algebraic theory, linear algebra, measure theory, and statistics, that can support the proofs of the properties. Developing such formal constructs and definitions in a machine-readable language is a challenging task since it requires domain knowledge of all aspects of the systems that need to be specified, knowledge of formal logic and reasoning techniques, and proficiency in the machine-readable language. For this reason, we have developed an open-source proof library in Athena[1] that can be used for reasoning about mathematical properties of data-driven aerospace systems [12,14]. Our specification and proof of the CRLB theorem (Fig. 1) rely on reusable formal constructs from our Athena library that are necessary for the proof to be mechanically verified. For example, we have created the Athena datatype `Estimator` to represent the domain of all estimators and a relation `biasedEstimator` over the class of all estimators to denote if an estimator is a biased estimator or not. Similarly, we have also created a datatype `Model` to denote the domain of all models. Our Athena statement of CRLB shown in Fig. 1 uses these constructs along with other constructs in our Athena library that we have developed for expressing concepts like expected value (`Random.expected`) and variance (`Random.var`) of random variables.

---

[1] The library can be found at https://wcl.cs.rpi.edu/assure.

```
datatype Estimator := (consEst Model.Model) #consEst(X ~ P_theta) -> theta_hat(X)

declare biasedEstimator : [Estimator Real] -> Boolean

declare estOut : [Estimator] -> Random.RandVar

datatype Model := (consMod Real Vector.VectRandom)
```

Fig. 2. Formal Athena constructs we have developed for the CRLB proof.

4   Experimental Results and Discussion

Structural vibration response data is obtained through a series of wind-tunnel experiments. A "fly-by-feel" capable modular wing (NACA 4412) outfitted with accelerometers and strain gauges is tested under multiple flights states characterized by airspeed ($8 \sim 20$ m/s with 2 m/s increment) and angle of attack ($1 \sim 15$ deg with 2 deg increment). Acceleration and strain signals from PCB ICP sensors are recorded at a sampling frequency of 512 Hz for 128 s ($N = 65536$). The flight state vector $\boldsymbol{k}$ is formed as $\boldsymbol{k} = [k_1, \ k_2]$, where k_1 designates the airspeed value and k_2 the angle of attack. A data segment of $N = 2048$ samples from each flight state is used for the VFP-AR model identification. The testing data set consists of non-overlapping, and different than the training, signal segments of $N = 2048$ samples collected from the same flight states. Simulated data corruption is applied to violate the data normality assumption emphasized by the machine-checked CRLB theorem. "Notching" is also created by uniformly replacing 10% of data samples with zeros, thus replicating sensor faults.

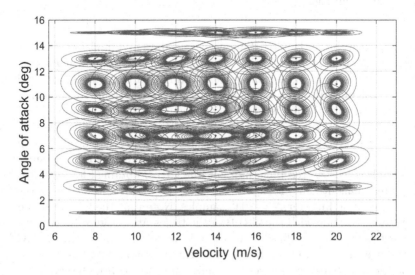

Fig. 3. CRLB 99% confidence interval ellipsoids obtained for the true flight state via the VFP-AR(20)$_{19}$ model for $N = 500$ (blue) to $N = 10000$ (red) samples.

The flight state estimation CRLB based on the selected VFP-AR model is obtained for increasing data length, from $N = 500$ to $N = 10000$ samples with 500 samples increment. Figure 3 presents the CRLB for the considered data sets as 2-D ellipsoids. Notice the state estimation bivariate lower uncertainty bounds as a function of increasing data size.

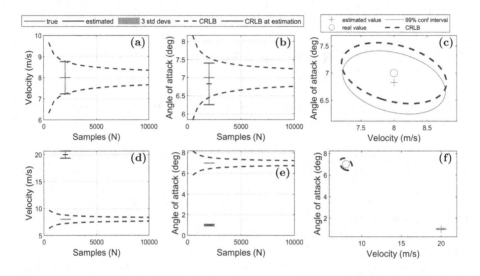

Fig. 4. Indicative CRLB estimation results comparing (a)-(c) pristine data and (d)-(f) corrupted via "notching" data for a velocity of 10 m/s and AoA of 1 deg. (c), (f) present the CRLB 99% ellipsoids.

Aiming to evaluate the CRLB under various data and model conditions, four different simulated data corruption cases and one sub-optimal model are used for flight state estimation, and the corresponding CRLB at the estimated flights states are calculated. The CRLB obtained from the corrupted conditions are then compared with the model residual covariance and CRLB obtained from the true flight states. Figure 4(a)(b) shows the CRLB baseline using pristine data and optimal model VFP-AR(20)$_{19}$ at flight states $AoA = 1$ deg, $V = 10$ m/s and $AoA = 7$ deg, $V = 8$ m/s. The CRLB convergence for increasing samples N is observed (dotted blue line) and the CRLB obtained for the estimated flight states (magenta line) is closely matching the CRLB obtained from the true state (dotted blue line) for $N = 2048$ samples. Note that the CRLB at the estimated states coincide with the CRLB of the true states. Figure 4(c) shows good agreement between the ellipsoids of the estimated (magenta) and true CRLB states (dotted blue line). For the data-corrupted cases, "notching" the signals provides poor state estimation results, shown in Fig. 4(d)-(e). Observe that both the estimated states and their CRLB do not coincide with the corresponding ones obtained under the true state.

A flight state runtime monitoring scheme is implemented and tested by introducing a "notching" corruption starting at $t = 26$ s for $\{AoA = 13\,\text{deg}, V = 14\ m/s\}$. The test data window ($N = 2000$ samples) starts from $t = 0\ s$ and advances with a $N = 1000$ increment. Figure 5 presents the flight state estimation results; observe the deviation of the CRLB from the true state when testing data begins to include corrupted data at $t = 26$ s.

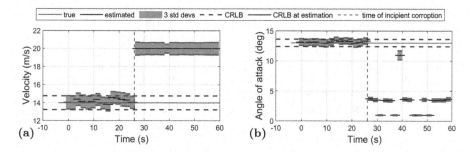

Fig. 5. Indicative CRLB runtime monitoring results for data corrupted at $t = 26$ s for a flight state corresponding to $14\,\text{m/s}$ velocity and AoA of 13 deg.

5   Conclusions

This work presented the introduction and application of a formally verified flight state estimation framework based on the Cramér-Rao lower bound theorem pertaining to stochastic VFP-AR time series models. The experimental assessment was presented via data collected from a wing structure under multiple flight states in wind tunnel experiments. The VFP-AR model enables flight state awareness and provides accurate and formally verified state estimation. The comparison between the obtained CRLB for pristine and corrupted data sets shows that the violation of data normality and model assumptions asserted by the formally verified CRLB theorem causes errors in the state estimation that are reflected on the obtained CRLB. The CRLB-based formal verification framework has great potential for verified state estimation via stochastic time series models.

Acknowledgment. This work is supported by the U.S. Air Force Office of Scientific Research (AFOSR) grant "Formal Verification of Stochastic State Awareness for Dynamic Data-Driven Intelligent Aerospace Systems" (FA9550-19-1-0054) with Program Officer Dr. Erik Blasch.

References

1. Arkoudas, K.: Simplifying proofs in fitch-style natural deduction systems. J. Autom. Reason. **34**(3), 239–294 (2005). https://doi.org/10.1007/s10817-005-9000-3
2. Arkoudas, K., Musser, D.: Fundamental Proof Methods in Computer Science: A Computer-Based Approach. MIT Press, Cambridge, MA (2017). https://doi.org/10.1017/s1471068420000071
3. Blasch, E., Ashdown, J., Kopsaftopoulos, F., Varela, C., Newkirk, R.: Dynamic data driven analytics for multi-domain environments. In: Artificial Intelligence and Machine Learning for Multi-Domain Operations Applications, vol. 11006, p. 1100604. International Society for Optics and Photonics (2019)
4. Breese, S., Kopsaftopoulos, F., Varela, C.: Towards proving runtime properties of data-driven systems using safety envelopes. In: Proceedings of the 12th International Workshop on Structural Health Monitoring (IWSHM 2019), pp. 1748–1757. Palo Alto, CA, USA (2019). https://doi.org/10.12783/shm2019/32302
5. Cruz-Camacho, E., Amer, A., Kopsaftopoulos, F., Varela, C.A.: Formal safety envelopes for provably accurate state classification by data-driven flight models. J. Aerospace Inf. Syst. **20**(1), 3–16 (2023). https://doi.org/10.2514/1.I011073
6. Kopsaftopoulos, F.P., Fassois, S.D.: Vector-dependent functionally pooled ARX models for the identification of systems under multiple operating conditions. In: Proceedings of the 16th IFAC Symposium on System Identification, (SYSID). Brussels, Belgium (2012)
7. Kopsaftopoulos, F., Chang, F.-K.: A dynamic data-driven stochastic state-awareness framework for the next generation of bio-inspired fly-by-feel aerospace vehicles. In: Blasch, E., Ravela, S., Aved, A. (eds.) Handbook of Dynamic Data Driven Applications Systems, pp. 697–721. Springer, Cham (2018). https://doi.org/10.1007/978-3-319-95504-9_31
8. Kopsaftopoulos, F., Nardari, R., Li, Y.H., Chang, F.K.: A stochastic global identification framework for aerospace structures operating under varying flight states. Mech. Syst. Signal Process. **98**, 425–447 (2018)
9. Kopsaftopoulos, F.P.: Advanced functional and sequential statistical time series methods for damage diagnosis in mechanical structures. Ph.D. thesis, Department of Mechanical Engineering & Aeronautics, University of Patras, Patras, Greece (2012)
10. Lamport, L.: How to Write a Proof. Am. Math. Mon. **102**(7), 600–608 (1995)
11. Manzano, M.: Extensions of First-Order Logic, vol. 19. Cambridge University Press, Cambridge, UK (1996)
12. Paul, S., Kopsaftopoulos, F., Patterson, S., Varela, C.A.: Dynamic data-driven formal progress envelopes for distributed algorithms. In: Dynamic Data-Driven Application Systems, pp. 245–252 (2020)
13. Paul, S., Kopsaftopoulos, F., Patterson, S., Varela, C.A.: Towards formal correctness envelopes for dynamic data-driven aerospace systems. In: Darema, F., Blasch, E. (eds.) Handbook of Dynamic Data-Driven Application Systems. Springer (2020). http://wcl.cs.rpi.edu/papers/DDDASHndbk2020_paul.pdf, preprint. To appear
14. Paul, S., Patterson, S., Varela, C.A.: Formal guarantees of timely progress for distributed knowledge propagation. In: Formal Methods for Autonomous Systems (FMAS). Electronic Proceedings in Theoretical Computer Science, vol. 348, pp. 73–91. Open Publishing Association, The Hague, Netherlands (2021)

Coupled Sensor Configuration and Path-Planning in a Multimodal Threat Field

Chase L. St. Laurent[1] and Raghvendra V. Cowlagi[2(✉)]

[1] AMETEK Inc., Peabody, MA 01960, USA
`clstlaurent@wpi.edu`
[2] Worcester Polytechnic Institute, Worcester, MA 01609, USA
`rvcowlagi@wpi.edu`
`https://labs.wpi.edu/ace-lab/`

Abstract. A coupled path-planning and sensor configuration method is proposed. The path-planning objective is to minimize exposure to an unknown, spatially-varying, and temporally static scalar field called the *threat* field. The threat field is modeled as a weighted sum of several scalar fields, each representing a *mode* of threat. A heterogeneous sensor network takes noisy measurements of the threat field. Each sensor in the network observes one or more threat modes within a circular field of view (FoV). The sensors are configurable, i.e., parameters such as location and size of field of view can be changed. The measurement noise is assumed to be normally distributed with zero mean and a variance that monotonically increases with the size of the FoV, emulating the FoV v/s resolution trade-off in most sensors. Gaussian Process regression is used to estimate the threat field from these measurements. The main innovation of this work is that sensor configuration is performed by maximizing a so-called task-driven information gain (TDIG) metric, which quantifies uncertainty reduction in the cost of the planned path. Because the TDIG does not have any convenient structural properties, a surrogate function called the self-adaptive mutual information (SAMI) is considered. Sensor configuration based on the TDIG or SAMI introduces coupling with path-planning in accordance with the dynamic data-driven application systems paradigm. The benefit of this approach is that near-optimal plans are found with a relatively small number of measurements. In comparison to decoupled path-planning and sensor configuration based on traditional information-driven metrics, the proposed CSCP method results in near-optimal plans with fewer measurements.

Keywords: sensor networks · trajectory- and path-planning · bayesian methods · sensor configuration

1 Introduction

Consider applications where an autonomous mobile agent learns about its unknown environment using data collected by an exteroceptive sensor network.

Supported by NSF grant #1646367 and #2126818.

E. Blasch et al. (Eds.): DDDAS 2022, LNCS 13984, pp. 53–68, 2024.
https://doi.org/10.1007/978-3-031-52670-1_6

For example, we envision a situation where the mobile agent - henceforth called an *actor* - needs to find a minimum-threat path in an adverse environment. The nature of the threat is multimodal and correlated, e.g., fire, smoke, and heat. A network of mobile sensors, e.g., unmanned aerial vehicles (UAVs), is available to collect data about the threat, but each sensor may be limited in its ability to detect the different threat modalities, e.g., one UAV may carry a camera that visually detect fire and smoke, whereas another UAV may carry a temperature sensor. In this situation, it may be: (1) possible to *configure* the sensors, e.g., send different types of sensors to different locations, and (2) crucial for the actor to find a plan with high confidence but with *as few measurements as possible.* To this end we ay ask: *how do we optimally configure sensors to find a near-optimal plan with a minimal number of measurements?*

We study the problem of finding a path of minimum threat exposure in an unknown environment. The unknown threat is a sum of multiple scalar fields representing modes. Information about the threat is gained through data collected by a heterogeneous sensor network, where each sensor is able to detect one or more modes of threat.

Related Work: Optimal path-planning and the related field of motion-planning are well studied for metrics such as minimum path length, maximum traversal utility, and obstacle avoidance [1,2]. Classical approaches include artificial potential fields, probabilistic roadmaps, and cell decomposition. Discrete-space methods such as Dijkstra's algorithm [3], A*, and its variants [4] are well-known for path-planning. Probabilistic techniques are used for planning under uncertainty [5], including dynamic programming and its variants. [6]. Partially observable Markov Decision process models are typically used in uncertain planning tasks when the agent has active onboard sensing [7].

The sensor configuration literature is focused on sensor *placement.* Optimal sensor placement addresses, for example, minimizing uncertainty, or maximizing spatial coverage and communication reachability [8]. Optimization metrics include set coverage and information theoretical criteria such as Kullback-Leibler divergence, Fisher information, and mutual information [9]. Sensor placement applications include placement for estimation of gaseous plumes [10], cooperative tracking of forest fires [11], and observing dynamics of volcanic ash [12]. Near-optimal sensor placement for linear inverse problems are studied by [13]. Clustering-based algorithms such as k-means [14] and density-based clustering [15] are studied for sensor placement in office spaces.

A comparison of *task-driven* versus *information-driven* sensor placement in tracking applications is discussed [16]. The literature cited above represents information-driven approaches. Task-driven approaches include recent works, for example, on optimal sensor selection for linear quadratic Gaussian feedback control systems [17], and hierarchical path-replanning concurrently with multi-agent data fusion [18]. Target tracking UAVs with limited field of view are studied in applications where the UAVs find optimal paths and optimal sensing position simultaneously [19].

The authors' previous work has addressed coupled sensor configuration and path-planning [20–22]. In all of these works, we assumed that the network of extroceptive sensors is homogeneous and that the underlying threat environment is unimodal. In practice, heterogeneous sensor types can be utilized to capture various modalities of an environment that can be measured observable by a particular sensor type. In this work, we address heterogeneous sensor networks that observe multimodal threat.

There are many situations in which a sensing agent (e.g., a UAV) could have a payload with multiple heterogeneous sensor types, allowing for simultaneous sensing of variably correlated threat modalities in an environment. In what follows, we address situations in which a sensor network can be comprised of heterogeneous sensing agents with the following scenarios: (1) every sensing agent payload is equipped with every sensor modality, (2) each the sensing agent's payload is deficient in at least one modality, and (3) the sensing agent payloads each contain a sensor modality.

The details of the threat and sensor models, and the proposed coupled sensing and path-planning technique are provided next.

2   Problem Formulation

We denote by $\mathbb{R}$ and $\mathbb{N}$ the sets of real and natural numbers, respectively, and by $\{N\}$ the set $\{1, 2, \ldots, N\}$ for any $N \in \mathbb{N}$. For any $\boldsymbol{a} \in \mathbb{R}^N$, $\boldsymbol{a}[i]$ is its ith element, $diag(\boldsymbol{a})$ is the $N \times N$ diagonal matrix with the elements of $\boldsymbol{a}$ on the principal diagonal, and $\boldsymbol{a}^{\circ(-1)}$ denotes the vector with reciprocal elements of $\boldsymbol{a}$. For any matrix $A \in \mathbb{R}^{M \times N}$, $A[i, j]$ is the element in the ith row and j^{th} column. For $A \in \mathbb{R}^{N \times N}$ and for the indicator vector $\boldsymbol{a} \in \{0, 1\}^N$, $diag(A)$ is the diagonal vector and $A[\boldsymbol{a}]$ is the submatrix of rows and columns indicated by $\boldsymbol{a}$. Similarly, $A[i, \boldsymbol{a}]$ denotes elements in the ith row and columns indicated by $\boldsymbol{a}$. $\boldsymbol{I}_{(N)}$ denotes the identity matrix of size N. For $\mu, \sigma \in \mathbb{R}$, $\mathcal{N}(\mu, \sigma^2)$ denotes the normal distribution with mean μ and variance σ^2.

The agent operates in a compact square planar region called the *workspace* $\mathcal{W} \subset \mathbb{R}^2$. Consider a uniformly-spaced square grid of points $i = 1, 2, \ldots, N_{\text{g}}$ and a graph $\mathcal{G} = (V, E)$ whose vertices $V = \{N_{\text{g}}\}$ are uniquely associated with these grid points. The set of edges E of this graph consist of pairs of geometrically adjacent grid points. In a minor abuse of notation, we label the vertices the same as grid points. We denote by $\boldsymbol{p}_i = (p_{ix}, p_{iy})$ the coordinates of the ith grid point and by Δp the distance between adjacent grid points. Without loss of generality we consider "4-way" adjacency of points (i.e., adajcent points are top, down, left, and right).

A *threat field* $c : \mathcal{W} \to \mathbb{R}_{>0}$ is a strictly positive temporally static scalar field. A *path* $\boldsymbol{\pi} = (\boldsymbol{\pi}[0], \boldsymbol{\pi}[1], \ldots, \boldsymbol{\pi}[\lambda])$ between prespecified initial and goal vertices $v_0, v_{\text{L}} \in V$ is a finite sequence, without repetition, of successively adjacent vertices such that $\boldsymbol{\pi}[0] = v_0$ and $\boldsymbol{\pi}[\lambda] = v_{\text{L}}$ for some $\lambda \in \mathbb{N}$. When the meaning is clear from the context, we also denote by $\boldsymbol{\pi}$ the unordered set of vertices in a path. A *path incidence vector* $\boldsymbol{v}_{\boldsymbol{\pi}} \in \{0, 1\}^{N_{\text{g}}}$ has $\boldsymbol{v}_{\boldsymbol{\pi}}[i] = 1$ if $i = \boldsymbol{\pi}[j]$ for

$j \in \{\lambda\}\backslash 0$ and $\boldsymbol{v}_\pi[i] = 0$ otherwise. The *cost* of a path $\boldsymbol{\pi}$ is the total threat exposure calculated as $\mathcal{J}(\boldsymbol{\pi}) := \Delta p \sum_{j=1}^{\lambda} c(\boldsymbol{p}_{\pi[j]})$. The main problem of interest is to find a path $\boldsymbol{\pi}^*$ of minimum cost.

We cannot solve this problem as stated because the threat field is unknown. The threat field can be measured by sensors, but each sensor may not be able to measure all of the modalities that constitute the threat. For the sake of a tangible example, sensor modalities may include (1) electro-optical (EO) imaging, (2) infrared (IR) imaging, and (3) a lidar (LI) point cloud. Fusion of these sensors could occur at the *sensor level* as raw data passes through circuitry, at the *data level* as the analog data acquisition becomes digitized, at the *feature level* where the data is combined through a latent embedded space, at the *decision level* which fuses independent output decisions from each sensor type, or at the *mission level* which fuses data with respect to spatial or task relevant correlations. In this work, we attempt to emulate mission level fusion via a statistical field estimation formulation. In terms of the aforementioned sensors, this would mean fusion occurs after the EO, IR, and LI sensor data is digitized, a context-based decision about the data is performed, including spatial association.

The *context* of the multimodal data fusion is of significant importance to how fusion is performed. We define each ith threat field modality as $c^{(i)} : \mathcal{W} \to \mathbb{R}_{>0}$ as a strictly positive temporally static scalar field, as illustrated in Fig. 1(a). We then define a *fused* threat field as $\check{c} := \boldsymbol{m}[1]c^{(1)} + \boldsymbol{m}[2]c^{(2)} + \cdots + \boldsymbol{m}[N_m]c^{(N_m)}$, where $\boldsymbol{m}$ is a user specified weighted fusion vector. The values prescribed to $\boldsymbol{m}$ define the context in which the fusion occurs. The path cost is then calculated as $\mathcal{J}(\boldsymbol{\pi}) := \Delta p \sum_{j=1}^{\lambda} \check{c}(\boldsymbol{p}_{\pi_j})$. Lastly, in what follows we make use of a multimodal vertex set $\tilde{V} := \{N_g N_m\}$ to represent the vertex indices scaled to N_m modalities.

Each sensor measures the threat in a circular FoV as shown in Fig. 1(b). The center $\boldsymbol{s}_k \in \mathcal{W}$ and radius $\varrho_k \in \mathbb{R}_{>0}$ of this circular FoV are parameters that we can choose for each $k \in \{N_s\}$. Maximum and minimum FoV radius constraints are specified as $\varrho^{\max}$ and $\varrho^{\min}$, respectively. The set of all sensor parameters is called a *configuration*, which we denote by $\mathcal{C} = \{\boldsymbol{s}_1, \varrho_1, \boldsymbol{s}_2, \ldots, \varrho_{N_s}\}$.

We introduce an observability incidence vector for each kth sensor where $\boldsymbol{o}_k \in \{0,1\}^{N_m}$, which characterizes if the kth sensor can view a threat modality and is specified by the user given each sensing agent's payload. We create an *observed cover incidence matrix* for each kth sensor as $\tilde{\boldsymbol{\nu}}_k := \text{vec}(\boldsymbol{\nu}_k \boldsymbol{o}_k) \in \mathbb{R}^{N_g N_m \times 1}$, where we define $\text{vec}(\cdot) := \mathbb{R}^{N_g \times N_m} \to \mathbb{R}^{N_g N_m \times 1}$. Finally, the combined observed covered incidence vector becomes $\tilde{\boldsymbol{\nu}} := (\tilde{\boldsymbol{\nu}}_1 \vee \tilde{\boldsymbol{\nu}}_2 \vee \ldots \vee \tilde{\boldsymbol{\nu}}_{N_s})$.

We update the sensor observations notation as follows. The collection of sensor data locations for a particular i^{th} sensor type is denoted as $\boldsymbol{X}^{(i)} = \{\boldsymbol{x}_{11}^{(i)}, \ldots, \boldsymbol{x}_{km}^{(i)}, \ldots, \boldsymbol{x}_{N_s M N_s}^{(i)}\} \,\forall\, i \in N_m$. We denote by, $\mathcal{X} = \{\boldsymbol{X}^{(1)}, \boldsymbol{X}^{(2)}, \ldots, \boldsymbol{X}^{(N_m)}\}$, the set of sensor data locations for each sensing modality which is aggregated or updated with each iteration ℓ. The training matrix of the training set augmented by the corresponding modality index is formulated as follows:

$$\tilde{\boldsymbol{X}} := \begin{bmatrix} \boldsymbol{X}^{(1)} & \boldsymbol{X}^{(2)} & \ldots & \boldsymbol{X}^{(N_m)} \\ \boldsymbol{0} & 1 & \ldots & N_m \end{bmatrix}^{\mathsf{T}} \tag{1}$$

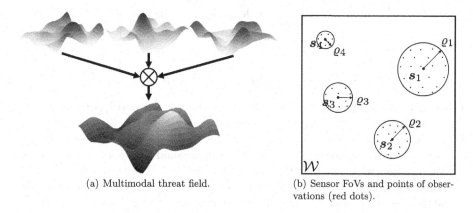

(a) Multimodal threat field.

(b) Sensor FoVs and points of observations (red dots).

Fig. 1. Illustration of the threat and sensor models considered in this work.

Similarly, we define the collection of vectorized noisy threat field observations and observation noise as $\tilde{z} = \begin{bmatrix} z_1^{\mathsf{T}} & z_2^{\mathsf{T}} & \ldots & z_{N_m}^{\mathsf{T}} \end{bmatrix}^{\mathsf{T}}$ and $\tilde{\sigma} = \begin{bmatrix} \sigma_1^{\mathsf{T}} & \sigma_2^{\mathsf{T}} & \ldots & \sigma_{N_m}^{\mathsf{T}} \end{bmatrix}^{\mathsf{T}}$, respectively. We say that each of the observations for each ith modality is modeled as $z_{km}^{(i)} = c^{(i)}(x_{km}) + \eta_{km}^{(i)}$. The measurement error η increases monotonically with the sensor FoV.

3   Multimodal Field Estimation

We use Gaussian Process regression (GPR) to construct an estimate of the multimodal threat field. Basic details regarding the GPR approach to unimodal threat estimation are provided in our previous work [22]. The joint posterior distribution is modified as follows to account for the multimodal vectorized output and multimodal threat field estimates:

$$\begin{bmatrix} \tilde{z} \\ \tilde{f} \end{bmatrix} \sim \mathcal{N} \left(0, \begin{bmatrix} K_{\tilde{z}} & K_* \\ K_*^{\mathsf{T}} & K_{**} \end{bmatrix} \right), \tag{2}$$

where K is the kernel function. We modify the kernel structure to enable learning of the cross-correlation between modalities. Namely, we utilize the intrinsic model of coregionalization (ICM) kernel K^X defined as:

$$K^X = (\Theta_w \Theta_w^{\mathsf{T}} + \mathrm{diag}(\theta_v)) \tag{3}$$

The ICM kernel is a matrix of size $N_m \times N_m$ and has learnable parameters Θ_w and θ_v. The matrix of parameters is of size $N_m \times r$, where $r \in \mathbb{N}$ is a small value to emulate low-rank positive definite correlation between modalities. The parameter vector θ_v is of size $N_m \times 1$ and models the independent scaling factor of each modality. For threat field modeling, we utilize the kernel $K_{ij} = K_{ij}^R \cdot K^X[i,j] \ \forall \ i,j \in N_m$. The resulting kernel has the form:

$$\boldsymbol{K} := K(\tilde{\boldsymbol{X}}, \tilde{\boldsymbol{X}}) = \begin{bmatrix} \boldsymbol{K}_{11} & \boldsymbol{K}_{12} & \cdots & \boldsymbol{K}_{1N_m} \\ \boldsymbol{K}_{21} & \ddots & \vdots & \vdots \\ \vdots & \vdots & \ddots & \vdots \\ \boldsymbol{K}_{N_m 1} & \cdots & \cdots & \boldsymbol{K}_{N_m N_m} \end{bmatrix}$$

The collection of hyperparameters to optimize are $\boldsymbol{\theta} = (\boldsymbol{\Theta}_r, \boldsymbol{\Theta}_w, \boldsymbol{\theta}_v)$. The multimodal input kernel with heteroscedastic noise vector is formulated as $\boldsymbol{K}_{\tilde{z}} := \boldsymbol{K} + diag(\tilde{\boldsymbol{\sigma}})$. The diagonal elements of $\boldsymbol{K}$ represent the auto-covariance of points within a modality, whereas the off-diagonal elements represent the cross-covariance which models the latent relationship between modalities.

From the joint distribution, we can obtain the current iteration multimodal threat field estimate and multimodal threat error covariance matrix as:

$$\tilde{\boldsymbol{f}}_\ell = \boldsymbol{K}_*^\mathsf{T} \boldsymbol{K}_{\tilde{z}}^{-1} \tilde{z}, \qquad\qquad \tilde{\boldsymbol{P}}_\ell = \boldsymbol{K}_{**} - \boldsymbol{K}_*^\mathsf{T} \boldsymbol{K}_{\tilde{z}}^{-1} \boldsymbol{K}_*.$$

We note that the multimodal threat field estimate and multimodal threat field error covariance matrix are constructed as:

$$\tilde{\boldsymbol{f}}_\ell = \left[\boldsymbol{f}_\ell^{(1)\mathsf{T}} \, \boldsymbol{f}_\ell^{(2)\mathsf{T}} \cdots \boldsymbol{f}_\ell^{(N_m)\mathsf{T}} \right]^\mathsf{T}, \quad \tilde{\boldsymbol{P}}_\ell = \begin{bmatrix} \boldsymbol{P}_\ell^{(11)} & \boldsymbol{P}_\ell^{(12)} & \cdots & \boldsymbol{P}_\ell^{(N_m)} \\ \boldsymbol{P}_\ell^{(21)} & \ddots & \vdots & \vdots \\ \vdots & \vdots & \ddots & \vdots \\ \boldsymbol{P}_\ell^{(N_m 1)} & \cdots & \cdots & \boldsymbol{P}_\ell^{(N_m N_m)} \end{bmatrix}.$$

The fused threat field estimate and fused threat field error covariance matrix can then be computed using the weighted fusion vector $\boldsymbol{m} \in \mathbb{R}^{N_m}$ as:

$$\check{\boldsymbol{f}}_\ell = \sum_{i=1}^{N_m} \boldsymbol{m}[i] \tilde{\boldsymbol{f}}_\ell^{(i)}, \quad \check{\boldsymbol{P}}_\ell = \sum_{i=1}^{N_m} \boldsymbol{m}^2[i] \tilde{\boldsymbol{P}}_\ell^{(ii)} + 2 \sum_{j=2}^{N_m} \sum_{k=1}^{j-1} \boldsymbol{m}[j] \boldsymbol{m}[k] \tilde{\boldsymbol{P}}_\ell^{(jk)}. \quad (4)$$

We may then use the fused threat field and fused threat error covariance matrix to find the estimated optimal path-plan and the estimated path-plan variance.

4   Coupled Sensor Configuration

The coupled sensor configuration and path-planning (CSCP) method for unimodal threats was introduced in our previous works, and is summarized in Fig. 2. The key step in the CSCP approach to configure sensors by maximizing what is called the task-driven information gain (TDIG). Informally, TDIG is a measure of entropy reduction in a region of interest "near" the currently optimal path. Further details are in our previous works. In [23], we introduced a function called *self-adaptive mutual information (SAMI)*, which approximates the TDIG, but has the added benefit of being submodular.

The introduction of heterogeneous sensors presents some additional nuances. Since each sensor network can have multiple modalities, we consider three separate situations in which the sensor configuration problem can be formalized.

1: Initialize: $\ell := 0$, set initial mean threat estimate to zero, and covariance arbitrarily large
2: Find initial estimated optimal path $\boldsymbol{\pi}_0^* := \arg\min \overline{\mathcal{J}}_0(\boldsymbol{\pi})$
3: **while** $\text{Var}_\ell(\boldsymbol{\pi}_\ell^*) > \varepsilon$ **do**
4:     Find optimal sensor configuration $\mathcal{C}_\ell^*$ (as discussed in this section)
5:     Record new measurements with the new sensor configuration
6:     Increment iteration counter $\ell := \ell + 1$
7:     Update mean threat estimate and estimation error covariance $\boldsymbol{P}_\ell$ (see §3)
8: Find $\boldsymbol{\pi}_\ell^* := \arg\min \overline{\mathcal{J}}_\ell(\boldsymbol{\pi})$

Fig. 2. Proposed iterative CSCP method.

In what follows, we describe situations in which every sensor is equipped with every modality, a mixture of modalities, or only a single modality.

First, we say that a multimodal region of interest is defined as the collection of region of interest vertices with index values biased by the appropriate mode index as $\tilde{\mathcal{R}} := \{\cup_{i=0}^{N_m-1}\mathcal{R} + iN_g\}$. The multimodal region of interest is used for indexing in the SAMI formulation. The entropy of any vertex $i \in \tilde{V}$ as:

$$h(i) := \frac{1}{2}\ln(2\pi e \tilde{\boldsymbol{P}}_\ell[i,i]) \tag{5}$$

In certain cases the conditional entropy computations can be batched and computed in parallel. For any $i \notin \mathcal{R}$, the batched conditional entropy vector $\boldsymbol{h}(\cdot|\tilde{\mathcal{R}}_{\backslash i})$ is computed as:

$$\boldsymbol{h}(\cdot|\tilde{\mathcal{R}}_{\backslash i}) = \frac{1}{2}\ln((2\pi e)diag(\tilde{\boldsymbol{P}}_\ell - \tilde{\boldsymbol{P}}_\ell[\cdot,\tilde{\mathcal{R}}]\tilde{\boldsymbol{P}}_\ell[\tilde{\mathcal{R}},\tilde{\mathcal{R}}]^{-1}\tilde{\boldsymbol{P}}_\ell[\tilde{\mathcal{R}},\cdot])) \tag{6}$$

We note that the matrix inverse only needs to be computed once. However, for any $i \in \mathcal{R}$, the conditional entropy of any vertex $i \in \{N_g N_m\}$ given $\tilde{\mathcal{R}}_{\backslash i}$ can be computed in parallel as:

$$h(i|\tilde{\mathcal{R}}_{\backslash i}) = \frac{1}{2}\ln((2\pi e)(\tilde{\boldsymbol{P}}_\ell[i,i] - \tilde{\boldsymbol{P}}_\ell[\tilde{\mathcal{R}}_{\backslash i},i]^\mathsf{T}\tilde{\boldsymbol{P}}_\ell[\tilde{\mathcal{R}}_{\backslash i},\tilde{\mathcal{R}}_{\backslash i}]^{-1}\tilde{\boldsymbol{P}}_\ell[\tilde{\mathcal{R}}_{\backslash i},i])) \tag{7}$$

Given these equations, we can calculate the mutual information I between the multimodal region of interest and any vertex for a particular mode $i \in \tilde{V}$ as:

$$I(\tilde{\mathcal{R}}_{\backslash i}; i) := h(i) - h(i|\tilde{\mathcal{R}}_{\backslash i}). \tag{8}$$

The SAMI reward term for multimodal threat fields is then:

$$R(i) := (1 - \alpha)I(\tilde{\mathcal{R}}_{\backslash i}; i) + \alpha I(\tilde{\mathcal{R}}_{\backslash i}^c; i) \tag{9}$$

In (9), the ROI complement is taken as $\tilde{\mathcal{R}}^c := \tilde{V} \backslash \tilde{\mathcal{R}}$ and $i \in \tilde{V}$. We also update the mutual information reward vector to be $\boldsymbol{\gamma} := \begin{bmatrix} \gamma(1) & \gamma(2) & \dots & \gamma(N_g) & \dots & \gamma(N_g N_m) \end{bmatrix}^\mathsf{T}$. The reward function given the sensor configuration is $\Gamma(\mathcal{C}_\ell) = \tilde{\boldsymbol{\nu}}^\mathsf{T}\boldsymbol{\gamma}$.

The SAMI penalty function is calculated as:

$$\Upsilon(\mathcal{C}_\ell) := -\frac{1}{2}\sum_{i \in \mathcal{F}}\left(\frac{1}{2}\ln(2\pi e) - \ln\sum_{k \in N_s}(\boldsymbol{\nu}_k \odot \boldsymbol{\sigma}_k^{-2})[\tilde{\mathcal{F}}]\right) \tag{10}$$

In (10), the $\odot$ operator is the Hadamard product, which is used to perform element-wise multiplication between the cover incidence vector and the inverse noise vector $\boldsymbol{\sigma_k}^{-2}$. The entries are indexed by $[\tilde{\mathcal{F}}]$ to ensure only covered vertices are accounted for prior to taking the natural logarithm of each element. The SAMI is the sum of the reward and penalty terms.

Sensors With All Modalities $o_k = 1 \, \forall \, k \in N_\mathrm{s}$: this problem reduces to maximizing the SAMI surrogate objective function directly over multimodal SAMI surrogate values. Because the SAMI is submodular, we can sequentially optimize sensor configuration for each set of kth sensor parameters $\{s_k, \varrho_k\}$ to find $\mathcal{C}_\ell^* := \arg\max S(\mathcal{C}_\ell)$.

Sensors With Overlapping Modalities: To illustrate this situation, suppose that we have available three sensors $\mathcal{S} = \{\mathcal{S}_1, \mathcal{S}_2, \mathcal{S}_3\}$ with the following modalities: $\mathcal{S}_1 = \{\mathrm{EO, IR}\}$, $\mathcal{S}_2 = \{\mathrm{EO, LI}\}$, and $\mathcal{S}_3 = \{\mathrm{IR, LI}\}$. To determine the sequence in which a sensing agent in the sensor network is optimized to adhere to the submodularity property, we need to calculate the total reward for each modality as $\bar{\gamma}_i = \sum_{j \in N_\mathrm{g}} (1 - \tilde{\nu}^{(i)}) \gamma^{(i)}$ prior to each sequential sensor configuration optimization. We denote the $\tilde{\nu}^{(i)}$ as the partition of the multimodal sensor cover incidence for the ith modality. Similarly, the ith modality partition of the SAMI reward vector is denoted $\gamma^{(i)}$. The total rewards is then $\bar{\gamma} = \begin{bmatrix} \bar{\gamma}_1 & \bar{\gamma}_2 & \dots & \bar{\gamma}_{N_m} \end{bmatrix}^\mathsf{T}$. We may then calculate the total potential reward for each kth sensor as $\bar{\gamma}^{(k)} = o_k \bar{\gamma}$. Therefore, each iteration of sequential sensor configuration, we choose $\mathcal{S}_k = \arg\max \bar{\gamma}^{(k)}$, optimize it with the SAMI objective function, and remove it from the set $\mathcal{S}$ (just for that round of optimization). In our example, if $\bar{R}^{(2)}$ was the maximum value, we would perform sensor configuration optimization with $\mathcal{S}_2$ and remove it from the set of optimizable sensing agents, leaving only $\mathcal{S}_1$ and $\mathcal{S}_3$ to be optimized. We proceed to score the sensing agents again until they have all been optimized.

Sensors With Unique Modalities: This is a special observability case which has unique sensor configuration optimization implications. We no longer require ranking the sensor modalities as they are entirely separable and additive. Due to this, the sensor configuration for uniquely observable sensor payloads allows for parallelization of sequential sensor configuration for each modality. In fact, each mode reduces to solving independent sensor configuration optimization N_m times in parallel, allowing for computational savings.

5   Results and Discussion

We conducted a study with four various sensor networks in an environment of area $9km^2$, workspace resolution of 21^2, and a desired termination threshold $\varepsilon = 1$. We considered a multimodal threat field with $N_m = 3$ correlated threat modalities. All mobile sensing agents were constrained to $\varrho^{\min} = 0.05$ and $\varrho^{\max} =$

0.5. The region of interest for the experiments was found with $N_a = 3$ alternate path plans.

We considered four various sensor network scenarios for the experiments. The first sensor network $\mathcal{S}_A$ was comprised of 3 sensors with all modalities. $\mathcal{S}_B$ had three sensors each with two modalities such that each threat modality was observable by at least 2 sensors. It also included one sensor with all modalities totaling $N_s = 4$. Sensor network $\mathcal{S}_C$ had the same pairs of modalities as $\mathcal{S}_B$, but instead of the single full-modal sensor, it had three unimodal sensors (one for each modality) totaling $N_s = 6$. Finally, we considered a sensor network $\mathcal{S}_D$ which was comprised of $N_s = 9$ unimodal sensors such that each modality equally had 3 sensors which could make observations. These networks were chosen to study the effect of modifying the degrees of freedom of the sensor network on convergence and sensor configuration optimization search time. The average iterations for convergence and the average sensor configuration optimization time for each sensor network are recorded in Table 1. We make the following observations from the collected results. In summary, the parallel optimization of sensor networks with unique modalities, by way of the separable nature of the multimodal formulation of the SAMI surrogate function, are able to quickly find sensor configurations whilst providing good convergence performance.

Table 1. Comparative study for various sensor payload configurations.

Sensor Network	Iterations	Configuration Time
Network $\mathcal{S}_A$	16.331 ± 4.163	2.313 ± 0.683
Network $\mathcal{S}_B$	16.033 ± 3.436	3.068 ± 0.995
Network $\mathcal{S}_C$	15.667 ± 4.619	4.900 ± 1.472
Network $\mathcal{S}_D$	14.330 ± 4.041	2.878 ± 1.004

Increasing Sensor Network Flexibility Improves Convergence Performance. We may characterize network flexibility in context of the constraint of having multiple modalities with identical FoVs to optimize on a single sensor. The flexibilities for the experiment sensor networks are in increasing order $\mathcal{S}_A < \mathcal{S}_B < \mathcal{S}_C < \mathcal{S}_D$. The results show that adding the increased flexibility by not confining multiple multimodal sensor payloads to a single sensing agent proportionally decreases the average iterations for path-plan convergence. This answers a typical quandary between, say, uying an expensive UAV equipped with many heterogeneous sensors versus buying multiple inexpensive UAVs each with a single modality. For coupled sensing and path-planning problems the additional flexibility by distributing the sensors across sensing agents within the network is more valuable.

Parallel Optimization of Unimodal Sensor Payloads Yields Low Configuration Times. Somewhat unsurprisingly, the parallel sequential optimization of sensor network $\mathcal{S}_D$ which had sensing agents with unimodal observability, was able to be optimized faster than sensor networks $\mathcal{S}_B$ and $\mathcal{S}_C$. We note that the reason it still takes longer to configure these sensors over sensor network $\mathcal{S}_A$ is that there are more potential configurations and globally optimizing is made more difficult. However, we note how close the runtime performance is for sensor configuration between $\mathcal{S}_A$ and $\mathcal{S}_D$ and note that the time difference is negligible in comparison to the average iterations until convergence performance gain.

5.1   Demonstrative Example

We demonstrate the MM-CSCP algorithm on an example randomly generated multimodal threat field with $N_m = 3$. The problem definition remains the same as the numerical experiments from the previous section, but we increase the environment area to $25km^2$. The individual threat field modalities and the fused threat field along with the optimal path-plan are shown in Fig. 3. We considered a sensor network with 3 sensing agents with unique pairings of observability and 2 sensing agents with unimodal observability for the first two threat modes. Therefore, the first two threat modes have up to three FoV covers per iteration while the third only has two. The initial sensor network configuration is depicted overlaying the SAMI reward function for each vertex in Fig. 4.

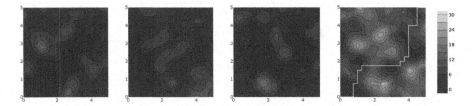

Fig. 3. Ground truth threat field modalities and the fused threat field (right) along with the optimal path-plan.

The data collected from the first iteration sensor configuration is then utilized to determine the multimodal threat field estimate and subsequently the fused

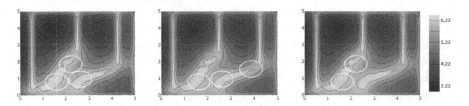

Fig. 4. Initial iteration sensor network configurations overlaying the multimodal SAMI reward values for each mode.

threat field estimate as shown in Fig. 5. The resultant estimated path-plan is shown in this figure, and Fig. 6 shows the corresponding multimodal threat error covariance values for each vertex and the fused covariance values. The algorithm proceeds until convergence at iteration $\ell = 22$. Figures 7, 8, 9, 10, 11, 12, 13, 14, 15, 16, 17 and 18 show the SAMI reward functions and sensor configurations, the multimodal and fused threat field estimates with the estimated path-plan, and the multimodal and fused threat variances at select iterations.

The SAMI reward examples show that in some cases sensing agents with paired observability are configured such that the reward of one modality is considered in conjunction with a potentially low reward modality. We also note that the multimodal threat field estimation and fusion visually does a good job at learning the environment. We also observe from the final fused threat error variance in Fig. 18 that the MM-CSCP still emphasizes learning task-driven regions of interest, rather than the entire environment, in order to plan the path. Ultimately, the estimated path plan is near-optimal.

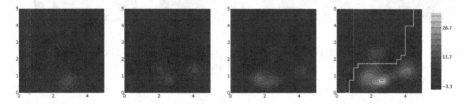

Fig. 5. Initial multimodal threat estimates and fused threat field estimate (right) along with the estimated optimal path plan (green) and the true optimal path plan (yellow). (Color figure online)

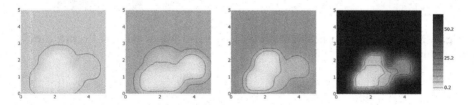

Fig. 6. Initial multimodal threat field error covariance vertex values and fused vertex covariance values.

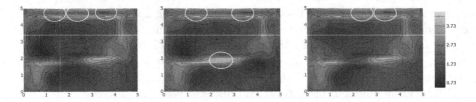

Fig. 7. Sensor network configurations overlaying the multimodal SAMI reward values for each threat mode at $\ell = 3$.

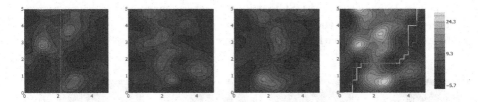

Fig. 8. Multimodal threat field estimates and fused threat field estimate (right) along with the estimated optimal path-plan (green) and the true optimal path-plan (yellow) at iteration $\ell = 3$. (Color figure online)

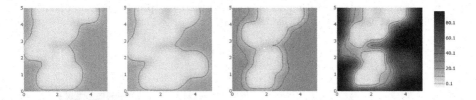

Fig. 9. Multimodal threat field error covariance vertex values and fused vertex covariance values at iteration $\ell = 3$.

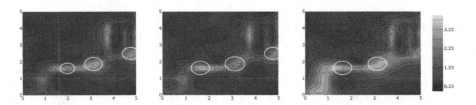

Fig. 10. Sensor network configurations overlaying the multimodal SAMI reward values for each threat mode at $\ell = 7$.

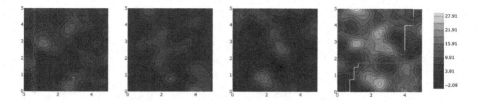

Fig. 11. Multimodal threat field estimates and fused threat field estimate (right) along with the estimated optimal path-plan (green) and the true optimal path-plan (yellow) at iteration $\ell = 7$. (Color figure online)

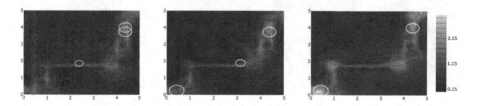

Fig. 12. Multimodal threat field error covariance vertex values and fused vertex covariance values at iteration $\ell = 7$.

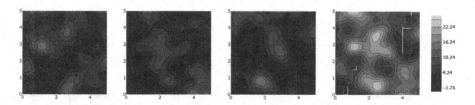

Fig. 13. Sensor network configurations overlaying the multimodal SAMI reward values for each threat mode at $\ell = 13$.

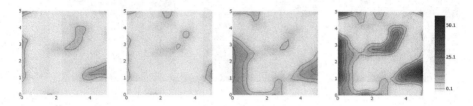

Fig. 14. Multimodal threat field estimates and fused threat field estimate (right) along with the estimated optimal path-plan (green) and the true optimal path-plan (yellow) at iteration $\ell = 13$. (Color figure online)

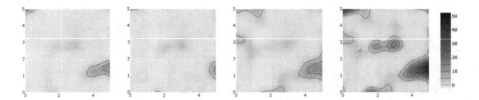

Fig. 15. Multimodal threat field error covariance vertex values and fused vertex covariance values at iteration $\ell = 13$.

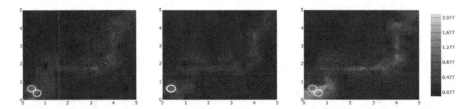

Fig. 16. Sensor network configurations overlaying the multimodal SAMI reward values for each threat mode at $\ell = 22$, the final iteration.

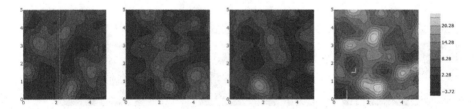

Fig. 17. Multimodal threat field estimates and fused threat field estimate (right) along with the estimated optimal path-plan (green) and the true optimal path-plan (yellow) at iteration $\ell = 22$, the final iteration. (Color figure online)

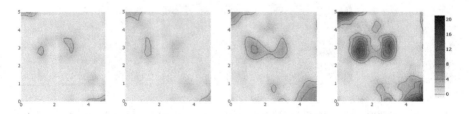

Fig. 18. Multimodal threat field error covariance vertex values and fused vertex covariance values at iteration $\ell = 22$.

6 Conclusion

A coupled path-planning and sensor configuration method was proposed. This work was based on the authors' previous work on CSCP in unimodal threats. To extend CSCP to multimodal threats, we introduced different ways of calculating and optimizing the SAMI for different cases where each sensor in the network is able to all, some, or one unique modality of the threat. Numerical simulation results were discussed for one demonstrative example and for more thorough studies comparing sensor networks with different levels of flexibilities. Future work includes extension to time-varying multimodal threat fields.

References

1. LaValle, S.M.: Planning Algorithms. Cambridge University Press, Cambridge (2006)
2. Aggarwal, S., Kumar, N.: Path planning techniques for unmanned aerial vehicles: a review, solutions, and challenges. Comput. Commun. **149**, 270–299 (2020)
3. Dijkstra, E.W.: A note on two problems in connexion with graphs. Numer. Math. **1**(1), 269–271 (1959)
4. Hart, P., Nilsson, N., Raphael, B.: A formal basis for the heuristic determination of minimum cost paths. IEEE Trans. Syst. Sci. Cybern. **4**(2), 100–107 (1968). https://doi.org/10.1109/tssc.1968.300136
5. Mohanan, M., Salgoankar, A.: A survey of robotic motion planning in dynamic environments. Robot. Auton. Syst. **100**, 171–185 (2018)
6. Esfahani, P.M., Chatterjee, D., Lygeros, J.: Motion planning for continuous-time stochastic processes: a dynamic programming approach. IEEE Trans. Autom. Control **61**(8), 2155–2170 (2016)
7. Kurniawati, H., Bandyopadhyay, T., Patrikalakis, N.M.: Global motion planning under uncertain motion, sensing, and environment map. Auton. Robot. **33**(3), 255–272 (2012)
8. Ramsden, D.: Optimization Approaches To Sensor Placement Problems. Ph.D. dissertation (2009)
9. Cochran, D., Hero, A.O.: Information-driven sensor planning: navigating a statistical manifold. In: 2013 IEEE Global Conference on Signal and Information Processing, GlobalSIP 2013 - Proceedings, pp. 1049–1052 (2013)
10. Demetriou, M., Gatsonis, N., Court, J.: Coupled controls-computational fluids approach for the estimation of the concentration from a moving gaseous source in a 2-d domain with a Lyapunov-guided sensing aerial vehicle. IEEE Trans. Control Syst. Technol. **22**(3), 853–867 (2013)
11. Merino, L., Caballero, F., Martínez-de Dios, J.R., Ferruz, J., Ollero, A.: A cooperative perception system for multiple UAVs: application to automatic detection of forest fires. J. Field Rob. **23**(34), 165–184 (2006)
12. Madankan, R.: Computation of probabilistic hazard maps and source parameter estimation for volcanic ash transport and dispersion. J. Comput. Phys. **271**, 39–59 (2014)
13. Ranieri, J., Chebira, A., Vetterli, M.: Near-optimal sensor placement for linear inverse problems. IEEE Trans. Signal Process. **62**(5), 1135–1146 (2014)
14. Li, S., Zhang, H., Liu, S., Zhang, Z.: Optimal sensor placement using FRFs-based clustering method. J. Sound Vib. **385**, 69–80 (2016)

15. Yoganathan, D., Kondepudi, S., Kalluri, B., Manthapuri, S.: Optimal sensor placement strategy for office buildings using clustering algorithms. Energy Build. **158**, 1206–1225 (2018)
16. Kreucher, C., Hero, A.O., Kastella, K.: A comparison of task driven and information driven sensor management for target tracking. In: Proceedings of 44th IEEE Conference on Decision and Control, pp. 4004–4009 (2005)
17. Tzoumas, V., Carlone, L., Pappas, G.J., Jadbabaie, A.: LQG control and sensing co-design. IEEE Trans. Autom. Control **66**(4), 1468–1483 (2021)
18. Allen, T., Hill, A., Underwood, J., Scheding, S.: Dynamic path planning with multi-agent data fusion - the parallel hierarchical replanner. In: 2009 IEEE International Conference on Robotics and Automation, pp. 3245–3250 (2009)
19. Skoglar, P., Nygards, J., Ulvklo, M.: Concurrent path and sensor planning for a UAV - towards an information based approach incorporating models of environment and sensor. In: 2006 IEEE/RSJ International Conference on Intelligent Robots and Systems, pp. 2436–2442 (2006)
20. St. Laurent, C.L., Cowlagi, R.V.: Breadth-first coupled sensor configuration and path-planning in unknown static environments. In: Proceedings of the 60th IEEE Conference on Decision & Control (2021)
21. Laurent, C.S., Cowlagi, R.V.: Depth-first coupled sensor configuration and path-planning in unknown static environments. In: Proceedings of the 2021 European Control Conference (2021)
22. Laurent, C.S., Cowlagi, R.V.: Coupled sensor configuration and path-planning in unknown static environments. In: Proceedings of the 2021 American Control Conference (2021)
23. St. Laurent, C.L.: Coupled sensor configuration and path-planning in uncertain environments using multimodal sensors. Ph.D. dissertation, Worcester Polytechnic Institute, Worcester, MA, USA (2022). https://digital.wpi.edu/show/n296x2271

Main-Track Plenary Presentations - Space Systems

Geometric Solution to Probabilistic Admissible Region Based Track Initialization

Utkarsh Ranjan Mishra[1]([⊠]), Weston Faber[2], Suman Chakravorty[1], Islam Hussein[3], Benjamin Sunderland[4], and Siamak Hesar[4]

[1] Texas A&M, College Station, TX 77840, USA
utkarshranjanmishra@tamu.edu
[2] L3 Applied Defence Solutions, Colorado Springs, CO 80921, USA
[3] Trusted Space, Columbia, MD 21044, USA
[4] Kayhan Space Corporation, Boulder, CO 80301, USA

Abstract. Probabilistic Admissible Region (PAR) is a technique to initialize the probability density function (*pdf*) of the states of a Resident Space Object (RSO). It combines apriori information about some of the orbital elements and a single partial-state observation to initialize the *pdf* of the RSO. A unified, geometrical solution to Probabilistic Admissible Region, G-PAR, is proposed. The proposed scheme gives a closed-form, clearly explainable solution for PAR particle mapping for the first time.

It is shown that the G-PAR can be posed as a Bayesian measurement update of the very diffuse *pdf* of the states given by the postulated statistics. The effectiveness of the proposed G-PAR will be shown on diverse combinations of sensors and apriori knowledge. Its unique advantages in resolving the data association problem inherent in initializing the *pdf* of the objects when tracking multiple objects will also be presented.

1 Introduction

Space Domain Awareness (SDA) is critical for safe and sustainable operations in the Earth's orbit. Cataloging of Resident Space Objects (RSO) is the most fundamental aspect of SDA. This involves initializing and recursively updating the uncertainty in the states of the RSOs in a catalog. The data required to initialize and update the catalog is gathered via sensors like SDA radars and telescopes.

A single short-arc measurement from an SSA sensor like a radar or a telescope gives only partial-state information and is insufficient to initialize all the states of an RSO. In an ideal single object tracking scenario, Initial Orbit Determination (IOD) [3] schemes piece together multiple measurements from a single object, over time, and try to fit a state vector. This method assumes that the series of observations came from the same Resident Space Object. But in the Multiple Object Tracking (MOT) scenarios, with multiple observations in each

---

Supported by Air Force Office of Scientific Research.

frame, there would be combinatorial growth in the possible 'hard' observation-to-observation (obs-to-obs) associations [2].

It is easy to mathematically evaluate the likelihood of an observation associated with a cataloged object compared to keeping track and evaluating the likelihood of associating each observation (among many) in one pass with each observation (among many) in a second pass. In fact, for the overwhelming majority of multiple space-object tracking scenarios, a mathematically rigorous formulation for weighing one observation-to-observation (obs-to-obs) association, between two sensor passes, versus another obs-to-obs association is impossible. This is because most objects appear physically the same to the sensor and even if identifying features like reflectivity and radar cross-sections could be measured making associations using them remains ad-hoc in most cases.

A 'target' is defined as an RSO whose *pdf* is available in the space object catalog. The problem of finding correct obs-to-obs association can be avoided if it can be posed as a target-to-observation (sometimes also called track-to-observation, or track-to-obs) association problem. This can be done by initializing the *pdf* of the states of the RSO that generated the observation, forward propagating the said *pdf* using the equations of motion, and calculating the measurement-likelihood of the measurements at the following passes. The *pdf* can be initialized using Constrained Admissible Region (CAR) or Probabilistic Admissible Region (PAR) algorithms [6,7].

This paper presents a geometrical solution to the PAR-based track initialization, called G-PAR. The proposed scheme can initialize the *pdf* of the states of the RSO using a single partial state measurement augmented with postulated statistics on a few states. The rest of the paper is laid out as follows: Sect. 2 gives greater background into admissible regions, Subsect. 2.2 details the G-PAR algorithm, Sects. 4 and 5 cover the specifics of G-PAR and present examples for radar and telescope cases, and Sect. 6 summarises the results.

2   Geometric Solution to Probabilistic Admissible Region

2.1   Probabilistic Admissible Region

The admissible region [10] is the set of physically acceptable orbits that can be constrained even further if additional constraints on some orbital parameters like semi-major axis, eccentricity, etc., are present [6–9]. This results in the constrained admissible region (CAR) [1]. If hard constraints are replaced, based on known statistics of the measurement process, with a probabilistic representation of the admissible region, it results in the probabilistic admissible region (PAR) [5]. PAR can be used for initializing the *pdf* in Bayesian tracking [4].

We look at some fundamental equations relevant to PAR. Let $\mathbf{r}$ be the inertial position of the RSO, $\mathbf{q}$ be the inertial position of the sensor site, and $\boldsymbol{\rho}$ be the position of the RSO with respect to the sensor site. This gives the following position and velocity relationships:

$$\mathbf{r} = \mathbf{q} + \boldsymbol{\rho} \tag{1}$$

$$\dot{\mathbf{r}} = \dot{\mathbf{q}} + \dot{\boldsymbol{\rho}} \tag{2}$$

The Vis-viva equation gives a relationship between the position and velocity, $\mathbf{r}$ and $\dot{\mathbf{r}}$,

$$\frac{\mu}{|\mathbf{r}|} - \frac{|\dot{\mathbf{r}}|^2}{2} = \frac{\mu}{2a} \tag{3}$$

where μ is the standard gravitation parameter and a is the semi-major axis (Fig. 1).

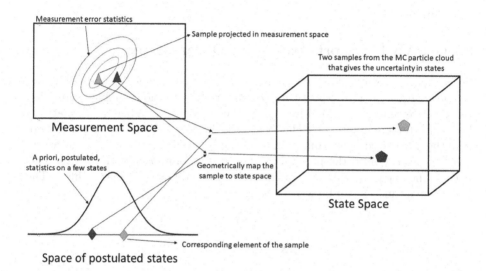

Fig. 1. Mapping samples from measurements and postulated states to the state space

2.2   G-PAR Algorithm

G-PAR maps uncertainty in measurement and uncertainty in a priori known states to uncertainty in the states of the RSO. It does so by MC sampling from the joint distribution on measurement and a priori known states and mapping each sample to state space. The resulting particle cloud in state space is a representation of the uncertainty in states.

A general template to get a geometric map between a single sample from the measurements and apriori-known states to state space for many different sensor modalities (angles-only telescope, angles-angle-rates telescope, or radar) is given below:

1. Pick a sample from the joint distribution on the measurement and the a priori known states.

2. Using the sample geometrically find out the relative position vector of the RSO with respect to the sensor site. This gives the inertial position vector corresponding to the sample.
3. Find the orbital plane(s) corresponding to the sample. This is usually fixed by sampled elements from the postulated uncertainty on either inclination or right ascension of ascending node or both.
4. Find the elements of the velocity vector using relevant constraints. This varies depending on the measurement model and postulated a priori states but some examples of constraints that can be used include the velocity vector being parallel to the orbital plane, the vis-viva equation, and the eccentricity vector equation.

3   G-PAR for Short-Arc Radar Observations

Space Domain Awareness (SDA) radars can take good range (ρ) and range-rate ($\dot{\rho}$) observations. It is also possible to extract right ascension (α) and declination (δ) measurements, from the direction of the radar beam, albeit with relatively high uncertainty due to the radar beam width. The G-PAR algorithm tries to map the postulated uncertainties in two orbital parameters (a, Ω) and measurements ($\rho, \dot{\rho}, \alpha, \delta$) into the probability density function (pd) of the states. This *pdf* can then be propagated and recursively updated using a Bayesian filter. The measurement model can be written as follows:

$$\mathbf{y} = h(\mathbf{x}) + \nu \tag{4}$$

$$h(\mathbf{x}) = [\rho, \dot{\rho}, \alpha, \delta]^T \tag{5}$$

where,

$$\rho = \sqrt{\boldsymbol{\rho} \cdot \boldsymbol{\rho}} \tag{6}$$

$$\dot{\rho} = \frac{\mathbf{v} \cdot \boldsymbol{\rho}}{\rho} \tag{7}$$

$$\frac{\boldsymbol{\rho}}{\rho} = \hat{\boldsymbol{\rho}} = [cos(\alpha)cos(\delta), \sin(\alpha)cos(\delta), sin(\delta)]^T \tag{8}$$

which gives

$$\alpha = tan^{-1}\left(\frac{\hat{\rho}(2)}{\hat{\rho}(1)}\right) \tag{9}$$

$$\delta = tan^{-1}\left(\frac{\hat{\rho}(3)}{\hat{\rho}(1) \times (1/cos(\alpha))}\right) \tag{10}$$

Equation 10 is written in a somewhat unconventional form as a hint to use four quadrant inverse (atan2) when implementing these equations on a computer.

Detailed modelling of the complexities of a modern SSA radar system is beyond the scope of this work. This includes details about the complex field of view of the radars, how multiple radar beams cover this field of view, modelling individual radar beams and coming up with associated measurement error statistics.

Unrealistic sensor models will inevitably bias the tracking performance. Therefore, efforts have been made to recreate scenarios and model sensor data realistically. Modern SSA radars have a beam-width of $0.1° \times 0.2°$, and can measure the range ρ with range residuals in the order of 15 m and range-rate with range-rate residuals in the order of 15 cm/s. In order to roughly recreate this for the simulation presented in Subsect. 3.2, it is assumed that the errors in the measurements can be modelled as zero mean Gaussian white noise with a measurement error covariance, R, given below:

$$R = \begin{bmatrix} (15\,m)^2 & 0 & 0 & 0 \\ 0 & (15\,cm)^2 & 0 & 0 \\ 0 & 0 & (0.05°/\sqrt{3})^2 & 0 \\ 0 & 0 & 0 & (0.1°/\sqrt{3})^2 \end{bmatrix} \tag{11}$$

The standard deviations of the measurement errors in α and δ being used here are just the standard deviation of a uniform distribution corresponding to the beam-width along α and δ. This measurement error covariance matrix is used on the simulation (to generate measurements) as well as on the model side. Note that the standard deviation of the errors in the angle measurements made by radars are still a couple of orders of magnitude higher than the typical values for SSA telescopes ($\sim 10^{-3}$ radians vs $\sim 10^{-5}$ radians).

3.1   Initialization: G-PAR MC Particle Cloud

Sample from : $p(a, \Omega, \rho, \dot{\rho}, \alpha, \delta)$ to get $(a^j, \Omega^j, \rho^j, \dot{\rho}^j, \alpha^j, \delta^j)$.
Use $(\rho^j, \alpha^j, \delta^j)$ to get the range vector i.e. the relative position vector of the space object with respect to the sensor site, ρ^j, using:

$$\rho^j = \rho^j [cos(\alpha^j)cos(\delta^j), sin(\alpha^j)cos(\delta^j), sin(\delta^j)]^T \tag{12}$$

Use ρ^j and sensor position vector $\mathbf{q}$ to get inertial position vector, $\mathbf{r}^j$, using

$$\mathbf{r}^j = \mathbf{q} + \rho^j \tag{13}$$

The unit vector perpendicular to the orbital plane, $\hat{\mathbf{h}}$ is completely defined by the inclination i and right ascension of ascending node Ω

$$\hat{\mathbf{h}}^j = [sin(i^j)sin(\Omega^j), -sin(i^j)cos(\Omega^j), cos(i^j)]^T \tag{14}$$

The sampled inertial position vector must lie on the orbital plane, which gives

$$\mathbf{r}^j \cdot \hat{\mathbf{h}}^j = 0 \tag{15}$$

Eqs. 14 and 15 can be used to solve for the appropriate values of i^j. This concludes the first two steps listed in the PAR algorithm, i.e., getting the position vector and the orbital plane corresponding to the sample.

Next we find the elements of the inertial velocity vector $\dot{\mathbf{r}}^j$ (expressed in the ECI reference frame), corresponding to the sample, using the following equations:

$$\dot{\mathbf{r}}^j \cdot \hat{\mathbf{h}}^j = 0 \tag{16}$$

$$\dot{\mathbf{r}}^j \cdot \hat{\boldsymbol{\rho}}^j = \dot{\rho}^j \tag{17}$$

Equations 17 and 16 are two linear equations with three unknowns, the third equation relating the elements of the vector $\dot{\mathbf{r}}^j$, is the Eq. 3 which is quadratic in the unknown and thus results in two possible solutions to the $\dot{\mathbf{r}}^j$. At this point we check if the solution satisfies the following

$$\frac{\mathbf{r}^j \times \dot{\mathbf{r}}^j}{|\mathbf{r}^j \times \dot{\mathbf{r}}^j|} = [sin(i^j)sin(\Omega^j), -sin(i^j)cos(\Omega^j), cos(i^j)]^T \tag{18}$$

Most of the time only one out of the two solutions of $\dot{\mathbf{r}}^j$ will pass this check but in some cases, depending on the geometry, it is possible that both might be correct. If such a situation occurs the MC particles in state space will have to be weighted.

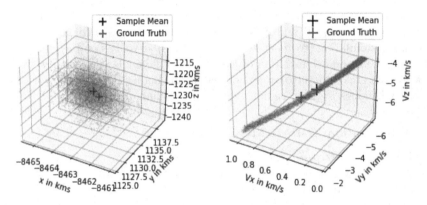

Fig. 2. Left: PAR Particle cloud in x-y-z space Right: PAR Particle cloud for radar measurements in $\dot{x}$-$\dot{y}$-$\dot{z}$ space

Figure 2 shows the particle cloud representing the uncertainty in position $\mathbf{r}$, velocity $\dot{\mathbf{r}}$. This was generated by running the G-PAR algorithm for radar measurements. Measurements were generated by a sensor with the measurement model given by Eq. 5. The measurement error covariance matrix used was listed in Eq. 11. The ground truth was given by,

$$[a, e, i, \Omega, \omega, \nu]^T = [8500\,\text{km},\ 0.05,\ 55°,\ -13.4°,\ -60°,\ 250°]^T \tag{19}$$

The following pieces of a priori knowledge were used:

$$a \sim U(8200\,\text{km}, 9700\,\text{km}) \tag{20}$$

$$\Omega \sim U(-20°, -10°) \tag{21}$$

Instead of p(a,Ω), with small modifications to what was presented here, G-PAR can be made to adapt to the following pairs of a priori knowledge: p(a, i), p(e,Ω), p(e, i).

4 G-PAR for Angles-Only, Short-Arc, Telescope Observation

SDA telescopes can take good right ascension (α) and declination (δ) measurements. The G-PAR algorithm tries to map the postulated uncertainties in four orbital parameters (a, e, i, Ω) and measurements (α, δ) into the probability density function (*pdf*) of the states. The measurement model can be written as follows:

$$\mathbf{y} = h(\mathbf{x}) + \nu \tag{22}$$

$$h(\mathbf{x}) = [\alpha, \delta]^T \tag{23}$$

Eq. 1 defines ρ which is the relative position vector of the RSO with respect to the sensor site. α and δ can be expressed in terms of ρ as following:

$$\frac{\boldsymbol{\rho}}{\rho} = \hat{\boldsymbol{\rho}} = [cos(\alpha)cos(\delta), sin(\alpha)cos(\delta), sin(\delta)]^T \tag{24}$$

$$\alpha = tan^{-1}\left(\frac{\hat{\rho}(2)}{\hat{\rho}(1)}\right) \tag{25}$$

$$\delta = tan^{-1}\left(\frac{\hat{\rho}(3)}{\hat{\rho}(1) \times (1/cos(\alpha))}\right) \tag{26}$$

Eq. 26 is written in a somewhat unconventional form as a hint to use four quadrant inverse (atan2) when implementing these equations on a computer.

4.1 Initialization: G-PAR MC Particle Cloud

Sample from : $p(a, e, i, \Omega, \alpha, \delta)$ to get $(a^j, e^j, i^j, \Omega^j, \alpha^j, \delta^j)$.
Use i^j, Ω^j to fix an orbital plane. The unit vector perpendicular to the orbital plane is given by:

$$\hat{\mathbf{h}}^j = [sin(i^j)sin(\Omega^j), -sin(i^j)cos(\Omega^j), cos(i^j)]^T \tag{27}$$

Use (α^j, δ^j) to get the line of sight vector i.e. the unit relative position vector of the RSO with respect to the sensor site, $\boldsymbol{\rho}^j$, using:

$$\hat{\boldsymbol{\rho}}^j = [cos(\alpha^j)cos(\delta^j), sin(\alpha^j)cos(\delta^j), sin(\delta^j)]^T \tag{28}$$

Use the fact that the RSO must lie on the orbital plane and it must also lie along the line of sight vector to write the following ,

$$\mathbf{r}^j = \mathbf{q} + k\hat{\boldsymbol{\rho}}^j \tag{29}$$

$$\mathbf{r}^j \cdot \hat{\mathbf{h}}^j = 0 \tag{30}$$

therefore,

$$(\mathbf{q} + k\hat{\boldsymbol{\rho}}^j) \cdot \hat{\mathbf{h}}^j = 0 \tag{31}$$

Solve for k using Eq. 31 and replace k it in Eq. 29 to get inertial position vector r^j. The left-hand side of the Fig. 3 visualizes the Eq. 31.

Now, use the sampled value of the semi-major axis a^j and eccentricity e^j to get the shape of the orbital ellipse. Consider the right half of the Fig. 3 that looks at the 2-D picture in the orbital plane. It can be shown that with an ellipse with a fixed shape (a^j, e^j) and a fixed focus (center of the Earth) can have at most two possible intersections with a point (the position of the RSO, given by r^j) on the orbital plane. The velocity vector corresponding to these points can be found. Details about the procedure are being skipped due to paucity of space but are very straightforward.

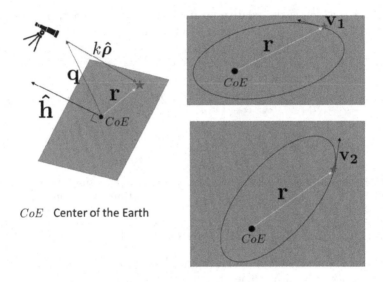

CoE Center of the Earth

Fig. 3. Left: Line of sight intersecting the orbital plane. Right: The velocity vectors corresponding to possible intersections between the orbital ellipse (with fixed shape) and the position vector

Figure 4 shows the particle cloud representing the uncertainty in position $\mathbf{r}$, velocity $\dot{\mathbf{r}}$. This was generated by running the G-PAR algorithm for telescope measurements. Measurements were generated by a sensor with the measurement model given by Eq. 23. The measurement error covariance matrix used is given by:

$$R = \begin{bmatrix} (2 \text{ arc second})^2 & 0 \\ 0 & (2 \text{ arc second})^2 \end{bmatrix} \tag{32}$$

The ground truth was given by,

$$[a, e, i, \Omega, \omega, \nu]^T = [27500 \text{km}, 0.05, 55°, -13.4°, -60°, 116°]^T \tag{33}$$

The following pieces of a priori knowledge were used:

$$a \sim U(24000 \text{km}, 30000 \text{km}) \tag{34}$$

$$\Omega \sim U(-25°, -5°) \tag{35}$$

$$i \sim U(50°, 60°) \tag{36}$$

$$e \sim U(0, 0.1) \tag{37}$$

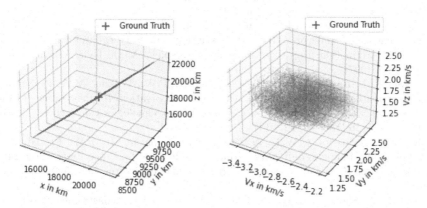

Fig. 4. Left: PAR Particle cloud in x-y-z space Right: PAR Particle cloud for telescope measurements in ẋ-ẏ-ż space

Many of the samples drawn from $p(a, e, i, \Omega, \alpha, \delta)$ will not map into state space because either there will be no appropriate (i.e. on the right side of the sensor) intersection between line o sight and the orbital plane or there will be no intersection between the orbital ellipse corresponding to the sample and the position vector corresponding to the sample.

5 Conclusion

A geometric solution to Probabilistic Admissible Regions was introduced for radar and optical telescope observations. The proposed scheme maps the uncertainty in measurements and postulated states to uncertainty in states. The resulting G-PAR particle cloud initializes the *pdf* of the states of the RSO with just a single measurement. Future work will show recursive updates of G-PAR *pdf* and its usage in single and multiple space object tracking.

Acknowledgements. The authors are thankful to Dr. Erik Blasch, AFOSR, and AFWERX for providing generous funding to support this research work via contract no. FA9550-21-P-0008.

References

1. DeMars, K.J., Jah, M.K.: Initial orbit determination via gaussian mixture approximation of the admissible region. Adv. Astron. Sci. **143** (2012)
2. Faber, W., Chakravorty, S., Hussein, I.: Multi-object tracking with multiple birth, death, and spawn scenarios using a randomized hypothesis generation technique (RFISST). In: Proceedings of the 19th International Conference on Information Fusion, pp. 154–161. IEEE, Darmstadt (2016)
3. Gooding, R.H.: A new procedure for the solution of the classical problem of minimal orbit determination from three lines of sight. Celest. Mech. Dyn. Astron. **66**(4), 387–423 (1997)
4. Hussein, I.I., Roscoe, C.W.T., Schumacher, Jr., P.W., Wilkins, M.P.: Probabilistic admissible region for short-arc angles-only observation. In: Proceedings of the Advanced Maui Optical and Space Surveillance Technologies Conference, Wailea, HI, 9–12 September 2014 (2014)
5. Hussein, I.I., Roscoe, C.W.T., Wilkins, M.P., Schumacher, Jr., P.W.: Probabilistic admissibility in angles-only initial orbit determination. In: Proceedings of the 24th International Symposium on Space Flight Dynamics, Laurel, MD, 5–9 May 2014 (2014)
6. Kelecy, T., Shoemaker, M., Jah, M.: Application of the constrained admissible region multiple hypothesis filter to initial orbit determination of a break-up. In: Proceedings of the 6th European Conference on Space Debris, Darmstadt, Germnay (2013)
7. Roscoe, C.W.T., Hussein, I.I., Wilkins, M.P., Schumacher, P.W., Jr.: The probabilistic admissible region with additional constraints. Adv. Astron. Sci. **156**, 117–130 (2015)
8. Roscoe, C.W.T., Schumacher, P.W., Jr., Wilkins, M.P.: Parallel track initiation for optical space surveillance using range and range-rate bounds. Adv. Astron. Sci. **150**, 989–1008 (2014)
9. Roscoe, C.W.T., Wilkins, M.P., Hussein, I.I., Schumacher, Jr., P.W.: Uncertain angles-only track initiation for SSA using different iod methods. Adv. Astron. Sci. **158** (2016)
10. Tommei, G., Milani, A., Rossi, A.: Orbit determination of space debris: admissible regions. Celest. Mech. Dyn. Astron. **97**, 289–304 (2007)

Radar Cross-Section Modeling of Space Debris

Justin K. A. Henry[1], Ram M. Narayanan[1(⊠)], and Puneet Singla[2]

[1] Department of Electrical Engineering, Pennsylvania State University, University Park, PA 16802, United States
rmn12@psu.edu
[2] Department of Aerospace Engineering, Pennsylvania State University, University Park, PA 16802, United States

Abstract. Space domain awareness (SDA) has become increasingly important as industry and society seek further interest in occupying space for surveillance, communication, and environmental services. To maintain safe launch and orbit-placement of future satellites, there is a need to reliably track the positions and trajectories of discarded launch designs that are debris objects orbiting Earth. In particular, debris with sizes on the order of 20 cm or smaller travelling at high speeds maintain enough energy to pierce and permanently damage current, functional satellites. To monitor debris, the Dynamic Data Driven Applications Systems (DDDAS) paradigm can enhance accuracy with object modeling and observational updates. This paper presents a theoretical analysis of modeling the radar returns of space debris as simulated signatures for comparison to real measurements. For radar modeling, when the incident radiation wavelength is comparable to the radius of the debris object, Mie scattering is dominant. Mie scattering describes situations where the radiation scatter propagates predominantly, i.e., contains the greatest power density, along the same direction as the incident wave. Mie scatter modeling is especially useful when tracking objects with forward scatter bistatic radar, as the transmitter, target, and receiver lie along the same geometrical trajectory. The Space Watch Observing Radar Debris Signatures (SWORDS) baseline method involves modeling the radar cross-sections (RCS) of space debris signatures in relation to the velocity and rotational motions of space debris. The results show the impact of the debris radii varying from 20 cm down to 1 cm when illuminated by radiation of comparable wavelength. The resulting scattering nominal mathematical relationships determine how debris size and motion affects the radar signature. The SWORDS method demonstrates that the RCS is proportional to linear size, and that the Doppler shift is predominantly influenced by translation motion.

Keywords: space domain awareness · space debris modeling · radar cross section · target tracking · bistatic radar · micro-Doppler · low earth orbit

© The Author(s), under exclusive license to Springer Nature Switzerland AG 2024
E. Blasch et al. (Eds.): DDDAS 2022, LNCS 13984, pp. 81–94, 2024.
https://doi.org/10.1007/978-3-031-52670-1_8

1 Introduction

Traditionally, Space Situational Awareness (SSA) focused on monitoring the human-made objects in space [1]. In 2019, the defense community changed the focus to Space Domain Awareness (SDA) [2] to address the growing interest in the industrial, commercial, and societal impacts of space [3]. SDA includes the identification, characterization, and understanding of any factor, passive or active, associated with the space environment that could affect space operations and thereby impact communications security (e.g., TV data integrity), navigational safety (e.g., vehicle autonomous positioning), environmental resiliency (e.g., space weather events), or the economy (e.g., timely financial transactions) [4–6]. Consequently, while SDA incorporates a holistic approach to space domain management, a key element remains of SSA for knowing the presence, characteristics and effects manned and unmanned objects.

From the SDA definition which includes tracking resident space objects (RSOs), there are numerous elements of "passive" human-made space object debris that require monitoring, assessment, maneuvering around, mitigation, or remediation. In particular, debris objects with size on the order of 20 cm or smaller travelling at high speeds maintain enough energy to pierce and permanently damage current, functioning satellites. According to the National Aeronautics and Space Administration (NASA), orbital debris can travel at speeds between 7 and 8 km/s and have impact velocities of up to 10 km/s [7]. Therefore, the debris objects have enough momentum to induce reasonable threat towards current and future satellites. Modern attempts to track space debris include electro-optical [8] and astronomical [9] sensing. Recent efforts include radar sensing due to the ability to remotely observe objects during any time of day and season. Also, there are many radar systems (transmitters and receivers) readily available for space sensing missions [10].

From the SDA definition, the Dynamic Data Driven Applications Systems (DDDAS) paradigm inspires an information management systems architecture. Wong, et al. [11] developed a DDDAS-inspired INFOrmation and Resource Management (INFORM) framework for SSA applications, shown in Fig. 1. Using the notional INFORM architecture that was designed for resident space object (RSO) detection (e.g., satellites, asteroids) from stereo vision, laser scanning, and photoclinometry data, there is an opportunity to utilize radar measurements to complement the optical sensing data. Figure 2 highlights the Space Watch Observing Radar Debris Signatures (SWORDS) approach. The approach is similar to the INFORM method aligned with the DDDAS paradigm that includes the instrumentation reconfiguration loop (IRL) and the data augmentation loop (DAL). While this paper focuses on radar, the concept is akin to the original optical approach for Space Watch Observing Optical Debris Signatures from Ref. [11].

SWORDS has three attributes for SDA: modeling events, sensing situations, and controlling awareness. The need for SDA includes modeling the operating conditions of the sensors, environment, and targets (SET). Specifically, this paper models a radar sensor's perspective in relation to the debris target. The operating conditions are utilized for the space scenario to feed the propagation models of the debris analysis. Sensing situations includes the observations from space sensors as well as using the propagation models for data fusion and data assimilation. Finally, the output of the results supports SDA products and subsequent instrumentation control for a User Defined Operating

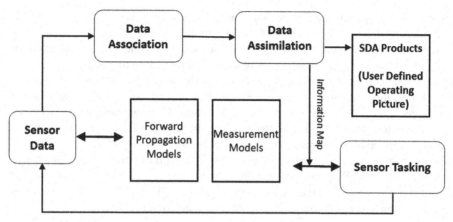

Fig. 1. INFORM Framework: A DDDAS for SSA applications (from [11]).

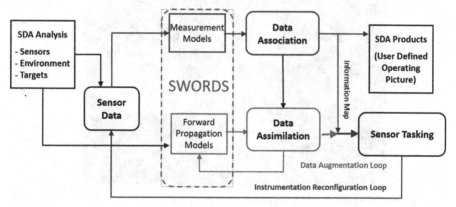

Fig. 2. SWORD Framework: A DDDAS for SSA applications for debris detection (from [11]).

Picture (UDOP). The modeling is based on the Mie scattering characteristics of the debris scattering signature.

The rest of the paper is structured as follows: Section 2 motivates Mie scattering analysis while Sect. 3 describes the bistatic radar analysis. Section 4 provides experiments with conclusions being presented in Sect. 5.

2 Mie Scattering Modeling Motivation

When the incident radiation wavelength is comparable to the radius of the debris object, *Mie scattering* is dominant [12]. Mie scattering describes situations where the radiation scatter propagates predominantly, i.e., contains the greatest power density, along the same direction as the incident wave as shown in Fig. 3 [13]. Two configurations are used in radar sensing, i.e., monostatic and bistatic radar.

Mie scatter modeling is especially useful when tracking objects with forward scatter bistatic radar, as the transmitter, target, and receiver lie along the same baseline, i.e., the line describing the distance between the transmitter and receiver in a particular bistatic pair. Additionally, for a monostatic radar operating at the radar wavelength, which could be receiving the backscatter, the least powerful section of the scattered radiation is measured. Monostatic radar is the more conventional radar configuration where the transmit antenna, or transmitter, is co-located with the receiver antenna, or receiver. Most times, these antennae are designed together. In contrast, bistatic radar involves transmitter and receiver antennae at separate locations. Consequently, there will be two range values to consider: the transmitter-to-target range, and the receiver-to-target range. The angle formed between these range vectors is known as the bistatic angle. The bistatic angle plays a crucial role in determining the Doppler shift of a travelling debris target. Once the bistatic angle is extended to 180°, the bistatic radar is said to be in Forward Scatter (FS) geometry, and the radar cross section (RCS) is determined as a function of Babinet's principle which can greatly simplify computation [14].

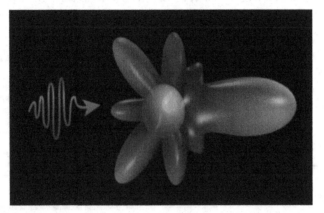

Fig. 3. Mie scattering pattern of an electromagnetic plane wave by a homogeneous sphere (from [13]).

Modern methods of space debris and satellite tracking involve the use of very large, quasi-monostatic phased-array radars operating in S-band [15, 16]. Both Refs. [15] and [16] describe the ability to detect debris as small as 10 cm, while recently, LeoLabs has claimed to be able to track debris as small as 2 cm [17]. However, these radar arrays are very expensive and require more constraints on the frequency spectrum available for radar services while potentially increasing clutter signatures. The motivation for space situational awareness (SSA) is based on passive bistatic radar (PBR). The passive nomenclature represents radar systems that employ transmitters of opportunity which are transmitters that are already available and could be optimized for the receiver unit. Hence, PBR is commonly called "green radar" [18] as it does not add to the clutter of signals in the Earth's atmosphere.

Current mathematical models show that Doppler shift goes to zero once the target crosses the baseline in a PBR system [19]. Therefore, to obtain non-zero values of Doppler, observation needs to occur over a range of angles before and after the baseline [20]. Utilizing PBR systems with sensing angle constraints, this paper focuses on the variation of the RCS with respect to Doppler and micro-Doppler shifts that stem from the debris altitude, debris velocity, debris rotation, and the illuminating frequency.

3 Bistatic Radar Characteristics and RCS

Three elements of radar processing include forward scatter geometry, radar cross section, and velocity-based Doppler shift, all of which are included in developments in deep learning radar methods [20].

3.1 Forward Scatter Geometry

Figure 4 depicts the geometry of the forward scatter bistatic radar as the bistatic angle approaches 180°, as well as, the corresponding vectors and angles; where Tx is the transmitter, Rx is the receiver, and r_T and r_R are the transmitter-to-target and receiver-to-target ranges, β is the bistatic angle, L is the baseline which is equal to $r_T + r_R$ at $\beta = 180°$, V is the target's velocity vector, and δ is the angle between the target's velocity vector and the bistatic bisector. In the FS configuration, δ is equal to 0° when approaching the baseline and 180° when leaving the baseline. Although not drawn to scale, it should be noted that this paper considers Rx to be a receiver on Earth while Tx is a GPS transmitter located 20,200 km away in medium earth orbit (MEO). Therefore, $|r_T| \gg |r_R|$.

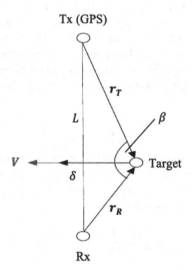

Fig. 4. Passive bistatic radar (PBR) in Forward Scatter (FS) configuration.

3.2 Radar Cross Section (RCS)

According to Balanis [12], in reference to a spherical object, the Mie scattering region is defined by values of radius, a, when $0.1\lambda < a < 2\lambda$, where λ is the illuminating wavelength. The GPS signal operates at a frequency of 1.575 GHz or at a wavelength of 19.05 cm, which would make it suitable for FS PBR when the target's radius is between 1.91 cm and 38.10 cm. For the purpose of this paper, we will look at targets of radii between 1 cm and 20 cm.

For the FS geometry, the RCS is dependent on Babinet's principle, which states that a signal diffracted around the target will produce an equal, but opposite signal compared to when the signal passes through a hole in a perfectly electrically conducting plane where the hole area is equivalent to the target's silhouette area [14]. Therefore, the RCS can be calculated from the area and dimension of a silhouette area which presents itself as a 2-dimensional (2D) projection of the 3-dimensional target. The paper motivation is to use a sphere and a cube as fundamental target shapes to resolve the geometrical configuration. Therefore, the RCS will depend on the circular and square surface areas as shown in the following relationship:

$$\sigma_{FS} = \frac{4\pi A^2}{\lambda^2} \tag{1}$$

where σ_{FS} is the forward scatter RCS, A is the silhouette area, and λ is the illuminating wavelength.

3.3 Velocity-Induced Doppler Shift

Since the target is in motion relative to the transmitter and receiver, the reflected (or diffracted) signal will be of a frequency that is slightly different from the transmitter frequency [19]. The frequency difference is widely known as a *Doppler shift* [21]. Similar to the RCS, the Doppler shift has a unique representation for bistatic geometry. In fact, there are two representations – both of which will be employed for the purpose of gauging the effects of bistatic Doppler shift on the RCS. The first expression is:

$$f_{D_{Trans}} = f_{D_{Bi}} = \frac{2f}{c}|V|\cos\left(\frac{\beta}{2}\right)\cos(\delta) \tag{2}$$

where $f_{D_{Trans}}$ is the translation-induced Doppler shift, $f_{D_{Bi}}$ is the bistatic Doppler shift and f is the illuminating frequency. The other variables were previously defined with β as the bistatic angle, and δ as the angle between the target's velocity vector and the bistatic bisector. The second expression is:

$$f_{D_{Bi}} = \frac{1}{2\pi}\frac{1}{\lambda}(v_T + v_R) \tag{3}$$

where v_T and v_R are the target's radial velocity towards the transmitter and receiver, respectively, and are defined as:

$$v_T = V\frac{r_T}{|r_T|} \tag{4}$$

$$v_R = V\frac{r_R}{|r_R|} \tag{5}$$

Equation (3) can be solved for wavelength, and substituted into Eq. (1) to produce a Doppler-shifted RCS given by

$$\sigma_{FS} = \frac{4\pi A^2}{\left[\frac{1}{2\pi}\frac{1}{f_{D_{Bi}}}(v_T + v_R)\right]^2}$$

(6)

3.4 Rotation-Induced Micro-doppler Shift

When objects have features on their bodies that move separately from the translation of the bulk center of mass, micro-Doppler shifts will be produced around the main Doppler shift. These micro-Doppler features can be points, edges, or other protrusions from the main body of the object. The most pronounced of these protrusions are titled "dominant scatterers," and produce the additional Doppler shifts. The motions are classified as rotations or vibrations, and the objects or bodies are classified as rigid body or nonrigid body [19]. This paper assumes that the target debris is solid and of finite size, and is therefore subject to rigid body motion [22]. Since the sphere has a fairly uniform shape with no special protrusions, only the cube will be considered to be subject to rotation-induced Doppler shifts. This phenomenon is defined by:

$$f_{mD_{Bi}} = \frac{2f}{c}\left\| \boldsymbol{\Omega} \times \boldsymbol{r_p}(t) \right\| \cos\left(\frac{\beta}{2}\right)$$

(7)

where $f_{mD_{Bi}}$ is the bistatic micro-Doppler shift, $\boldsymbol{\Omega}$ is the rotation vector (see Fig. 5) represented as $\boldsymbol{\Omega} = \left(\omega_x, \omega_y, \omega_z\right)^T$ and $\boldsymbol{r_p}(t)$ is the position vector describing the location of a point, P, a scatterer on the surface of the debris target. Assuming the rotating body (cube) has its own coordinate system centered at (x, y, z), one can describe a point P as a vertex located at $\boldsymbol{r_p} = \left(\frac{d}{2}, \frac{d}{2}, \frac{d}{2}\right)^T$ when $t = 0$ and d is the linear dimension or the length of an edge as the average distance $\left(\frac{d}{2}\right)$ for each dimension.

Finally, after considering the translation velocity of the bulk object and the rotation vector of the dominant scatterer, the total Doppler shift is:

$$f_{D_{Bi}} = f_{D_{Trans}} + f_{mD_{Bi}} = \frac{2f}{c}\cos\left(\frac{\beta}{2}\right)\left\{ |V|\cos(\delta) + \left\| \boldsymbol{\Omega} \times \boldsymbol{r_p}(t) \right\| \right\}$$

(8)

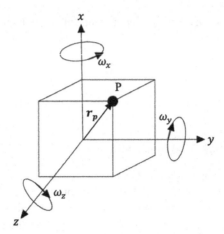

Fig. 5. Cube geometry showing body-centered axes with rotation vectors and dominant scatterer, P.

4 Simulation Results

4.1 RCS vs. Silhouette Area

The Mie scattering regime dictates that it is only present when $0.1\lambda < a < 2\lambda$. This paper looks at objects sized around $1\,\text{cm} \leq d \leq 20\,\text{cm}$, where d is the linear dimension of the silhouette which will be used in place of a. For the sphere and cube targets, d is the radius of the sphere and the magnitude or length of a cube edge.

Figure 6 shows that the RCS for the sphere target is greater than that of the cube target. However, the result is not likely due to the silhouette being larger $\left(\pi d^2 \text{vs.} d^2\right)$. The most significant takeaway is that both of the RCS values are rather low with the cube target being lower than 0 dBsm ($1\,\text{m}^2$) over the entire size range at the GPS wavelength. The sphere target RCS values began to increase past 0 dBsm when the ratio of the radius to the wavelength began to approach unity.

4.2 Translation (Velocity-Induced) Doppler Shift

Since the Doppler average is zero at the baseline, there is a need to account for the situation when the target is approaching and departing the baseline to gather any meaningful Doppler information [14]. Since Eq. (2) only takes into consideration the radar geometry, operating frequency and target velocity, the Doppler shifts will be the same irrespective of target shape.

Figure 7 simulates a case in which the velocity at a constant magnitude of 7.5 km/s is the average of the range provided by NASA [7]. Figure 7 displays the Doppler shift over the journey of the debris target starting from a bistatic angle of 160° as the target approached the baseline and finishing at an angle of 160° on the opposite side as the target departed the baseline. Not surprisingly, the constant velocity produced a very linear response. As 180° was approached, the Doppler shift approached zero as expected. The reverse is true for the second leg of the target's journey.

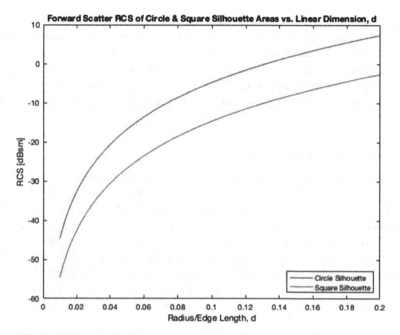

Fig. 6. RCS vs. Radius/Edge length for the circle and square silhouette areas.

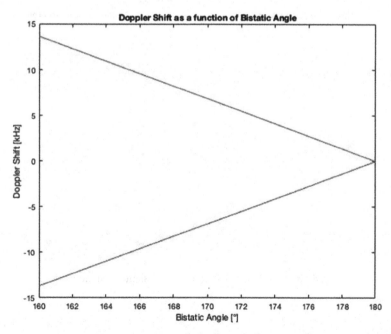

Fig. 7. Target Doppler when approaching and departing the FS PBR baseline.

Also, it is important to note that the Doppler range at these angles was only $\pm$ 13.67 kHz at an operating frequency of 1.575 GHz. Combining the observation with the fact that the RCS values are relatively low without noise and operational conditions, Ref. [23] emphasizes how much the radar gain, integration gain, and post-processing gain [24] could be beneficial in a practical system. The observations could be implemented as part of a data-driven system that would allow for further analysis and predictive modeling [25].

4.3 Translation-Doppler-Shifted RCS

Figure 8 displays ranges of RCS values for debris targets of linear dimensions: $d =$ 0.2 m, 0.1 m, 0.05 m, 0.025 m and 0.01 m, effectively covering the range of values from Fig. 6. The maximum RCS magnitudes occurred at the greatest Doppler shifts which correspond to the smallest bistatic angles. For example, when considering a 0.2 m target at $\pm$13.67 kHz of shift, the sphere had an RCS of 13.99 dBsm, and the cube had an RCS of 4.04 dBsm. In Fig. 6, the system accounted for only the static target case at 180°. Now that the targets have a high velocity, the RCS plunges to several hundred dB below 0 dBsm. The targets effectively disappear into noise as the time spent in that location (on the baseline) is extremely small. Again, long integration times and signal processing gain are a must.

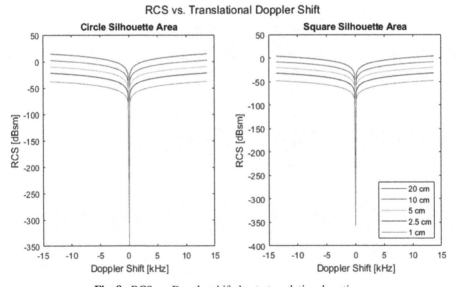

Fig. 8. RCS vs. Doppler shift due to translational motion.

4.4 Micro-doppler (Rotation-Induced) Shift

Only the cube targets were considered for the micro-Doppler calculation as only the cube shape has significant protrusions that may play a part in further diffracting the illuminating signal. Once again, the largest value of d was used so the point scatterer vector is $r_p = \left(\frac{d}{2}, \frac{d}{2}, \frac{d}{2}\right)^T$. The body-centered motion example that we used was a half-rotation in x and z and a full rotation in y. Therefore, the rotation vector, $\Omega = (\pi, 2\pi, \pi)^T$ in units of rad/sec. With these values, Fig. 9 shows the rotation-induced Doppler shift versus bistatic angle.

From the results, the same linear relationship between the micro-Doppler and the bistatic angle are resolved as from the velocity-induced Doppler shift in Fig. 7. However, one can note that for positive rotation values, the Doppler shift remains positive as from testing negative rotation vectors. It is worth noting that the micro-Doppler shifts are extremely small, but proportional – a 1:1 ratio between linear dimension and micro-Doppler shift can be seen. Previously, the Doppler shifts were on the order of kHz, and now, the micro-Doppler shifts are on the order of < 1 Hz, which will most likely not vary the overall Doppler shift by any significant amount. In fact, rotation of the very small targets ($d = 0.01$ m) is virtually invisible to the radiation.

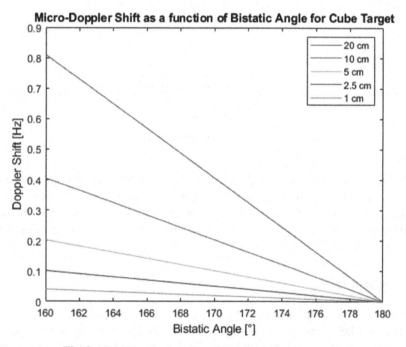

Fig. 9. Cube Target micro-Doppler due to point scatterer, P.

4.5 Translation- and Rotation-Shifted RCS

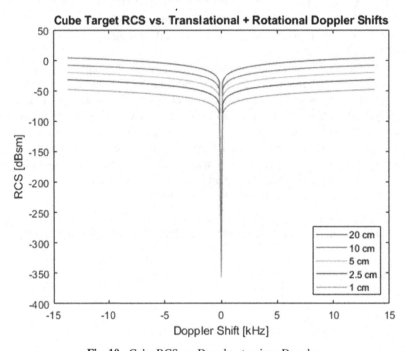

Fig. 10. Cube RCS vs. Doppler + micro-Doppler.

The RCS as a function of total Doppler shift resulting from translation and rotation motions is plotted in Fig. 10. As expected, the graph is virtually identical to that of Fig. 8. In fact, for the $d = 0.2$ m target, the RCS value at ± 13.67 kHz of shift remained the same due to the micro-Doppler shift being extremely small. Therefore, the micro-Doppler at these scales could be ignored.

5 Conclusions

The paper presents radar sensing updates to the DDDAS-inspired INFORM testbed for SDA as SWORDS. A preliminary analysis demonstrates the feasibility of using forward scatter passive bistatic radar (FS-PBR) for the detection and tracking of space debris objects at sizes on the order of 20 cm or smaller when using the L1 GPS signal as an illuminator of opportunity. It has been shown that the radar cross section (RCS) is proportional to its physical size, and that the translational velocity has virtually total influence on any Doppler shift as the magnitude of micro-Doppler signatures are extremely insignificant. Nonetheless, the RCS values especially when approaching 1 cm provide low impact. Thus, to compensate, large and/or many-element radar arrays along with sophisticated signal processing techniques will be required to efficiently separate the signal from the noise and observe the Doppler shift. Next steps could involve analyzing rough-surfaced and irregularly shaped objects and testing array designs. Additionally, motivation is to use sensor data fusion to combine optical and radar information for small space debris tracking to enhance the INFORMS testbed.

Acknowledgments. We thank Dr. Erik Blasch for concept development and paper editing. Partial research support through the Air Force Office of Scientific Research (AFOSR) Grant Number FA9550-20-1-0176 is acknowledged. The views and conclusions contained herein are those of the authors and should not be interpreted as necessarily representing the official policies or endorsements, either expressed or implied, of the Air Force Research Laboratory or the U.S. Government.

References

1. Chen, G., Pham, K.D., Blasch, E.: Special section guest editorial: sensors and systems for space applications. Opt. Eng. **58**(4), 041601 (2019). https://doi.org/10.1117/1.OE.58.4.041601

2. Erwin, S.: Air Force: SSA is no more; it's 'space domain awareness', SpaceNews, 14 November 2019 (2019). https://spacenews.com/air-force-ssa-is-no-more-its-space-domain-awareness/

3. Blake, T.: Space domain awareness (SDA) (2011). https://apps.dtic.mil/sti/pdfs/ADA550594.pdf

4. Holzinger, M.J., Jah, M.K.: Challenges and potential in space domain awareness. J. Guid. Control. Dyn. **41**(1), 15–18 (2018). https://doi.org/10.2514/1.G003483

5. Vasso, A., Cobb, R., Colombi, J., Little, B., Meyer, D.: Augmenting the space domain awareness ground architecture via decision analysis and multi-objective optimization. J. Defense Analytics Logistics **5**(1), 77–94 (2021). https://doi.org/10.1108/JDAL-11-2020-0023

6. Blasch, E., Shen, D., Chen, G., Sheaff, C., Pham, K.: Space object tracking uncertainty analysis with the URREF ontology. In: 2021 IEEE Aerospace Conference, Big Sky, MT (2021).https://doi.org/10.1109/AERO50100.2021.9438207

7. NASA: Frequently asked questions: Orbital debris (2011). https://www.nasa.gov/news/debris_faq.html

8. Sciré, G., Santoni, F., Piergentili, F.: Analysis of orbit determination for space based optical space surveillance system. Adv. Space Res. **56**(3), 421–428 (2015). https://doi.org/10.1016/j.asr.2015.02.031

9. Muntoni, G., et al.: Space debris detection in low earth orbit with the Sardinia radio telescope. Electronics **6**(3), 59 (2017). https://doi.org/10.3390/electronics6030059

10. Jia, B., Pham, K.D., Blasch, E., Wang, Z., Shen, D., Chen, G.: Space object classification using deep neural networks. In: 2018 IEEE Aerospace Conference, Big Sky, MT (2018). https://doi.org/10.1109/AERO.2018.8396567

11. Wong, X.I., Majji, M., Singla, P.: Photometric stereopsis for 3D reconstruction of space objects. In: Chapter 13 in Handbook of Dynamic Data Driven Applications Systems, vol. 1, 2nd Edn. (2022). https://doi.org/10.1007/978-3-319-95504-9_13

12. Balanis, C.A.: Advanced Engineering Electromagnetics, New York, NY. Wiley, pp. 655–665 (2012)

13. Wikipedia: Mie scattering. https://en.wikipedia.org/wiki/Mie_scattering

14. Baker, C.: PCL waveforms: NATO S&T organization educational note STO-EN-SET-243-02 (2018)

15. Stevenson, M., Nicolls, M., Park, I., Rosner, C.: Measurement precision and orbit tracking performance of the kiwi space radar. In: Advanced Maui Optical and Space Surveillance (AMOS) Technologies Conference (2020). https://amostech.com/technicalpapers/2020/optical-systems-instrumentation/stevenson.pdf
16. GlobalSecurity.org: Space fence (AFSSS S-Band). https://www.globalsecurity.org/space/systems/space-fence.htm
17. LeoLabs: Global phased-array radar network. https://leolabs.space/radars/
18. Ulander, L.M.H., Frölind, P.-O., Gustavsson, A. Ragnarsson. R., Stenström, G.: Airborne passive SAR imaging based on DVB-T signals. In: 2017 IEEE International Geoscience and Remote Sensing Symposium (IGARSS), Fort Worth, TX, pp. 2408–2411 (2017). https://doi.org/10.1109/IGARSS.2017.8127477
19. Chen, V.C.: The Micro-Doppler Effect in Radar, Norwood, MA. Artech House (2019)
20. Majumder, U., Blasch, E., Garren, D.: Deep Learning for Radar and Communications Automatic Target Recognition, Norwood, MA. Artech House (2020)
21. Niu, R., Zulch, P., Distasio, M., Blasch, E., Shen, D., Chen, G.: Joint sparsity based heterogeneous data-level fusion for target detection and estimation. In: SPIE Conference on Sensors and Systems for Space Applications X, Anaheim, CA, vol. 10196 (2017). https://doi.org/10.1117/12.2266072
22. Junkins, J.L., Singla, P.: How nonlinear is it? A tutorial on nonlinearity of orbit and attitude dynamics. J. Astron. Sci. 52(1–2), 7–60 (2004). https://doi.org/10.1007/BF03546420
23. Kahler, B., Blasch, E.: Sensor management fusion using operating conditions. In: 2008 IEEE National Aerospace and Electronics Conference (NAECON), Dayton, OH, pp. 281–288 (2008). https://doi.org/10.1109/NAECON.2008.4806559
24. Choi, E.J., et al.: A study on the enhancement of detection performance of space situational awareness radar system. J. Astron. Space Sci. 35(4), 279–286 (2018). https://doi.org/10.5140/JASS.2018.35.4.279
25. Darema, F., Blasch, E.P., Ravela, S., Aved, A.J. (Eds.), Handbook of Dynamic Data Driven Applications Systems, vol. 2. Springer, Cham (2023). https://doi.org/10.1007/978-3-031-27986-7

High-Resolution Imaging Satellite Constellation

Xiaohua Li[1]([envelope]), Lhamo Dorje[1], Yezhan Wang[1], Yu Chen[1],
and Erika Ardiles-Cruz[2]

[1] Binghamton University, Binghamton, NY 13902, USA
{xli,ldorje1,ywang516,ychen}@binghamton.edu
[2] The U.S. Air Force Research Laboratory, Rome, NY 13441, USA
erika.ardiles-cruz@us.af.mil

Abstract. Large-scale Low-Earth Orbit (LEO) satellite constellations
have wide applications in communications, surveillance, and remote sensing. Along with the success of the SpaceX Starlink system, multiple LEO
satellite constellations have been planned and satellite communications
has been considered a major component of the future sixth generation
(6G) mobile communication systems. This paper presents a novel LEO
satellite constellation imaging (LEOSCI) concept that the large number
of satellites can be exploited to realize super high-resolution imaging of
ground objects. Resolution of conventional satellite imaging is largely limited to around one meter. In contrast, the LEOSCI method can achieve
imaging resolution well below a centimeter. We first present the new imaging principle and show that it should be augmented with a Dynamic Data
Driven Applications Systems (DDDAS) design. Then, based on the practical Starlink satellite orbital and signal data, we conduct extensive simulations to demonstrate a high imaging resolution below one centimeter.

Keywords: THz imaging · LEO satellite constellation · Synthetic
aperture radar (SAR) · Integrated communication and sensing · Starlink

1 Introduction

Spaceborne imaging systems have wide applications, including surveillance and
remote sensing [11]. Typically, satellite-based sensing can use a synthetic aperture, e.g., for radar (SAR), to generate ground images, but the resolution is relatively low. The state of the art of the current satellite SAR imaging resolution
is around one meter. Increasing resolution has always been a highly-demanded,
but challenging objective of satellite-based imaging.

Along with the development of SpaceX's Starlink satellite constellation,
which has over 2000 satellites deployed in space and over 50,000 satellites
planned to deploy, it has become a promising research topic to use LEO satellite

Thanks to Erik Blasch and Alex Aved for concept development and co-authorship.
This work is supported by the U.S. Air Force Office of Scientific Research (AFOSR)
under Grant FA9550-20-1-0237.

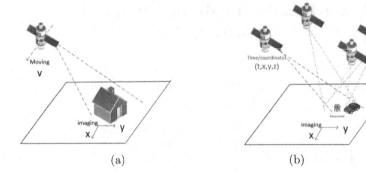

(a)   (b)

Fig. 1. (a) Conventional single-satellite SAR imaging. (b) Proposed LEO satellite constellation imaging (LEOSCI) with high resolutions to sense small targets.

constellations for global communications, including WiFi wireless access and the future sixth generation (6G) mobile communications [5]. Several large-scale LEO satellite constellations are deployed or in planning. They are designed mainly for communications because only the large mobile communications market can cover the huge cost of their deployment and maintenance. Reusing them as a sensing and imaging platform via the integrated communication and sensing design [9] is both interesting and attractive since it will create novel applications and generate new revenues.

This paper develops a novel LEO satellite constellation imaging (LEOSCI) method that uses a large number of satellites to create ground target images with a super-high resolution well below one centimeter. By comparison, as the state-of-the-art of existing commercial satellite imaging, optical camera imaging resolution is 0.3 meters and SAR imaging resolution is much lower, around one meter. To guarantee the performance of this complex imaging system in a dynamic environment, Dynamic Data Driven Applications Systems (DDDAS) design principles [3] will be adopted to address the challenging multi-objective parameterization among imaging resolution, imaging area, number of satellites, imaging time delay, multiple satellite data collection, and data processing. We will use the realistic data of the SpaceX Starlink to conduct simulations to demonstrate the effectiveness of the LEOSCI method.

The organization of this paper is as follows. Section 2 describes the LEO satellite constellation imaging model. Section 3 develops the imaging algorithm and Sect. 4 presents simulation results. Conclusions are given in Sect. 5.

2   Model of Satellite Constellation Imaging

Large-scale LEO satellite constellations such as Starlink create a cellular structure similar to mobile networks, with satellite antenna beams replacing terrestrial cellular towers [14]. At any given time, a cell on the ground is illuminated by one or more beams. The size of the cell is determined by the beam width.

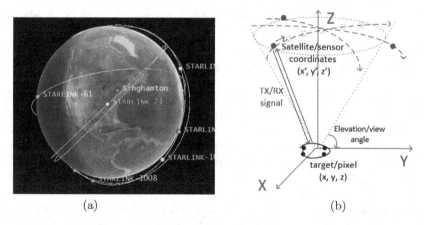

(a)   (b)

Fig. 2. (a) Orbits of some Starlink satellites and the illumination of the target Binghamton by satellite STARLINK-71 (green line). (b) Coordinates of satellites and target. A satellite (red dot) enters the sensing area (red dotted circle) of the target and starts collecting sensing data at various locations along its orbit. (Color figure online)

Each satellite beam only serves this cell for a short time. The cell is continuously served by the satellites flying over.

Existing works that use satellites for ground imaging [6,10] in general apply the synthetic aperture method from only one satellite, as shown in Fig. 1(a). The Y-direction (range) resolution is $\Delta Y = \frac{c}{2B}$, where c is the speed of light and B is the signal bandwidth. Starlink has beam signal bandwidth 250 MHz, so the ground resolution will be 0.85 meters (assuming a 45° incidence angle).

For high-resolution imaging, we propose to use many satellites instead of one only and use imaging techniques developed in millimeter-wave/Tera-Hz (THz) imaging such as airport screening systems [2,4]. As shown in Fig. 1(b), in an LEO satellite constellation, many satellites will fly over the imaging area sequentially. We call them *valid* satellites. When a valid satellite flies over the area, it keeps illuminating the target and recording the reflected echos. Either a communication signal or radar signal can be sent. Multiple satellites within this area can transmit and receive at the same time, just with different frequencies. These satellites' data will be collected and processed together to generate an image of the target. The satellites effectively form a virtual antenna array. The size of this array is the synthetic aperture and determines the imaging resolution.

This LEO satellite constellation imaging concept is a new opportunity with a lot of challenges. One of the major challenges is that only a small number of satellites within the constellation can sense a target, which can hardly form a dense and regular array. THz imaging requires a dense 2D antenna array with antennas evenly spaced at a half-wavelength distance. For example, an array used 736 transmit (Tx) antennas and 736 receive (Rx) antennas to electronically scan 25,600 antenna positions within a regular grid of 50 cm×50 cm aperture [1]. This kind of scan over a 2D regular grid is impossible for satellites. To address this challenge, we adopt the algorithm developed in [8] and will show by simulations that a few satellites are indeed enough.

3   Imaging Algorithm and Resolution

Consider the case that a satellite flies over the target area, transmits a sensing signal, and collects the target echos. As an illustration, Fig. 2(a) shows the STARLINK-71 satellite moves along its orbit and passes over the target located in Binghamton, NY. It keeps sensing the target until flying out of the sensing area, as illustrated in Fig. 2(b). The sensing area is determined by the elevation or viewing angle of the target, which further determines the number of valid satellites and imaging delay.

Assume the coordinates of the satellite and the target are $\mathbf{r}' = (x', y', z')$ and $\mathbf{r} = (x, y, z)$, respectively. The satellite transmits signal $p(t)$ with carrier frequency f_c, and receives the signal $\hat{s}_{\mathbf{r}'}(t)$ reflected by the target,

$$\hat{s}_{\mathbf{r}'}(t) = \int_{\mathbf{r}} \sigma_{\mathbf{r}} p(t - \tau_{\mathbf{r}'\mathbf{r}}) d\mathbf{r} + \hat{v}_{\mathbf{r}'}(t), \tag{1}$$

where $\tau_{\mathbf{r}'\mathbf{r}}$ is the propagation delay, $\sigma_{\mathbf{r}}$ is the target's reflection coefficient, and $\hat{v}_{\mathbf{r}'}(t)$ denotes noise, interference, and clutter [13]. The satellite can transmit either Frequency-Modulated Continuous Wave (FMCW) radar signal waveform [6] or communications signal such as Digital Video Broadcasting - Satellite - 2nd Generation (DVB-S2X) signal waveform [12]. Conventional pulsed radar signals may not be appropriate due to the limited peak transmission power of the communication satellites. Using communication signals directly for sensing is more attractive as it reduces hardware complexity and cost.

With the signal processing outlined in [8], we get a data sample at each sensing position $\mathbf{r}'$ which is

$$s_{\mathbf{r}'} = \int_{\mathbf{r}} \sigma_{\mathbf{r}} e^{j2\pi f_c \tau_{\mathbf{r}'\mathbf{r}}} d\mathbf{r} + v_{\mathbf{r}'}, \tag{2}$$

where $v_{\mathbf{r}'}$ denotes the processed $\hat{v}_{\mathbf{r}'}(t)$. For 2D imaging, the goal is to reconstruct an $I \times J$ target image $\mathbf{X}$ whose pixel is X_{ij}. We stack the columns of $\mathbf{X}$ into an N-dimensional column vector $\mathbf{x} = \text{vec}(\mathbf{X})$, where $N = IJ$. Assume we have collected M data samples from M satellite sensing positions $\mathbf{r}'_m = (x'_m, y'_m, z'_m)$, $m = 0, \cdots, M - 1$. We stack the data into an M-dimensional column vector $\mathbf{y}$, whose mth element is $y_m = s_{\mathbf{r}'_m}$. Then we have

$$\mathbf{y} = \mathbf{H}\mathbf{x} + \mathbf{v}, \tag{3}$$

where the element of the $M \times N$ matrix $\mathbf{H}$ is $H_{mn} = e^{j4\pi R_{mn}/\lambda}$, R_{mn} is the distance between the satellite and the target, and λ is the wavelength. The vector $\mathbf{v}$ includes noise, interference, and clutter.

Based on Eq. (3), one can apply the computationally efficient back-projection (BP) algorithm to reconstruct the image, which gives

$$\hat{\mathbf{x}} = \mathbf{H}^H \mathbf{y}. \tag{4}$$

where $(\cdot)^H$ is the Hermitian transpose. The computational complexity is $O(MN)$. From Eqs. (3) and (4), one can easily see that an extremely large number of antenna samples, i.e., large M, is needed to make $\mathbf{H}^H\mathbf{H}$ close to the identity. This is why a large-scale LEO satellite constellation is needed. Fortunately, each satellite is able to collect a large number of data samples along its orbit. So the BP may just need a few valid satellites for each sensing area.

The signal-to-noise ratio (SNR) of the received signal can be derived as

$$\text{SNR} = \frac{P_t G_t G_r \lambda^2 \theta \Delta_S \sigma_0}{(4\pi R)^3 kTBFL_s}. \tag{5}$$

where P_t is the transmit power, G_r is the receive antenna gain, G_t is the transmit antenna gain, θ is the beam width, Δ_S is the resolution, σ_0 is the radar cross section (RCS) of the target, R is the range, and L_s is the system loss. The noise power is $P_N = kTBF$ where T is the noise temperature, k is the Boltzmann constant, B is the signal bandwidth, and F is the noise factor of the system.

The imaging resolution is

$$\Delta X = \Delta Y = \frac{\lambda R}{2D} \tag{6}$$

where D is the array aperture in each dimension [13]. D is determined by the satellite height and the target elevation angle.

The imaging resolution and quality depend on a lot of factors in a complex way. Smaller elevation angle leads to more valid satellites, larger D, and better resolution, but suffers from longer imaging delay, lower SNR, etc. It is computationally complex to generate a high-resolution image over a large area. It is challenging to guarantee the performance of high resolution imaging in dynamic environments with uncertain propagation attenuation and target reflectivity. Following the DDDAS design principle, we can apply the dynamic feedback control to achieve desired trade-offs [3]. Specifically, starting from a coarse image of a large area, DDDAS leverages continuous sensor measurement data and imaging model to adaptively adjust satellite configuration, signal power/beam width, and image pixel size. This way, we can generate finer images over smaller areas with an affordable complexity [8,15]. Applying the DDDAS concept in the development of this complex system, the super-high resolution images obtained in real-time will enable an application to dynamically incorporate new information to make the working procedure be more adaptive to the environment accurately.

4   Simulation Results

We compared the proposed LEOSCI method with the conventional single-satellite SAR imaging based on the practical SpaceX Starlink satellite constellation data. Specifically, we used the Starlink's two-line element (TLE) file[1] to get the orbital information and coordinates of all the satellites. The July 7, 2022

---

[1] Downloaded from https://celestrak.org/.

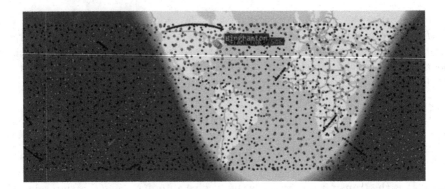

Fig. 3. Starlink satellite constellation on July 7 2022 13:41:11. Satellites are shown as black and red dots. Red dots are the valid satellites that can contribute to imaging the target located at Binghamton, NY (green dot). At this moment, two satellites (bigger red dots) can sense the target. (Color figure online)

TLE file consisted of 2469 Starlink satellites. We simulated a target located at the coordinates of Binghamton, NY. With an elevation angle of 30°, we had 435 valid satellites. All the satellites are shown in Fig. 3. Their distance to the target was around $R = 462$ km. This means we had an effective radar aperture of $1600 \times 1600 \, \text{km}^2$, i.e. $D = 1600 \, \text{km}$. The theoretical resolution is thus $\Delta X = \Delta Y = 0.004$ meter according to (6).

We applied the typical Starlink signal parameters [14]: Carrier frequency $f_c = 11.7$ GHz, bandwidth $B = 250$ MHz, beam width 2.5°, antenna gain $G_t = G_r = 53$ dB, noise power $P_N = -120$ dB, system loss $L_s = 6$ dB, transmit power $P_t = 37$ dBW. If the signal dechirping or cross-correlation provides 30 dB gain, then we have 19 dB SNR for target RCS $\sigma_0 = 100$ and 1 cm resolution.

To generate the simulation data, we created a target consisting of two circles as shown in Fig. 4(a): 4 points on the inner circle of radius 0.02 m and 8 points on the outer circle of radius 0.05 m. We simulated the satellite signal waveforms as either FMCW radar wave or DVB-S2X wave. Each of the valid satellites conducted a transmission and receiving at each of the 435 randomly selected locations on its orbit. The received signals were dechirped in the case of FMCW or cross-correlated with the transmitted signal in the case of DVB-S2X to find the maximum samples, which resulted in a 435×435 data matrix. So $M = 189225$. With the data matrices, we reconstructed images with size $I \times J = 30 \times 30$ (so $N = 900$) for the target area of $0.1 \times 0.1 \, \text{m}^2$ based on (4).

Figures 4(b) and (c) show the reconstructed images with the LEOSCI method when the satellite signal was FMCW and DVB-S2X waveforms, respectively. For comparison, Fig. 4(d) shows the reconstructed image with the conventional single-satellite SAR method. Image reconstruction quality was compared quantitatively in Table 1 in terms of peak-signal-to-noise ratio (PSNR), structural similarity method (SSIM) [7], and mean square error (MSE) between the reconstructed image and the true image of Fig. 4(a).

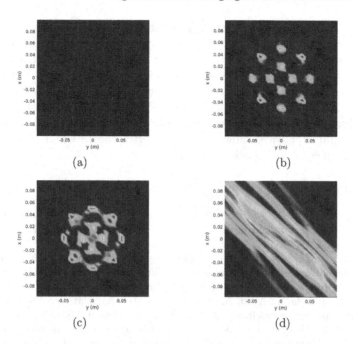

Fig. 4. (a) True target. (b) Target image reconstructed by the LEOSCI method with FMCW waveform. (c) Target image reconstructed by the LEOSCI method with DVB-S2X waveform. (d) Target image reconstructed by conventional single-satellite SAR.

It is easy to see that the LEOSCI method could provide a super high imaging resolution of less than 1 cm, while the conventional SAR failed completely due to insufficient resolution. The true image shown in Fig. 4(a) had a larger size of 200×200 pixels. It was resized to 30×30 pixels when calculating performance metrics. The reconstructed images in Figs. 4(b) and (c) in fact looked much more similar to the resized true image than to Fig. 4(a).

Table 1. Comparison of imaging quality and (theoretical) imaging resolution.

Imaging Method	Resolution (m)	PSNR (dB)	SSIM	MSE
LEOSCI (FMCW)	0.004	15.0818	0.0697	0.0310
LEOSCI (DVB-S2X)	0.004	16.3638	0.0840	0.0231
Single-Satellite SAR	0.85	8.6805	0.0081	0.1355

5   Conclusions

It has been well recognized to leverage the large number of satellites for global communications, but there is not sufficient attention given to their capability as a sensing and imaging constellation. The satellite constellation is potential to serve the compelling need of DDDAS in the space applications. This paper proposes the LEO satellite constellation imaging (LEOSCI) concept that exploits a few set of valid satellites to generate super high-resolution images. Simulations demonstrate that the method can provide ground target images with resolution lower than one centimeter, which is hardly achievable by existing satellite SAR imaging. As future work, we will verify the imaging algorithm with real-measured millimeter wave radar data and verify the LEOSCI concept using UAVs (unmanned aerial vehicles) to collect data.

References

1. Ahmed, S.S., Schiessl, A., Schmidt, L.P.: A novel fully electronic active real-time imager based on a planar multistatic sparse array. IEEE Trans. Microw. Theory Tech. **59**(12), 3567–3576 (2011)
2. Alexander, N.E., et al.: Terascreen: multi-frequency multi-mode terahertz screening for border checks. In: Passive and Active Millimeter-Wave Imaging XVII, vol. 9078, p. 907802. International Society for Optics and Photonics (2014)
3. Blasch, E., Ravela, S., Aved, A.: Handbook of Dynamic Data Driven Applications Systems. Springer, Heidelberg (2018). https://doi.org/10.1007/978-3-319-95504-9
4. Cooper, K.B., Dengler, R.J., Llombart, N., Thomas, B., Chattopadhyay, G., Siegel, P.H.: Thz imaging radar for standoff personnel screening. IEEE Trans. Terahertz Sci. Technol. **1**(1), 169–182 (2011)
5. Giordani, M., Zorzi, M.: Non-terrestrial networks in the 6g era: challenges and opportunities. IEEE Netw. **35**(2), 244–251 (2020)
6. Hoogeboom, P., et al.: Panelsar, an fmcw based x-band smallsat sar for infrastructure monitoring. In: The 27th Annual AIAA/USU Conference on Small Satellites, Logan, USA, pp. 1–5 (2013)
7. Hore, A., Ziou, D.: Image quality metrics: PSNR vs. SSIM. In: 2010 20th International Conference on Pattern Recognition, pp. 2366–2369. IEEE (2010)
8. Li, X., Chen, Y.: Lightweight 2d imaging for integrated imaging and communication applications. IEEE Signal Process. Lett. **28**, 528–532 (2021)
9. Liu, F., et al.: Integrated sensing and communications: towards dual-functional wireless networks for 6g and beyond. IEEE J. Sel. Areas Commun. **40**, 1728–1767 (2022)
10. Liu, Y., Deng, Y.K., Wang, R., Loffeld, O.: Bistatic FMCW SAR signal model and imaging approach. IEEE Trans. Aerosp. Electron. Syst. **49**(3), 2017–2028 (2013)
11. Majumder, U.K., Blasch, E.P., Garren, D.A.: Deep Learning for Radar and Communications Automatic Target Recognition. Artech House, Norwood (2020)
12. Pisciottano, I., Santi, F., Pastina, D., Cristallini, D.: DVB-S based passive polarimetric ISAR-methods and experimental validation. IEEE Sens. J. **21**(5), 6056–6070 (2020)
13. Richards, M.A., Scheer, J., Holm, W.A., Melvin, W.L.: Principles of modern radar. Citeseer (2010)
14. Sayin, A., Cherniakov, M., Antoniou, M.: Passive radar using starlink transmissions: a theoretical study. In: 2019 20th International Radar Symposium (IRS), pp. 1–7. IEEE (2019)
15. Wu, R., Liu, B., Chen, Y., Blasch, E., Ling, H., Chen, G.: A container-based elastic cloud architecture for pseudo real-time exploitation of wide area motion imagery (WAMI) stream. J. Signal Process. Syst. **88**(2), 219–231 (2017)

Main-Track Plenary Presentations - Network Systems

Reachability Analysis to Track Non-cooperative Satellite in Cislunar Regime

David Schwab$^{(\boxtimes)}$ ⓘ, Roshan Eapen ⓘ, and Puneet Singla ⓘ

The Pennsylvania State University, University Park, PA 16802, USA
{dvs5558,rpe5185,psingla}@psu.edu

Abstract. Space Domain Awareness (SDA) architectures must adapt to overcome the challenges present in cislunar space. Dynamical systems theory provides tools which may be leveraged to address some of the many challenges associated with cislunar space. The PSS is an analysis tool used to reduce dimensionality and help study the properties of the system flow. Invariant manifolds have been combined with the PSS to prescribe trajectories through various cislunar regimes by other researchers. In this work, the PSS and the invariant manifolds are used to pose a set of boundary value problems which define the $\Delta \mathbf{v}$ from a nominal L_2 Lyapunov orbit through the PSS. By approximating the solutions through the PSS, the admissible controls onto these highways are approximated. One viable use for this formulation of a reduced reachable set will allow an SDA operator to intelligently task sensors to regain custody of a maneuver spacecraft. This paper examine uses concepts of a admissible region to intelligently reduce the reachability set for maneuver spacecraft and studies the efficacy for multiple maneuver windows and the affects of various user set parameters.

Keywords: Three-Body Problem · Space Domain Awareness · Reachable set · Admissible region

1 Introduction

As reflected in the Air Force Space Command's report on the *The Future of Space 2060 & Implications for U.S. Strategy*, cislunar space will likely be an important civil, commercial, and military domain in the near future [1]. The current Space Domain Awareness (SDA) architecture has focused on objects in Geostationary and below [2], many of which are not applicable to the cislunar regime. There are significant challenges in the cislunar regime that will require development of new SDA algorithms and platforms, such as the AFRL xGEO space domain awareness flight experiment launching in the near future [3]. These challenges include 1) Data-sparsity, 2) Low sensor signal to noise ratio (SNR), and 3) The sheer volume of the cislunar domain [2,13]. These difficult challenges are only exacerbated in the presence of non-cooperative, maneuvering spacecraft.

This material is based upon work supported jointly by the AFOSR grant FA9550-20-1-0176, FA9550-22-1-0092, as well as the DoD SMART Scholarship Program.

E. Blasch et al. (Eds.): DDDAS 2022, LNCS 13984, pp. 105–113, 2024.
https://doi.org/10.1007/978-3-031-52670-1_10

The reachability set search (RSS) method is a DDDAS-based method to regain custody of a non-cooperative, maneuvering spacecraft [4,5,14,15]. In this algorithm, a user-defined representation of the target's reachability set is used to guide a SDA sensor's search for lost spacecraft. As shown in previous work [5], the reachability set of a spacecraft given a simple bound on $\|\Delta\mathbf{v}\|$ may quickly grow too large to search in any practical sense. They will include unlikely/low-priority trajectories, such as those which leave the L_2 gateway. Therefore, the reachability set may be intelligently reduced by leveraging the concept of admissible regions. Similar work has been done in the GEO- and-below orbital domain via initial orbit determination where only tracklets returning pre-specified bounds, such as Earth-bounded orbits [6,7], are defined. By taking advantage of the invariant manifold structures in cislunar dynamics, there exists a framework to define a set of admissible impulsive controls which produces transit trajectories of interest for a cislunar SDA architecture as shown in [8]. In this way, the reachable set may be intelligently reduced; therefore providing a practical search space for the RSS algorithm. This paper intends to investigate that capabilities of that framework in various maneuver windows.

2 Methodology

Given that a simple bound on $\|\Delta\mathbf{v}\|$ could lead to large reachable set, the authors developed a way to approximate admissible control onto the cislunar highway between L_2 and the lunar plane in the planar circular restricted three-body problem (PCR3BP) in [8], which are governed by the equations

$$\ddot{x} = x + 2\dot{y} - \frac{1-\mu}{r_1^3}(x+\mu) - \frac{\mu}{r_2^3}(x - (1-\mu)) \tag{1}$$

$$\ddot{y} = y - 2\dot{x} - \frac{1-\mu}{r_1^3}y - \frac{\mu}{r_2^3}y. \tag{2}$$

in the synodic reference frame where μ is the mass ratio in the PCR3BP, the state is given by $\mathbf{x} = [x\ y\ \dot{x}\ \dot{y}]^T$, and r_1, r_2 are the magnitudes of distance between the third body and the first and second primaries, respectively. Dynamical systems theory for the PCR3BP defines invariant manifolds which act as separatrices to periodic orbits; therefore, trajectories that lie on the manifold, tend towards or away from the periodic orbit as $t \to \infty$. The intersections of these invariant manifolds on the Poincaré Surface of Section (PSS)–a tool for analysis simplifies the study of specific properties of the dynamics [9]–have been used to define trajectories which pass through the Lagrange point gateways and into the interior of the system [10]. The authors take advantage of this by posing a two point boundary value problem (TPBVP) between positions on a nominal L_2 Lyapunov orbit and a bounded region on this PSS to define admissible $\Delta\mathbf{v}$ which would send a spacecraft through the PSS.

A periodic orbit in a given family is defined by the Jacobi constant, C. Therefore, the state vector on the Lyapunov orbit is uniquely defined by a Jacobi constant, $\Psi_1 \in [C_{max}, C_{min}]$, and the ratio of time passed to the period, $\Psi_2 \in [0, 1]$. An unstable manifold may be generated from this point and propagated to the PSS. Therefore, a surface may be fit to the manifold curves on the PSS, such that

$$
\begin{bmatrix} y \\ \dot{y} \end{bmatrix} \approx \mathbf{P}\phi\left(\begin{bmatrix} \Psi_1 \\ \Psi_2 \end{bmatrix}\right),
\tag{3}
$$

where $\mathbf{P}$ is the coefficient matrix and $\phi(\cdot)$ is a user-defined library of basis functions. This allows for the direct generation of points which lie between the invariant manifold curves of the minimal and maximum C on the PSS. This subspace defines a conservative admissible region on the PSS as shown in [8].

By generating samples within the subspace on the PSS and samples of the target's position at maneuver time, a set of TPBVPs is formed. Given samples of Ψ_i and samples of target position at maneuver time, $\mathbf{r}_i$, the formulation of a single TPBVP is such that

$$
\left.\begin{matrix} x_{U_3} = 1 - \mu \\ \begin{bmatrix} y_{U_3} \\ \dot{y}_{U_3} \end{bmatrix} = \mathbf{P}\phi(\Psi_i) \\ \mathbf{r}(\beta_i) \end{matrix}\right\} \text{ is specified,} \qquad\qquad \left.\begin{matrix} \Delta v_x \\ \Delta v_y \\ \sigma := \text{Time of Flight} \end{matrix}\right\} \text{ is free.} \tag{4}
$$

where β is the angle with respect to L_2 on the corresponding Lyapunov orbit. Under these assumptions, a shooting method formulation where the trajectory is propagated from $\mathbf{r}_i$ with an initial guess for Δv may be used to solve each TPBVP. The particular formulation of the shooting method used in this work is known as *stabilized continuation*, which provides adaptive step-size selection through use of a numerical propagator, such as MATLAB's `ode45`, among other benefits [11, 12].

For computational efficiency, a library of polynomial basis functions are fit to the solution surface of the TPBVPs. Quadrature points, $\zeta \in [-1, 1]$, are generated and used for training under the assumption that

$$
\zeta_1 = \frac{\beta}{\pi}, \qquad \zeta_2 = \bar{\Psi}_1, \qquad \zeta_3 = R(\bar{\Psi}_2)
\tag{5}
$$

where $R(\bar{\Psi}_2)$ is a user defined map to ensure that the solution surface is continuous and that the resulting $\|\Delta v\|$'s are reasonable. This can be done because $\bar{\Psi}_2$ is a cyclic variable, therefore this mapping is analogous to unwrapping angle variables or adding a phase shift. The effects of the mapping are investigated later in this work. This mapping will allow for the direct approximation of Δv_x, Δv_y, and σ via a least squares approximation for $\mathbf{P}$ such that

$$
\begin{bmatrix} \Delta v_x \\ \Delta v_y \\ \sigma \end{bmatrix} \approx \mathbf{P}\phi\left(\begin{bmatrix} \zeta_1 \\ \zeta_2 \\ \zeta_3 \end{bmatrix}\right).
\tag{6}
$$

3 Results

The bounds on a maneuver window and the mappings in (5) are defined a priori. Therefore, different maneuver windows, i.e. a fixed range in β, define different approximations. Gauss-Legendre quadrature points are used with legendre polynomial basis functions. The number of quadrature points selected are 4, 6, and 23 in ζ_1, ζ_2, and ζ_3, respectively. The phase shift, $R(\bar{\Psi}_2)$ in (5), is determined by propagating a $\frac{20}{344000}$ perturbation in the velocity component of the eigenvector generated on the nominal Lyapunov orbit at the lower bound of the given maneuver window. The value of $\bar{\Psi}_2$ at that location determines $R(\bar{\Psi}_2)$. The Monte Carlo points are defined such that 10 random β are generated and then 500 points are generated on the PSS at each β, which helps ensure the convergence of the TPBVP solver used to determine the ground truth.

Figure 1 shows the TPBVP solution surfaces of the quadrature points used for training in (6) for the maneuver windows of $\beta \in [15°, -5°]$ and $\beta \in [-155°, -175°]$. The solution surface for both Δv_x and Δv_y varies almost linearly with respect to ζ_2, while ζ_1–mapped to the maneuver location via (5)–essentially adds thickness to the surfaces. This is especially pronounced as $\zeta_3 \to -1$. The solution surface for ToF is a nearly-flat plate increasing as $\zeta_3 \to 1$ and showing only slight variations with respect to ζ_1 and ζ_2. There is a near-exponential increase in Δv_x as $\zeta_3 \to -1$, which is especially pronounced in Fig. 1(a). The reason why Fig. 1(b) has a less pronounced change is because the selection of $R(\bar{\Psi}_2)$ captures less of the high Δv region.

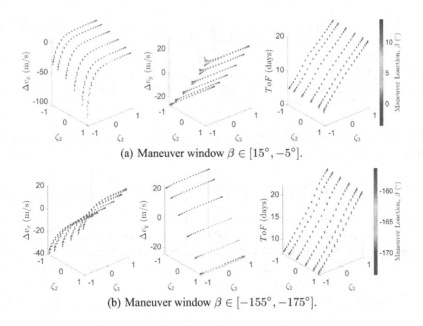

(a) Maneuver window $\beta \in [15°, -5°]$.

(b) Maneuver window $\beta \in [-155°, -175°]$.

Fig. 1. TPBVP solution surface for the quadrature points used for training.

The scale of the coefficients as a function of the polynomial order in each input dimension is shown in Fig. 2. By simply looking at the scale of the coefficients, the effects of capturing that exponential curve is evident. Keeping in mind the difference in scale, the number of significant coefficients is much larger in the $\beta \in [15°, -5°]$ maneuver window shown in Fig. 2(a). This is likely the effect of the region as $\zeta_3 \to -1$, where the variations due ζ_1 and ζ_3 are much larger. For both windows, the coefficient orders decrease as the order of ζ_1 and ζ_2 increases, but at a much more rapid pass for ζ_1. The strong dependence on ζ_3 is evident as the magnitude of the coefficients are significant until polynomial orders in the 10 s.

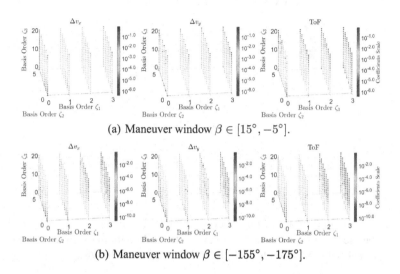

(a) Maneuver window $\beta \in [15°, -5°]$.

(b) Maneuver window $\beta \in [-155°, -175°]$.

Fig. 2. Scale of the coefficients for the TPBVP solution surfaces.

Figure 3 shows the $\|\Delta \mathbf{v}\|$, direction, and ToF errors versus the ground truth for the two maneuver windows studied. The ground truth $\|\Delta \mathbf{v}\|$ approaches 100 m/s and 50 m/s for $\beta \in [15°, -5°]$ and $\beta \in [-155°, -175°]$, respectively. Both methods produce errors consistently two orders of magnitude below that of the ground truth. The pointing errors for both methods are generally less than 0.1°. The few points greater than 0.1° error correspond to $\|\Delta \mathbf{v}\|$'s on the order of 10 s of cm/s, where the direction is more sensitive to small errors in the Δv_x and Δv_y approximations. The ground truth Cartesian $\Delta \mathbf{v}$ angles show a dense region near the extremes of $\sim 115°$ and $\sim 295°$ for $\beta \in [15°, -5°]$ and $\sim 130°$ and $\sim 320°$ for $\beta \in [15°, -5°]$. The ToF approximations are consistently on the order of 100 s while the true ToF is between 4 and 25 d. The largest errors are on the order of 1000 s in the low ToF , high $\|\Delta \mathbf{v}\|$ region shown in Fig. 3(a), which is again likely due to the large exponential region as $\zeta_3 \to -1$ in the $\beta \in [15°, -5°]$ window.

Figure 4 shows the $\|\Delta \mathbf{v}_{GT}\|$, $\|\Delta \mathbf{v}_{GT} - \Delta \mathbf{v}_{Approx}\|$, and $ToF_{GT} - ToF_{Approx}$ on a 3D plot of the PSS expended in the z-axis by the ground truth ToF. The $\|\Delta \mathbf{v}_{GT}\|$ grows as a function of distance away from the nominal orbit's manifold curve on the PSS and as the ToF decreases. The $\Delta \mathbf{v}$ and ToF errors are lower along lines within the spiral associated with the locations of the quadrature points used for defining the coefficients. They are highest in regions of high $\Delta \mathbf{v}$ and in the thickest regions of the PSS, where there is more sparsity in the training points. The position of the discontinuity in the ToF

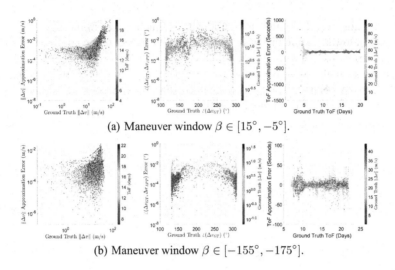

(a) Maneuver window $\beta \in [15°, -5°]$.

(b) Maneuver window $\beta \in [-155°, -175°]$.

Fig. 3. Approximation errors of $\Delta\mathbf{v}$ and ToF versus their ground truth for both maneuver windows.

on Fig. 4 is a function the a priori user selection of $R(\bar{\boldsymbol{\Psi}}_2)$ and β. Given a fixed β, this spiral can be viewed as a snapshot a larger, finite, and continuous spiral between the PSS and the ToF with the bounds of the snapshot defined by $R(\bar{\boldsymbol{\Psi}}_2)$.

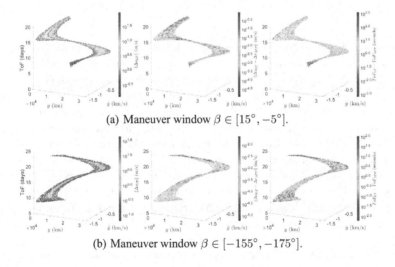

(a) Maneuver window $\beta \in [15°, -5°]$.

(b) Maneuver window $\beta \in [-155°, -175°]$.

Fig. 4. The $\|\Delta\mathbf{v}_{GT}\|$, $\|\Delta\mathbf{v}_{GT} - \Delta\mathbf{v}_{Approx}\|$, and $ToF_{GT} - ToF_{Approx}$ with respect to position on the 3D plot of the PSS with the ground truth ToF as the third axis.

Figure 5 shows the effects of the approximation errors in propagation towards the PSS for both the full approximation–$\Delta\mathbf{v}$ and ToF–and for propagating the $\Delta\mathbf{v}$ approximation to the intersection with the PSS, eliminating any error caused by the ToF approximation. The errors for the full approximation are large given the the sensitivity of errors in ToF at the PSS. There is a two to three orders of magnitude increase in accuracy when the approximated $\Delta\mathbf{v}$ is propagated to the PSS. The errors do remain largest in the regions of high ToF as the small errors in $\Delta\mathbf{v}$ have a much longer time to accumulate. The ToF errors between the approximated ToF and the true ToF to the PSS propagating the approximated $\Delta\mathbf{v}$ starts on the order of seconds in the low ToF region and increases to the order of 1 h in the longer ToF region.

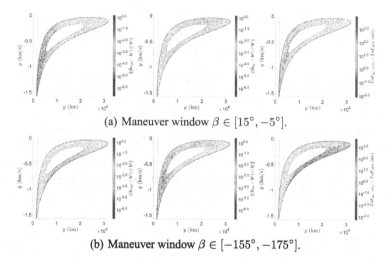

(a) Maneuver window $\beta \in [15°, -5°]$.

(b) Maneuver window $\beta \in [-155°, -175°]$.

Fig. 5. Propagation error for the full approximated $\Delta\mathbf{v}$ and ToF (left), error of the approximated $\Delta\mathbf{v}$ propagated to the PSS (middle), and ToF error (right).

Figure 6 shows the propagation 5000 random samples of the TPBVP solution surface at 96, 192, and 288 h. The point cloud emanates from a curve containing the nominal orbit and depart as time increases. Notice that in these point clouds, there are large $\Delta\mathbf{v}$ which still take a long time to depart the vicinity of the nominal trajectory. These $\Delta\mathbf{v}$ are associated with the boundaries of the PSS and therefore relate to ground truth Cartesian angles at the extremes as shown in Fig. 3. Throughout propagation, the point cloud takes on the shape similar to a twisting ribbon. There are time steps with very condensed pinch points–exemplified at 192 h in Fig. 6(a)–and time steps with a great amount of sparsity–exemplified at 288 h in Fig. 6(b).

4   Limits of the Method

Multiple solutions exist for much of the PSS and it would not be difficult to capture more than one solution by simply extending this method. However, this is because the ToFs

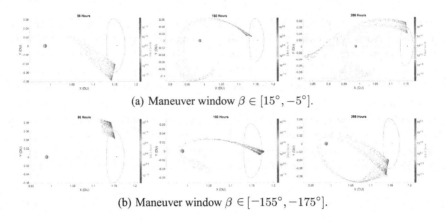

(a) Maneuver window $\beta \in [15°, -5°]$.

(b) Maneuver window $\beta \in [-155°, -175°]$.

Fig. 6. Propagation of 5000 randomly sampled approximations to the TPBVP solution surface.

of those multiple solutions are separated on the order of 10 d in the regions reported in this work. There are two instances where this is not true and the proposed method may fail 1) a poor selection of $R(\bar{\Psi}_2)$ which captures too much of the exponential region exemplified in Fig. 1(a) and 2) a maneuver region characterized by a low velocity. The issue in both of these scenarios is that the of the magnitude of the velocity in the nominal orbit is very close to the $\|\Delta \mathbf{v}\|$ of the TPBVP solutions, which results in an ill-posed TPBVP.

5   Conclusions

The authors have developed a method for defining admissible control onto a cislunar highway from the vicinity of L_2 through the lunar plane [8]. This work further explored the capabilities of this method through examination of multiple maneuver windows. The a priori, user-defined parameter $R(\bar{\Psi}_2)$ determines the snapshot of the continuous, finite TPBVP solution surface defined in the approximation. The method selecting $R(\bar{\Psi}_2)$ used in this work required no a priori knowledge of the solution surface. Through further a priori analysis, users may choose this parameter to limit the exponentially increasing $\Delta \mathbf{v}$ region in the solution surface. This will increase accuracy, reduce the capture of $\Delta \mathbf{v}$ outside the bounds of likely maneuver capabilities, and ensure a well-posed TPBVP so long as the maneuver location is also suitable. Monte Carlo samples are generated from the solution set are then propagated and the reachable sets analyzed. The shape of this region through propagation and the ToF output by the approximation may be leveraged by a cislunar SDA operator using the RSS method to prioritize sensors to regions of the reachable set with short ToFs and by taking observations of the dense pinch points.

References

1. The future of space 2060 & implications for U.S. strategy: report on the space futures workshop, Tech. rep., Air Force Space Command (2019)
2. Holzinger, M.J., Chow, C.C., Garretson, P.: A Primer on Cislunar Space. White Paper 2021–1271. AFRL (2021)
3. Perkins, J.: AFRL's Cislunar Highway Patrol System Seeks Industry Collaboration (2022)
4. Hall, Z., Singla, P., Johnson, K.: A particle filtering approach to space-based maneuvering satellite location and estimation. In: AAS Astrodynamics Specialist Conference (2020)
5. Hall, Z., et al.: Reachability-based approach for search and detection of maneuvering cislunar objects. In: AIAA SCITECH 2022 Forum. AIAA SciTech Forum. San Diego, CA & Virtual: American Institute of Aeronautics and Astronautics (2022). https://doi.org/10.2514/6.2022-0853
6. Tommei, G., Milani, A., Rossi, A.: Orbit determination of space debris: admissible regions. Celest. Mech. Dyn. Astron. **97**(4), 289–304 (2007). https://doi.org/10.1007/s10569-007-9065-x
7. Fujimoto, K., Scheeres, D.J.: Applications of the admissible region to space-based observations. Adv. Space Res. **52**(4), 696–704 (2013). https://doi.org/10.1016/j.asr.2013.04.020
8. Schwab, D., Eapen, R., Singla, P.: Approximating admissible control onto the cislunar highways for detection and tracking of spacecraft. In: 2022 AAS/AIAA Astrodynamics Specialist Conference. Charlotte, North Carolina (2022)
9. Villac, B.F., Scheeres, D.J.: Escaping trajectories in the hill three-body problem and applications. J. Guidance Control Dyn. **26**(2), 224–232 (2003). https://doi.org/10.2514/2.5062
10. Wang S.K., et al.: Dynamical Systems, the Three-Body Problem and Space Mission Design. Chapter 7: Invariant manifolds and end to end transfer. Useful information for trajectory planning. Marsden Books (2011). ISBN: 978-0-615-24095-4
11. Ohtsuka, T., Fujii, H.: Stabilized continuation method for solving optimal control problems. J. Guidance Control Dyn. **17**(5), 950–957 (1994). https://doi.org/10.2514/3.21295
12. Vedantam, M., Akella, M.R., Grant, M.J.: Multi-stage stabilized continuation for indirect optimal control of hypersonic trajectories. In: AIAA Scitech 2020 Forum. Orlando, FL: American Institute of Aeronautics and Astronautics (2020). ISBN: 978-1-62410-595-1. https://doi.org/10.2514/6.2020-0472
13. Bolden, M., Craychee, T., Griggs, E.: An evaluation of observing constellation orbit stability, low signal-to-noise, and the too-short-arc challenges in the cislunar domain. In: Advanced Maui Optical and Space Surveillance Technologies Conference (AMOS). Maui, HI (2020)
14. Hall, Z., Singla, P.: Reachability analysis based tracking: applications to non-cooperative space object tracking. In: Darema, F., Blasch, E., Ravela, S., Aved, A. (eds.) DDDAS 2020. LNCS, vol. 12312, pp. 200–207. Springer, Cham (2020). https://doi.org/10.1007/978-3-030-61725-7_24
15. Hall, Z.: A probabilistic framework to locate and track maneuvering satellites, Ph. D. thesis, University Park, PA: The Pennsylvania State University (2021)

Physics-Aware Machine Learning for Dynamic, Data-Driven Radar Target Recognition

Sevgi Zubeyde Gurbuz[(✉)] [iD]

Department of Electrical and Computer Engineering, The University of Alabama, Tuscaloosa,
AL 35487, USA
szgurbuz@ua.edu

Abstract. Despite advances in Artificial Intelligence and Machine Learning
(AI/ML) for automatic target recognition (ATR) using surveillance radar, there
remain significant challenges to robust and accurate perception in operational
environments. Physics-aware ML is an emerging field that strives to integrate
physics-based models with data-driven deep learning (DL) to reap the benefits of
both approaches. Physics-based models allow for the prediction of the expected
radar return given any sensor position, observation angle and environmental scene.
However, no model is perfect and the dynamic nature of the sensing environment
ensures that there will always be some part of the signal that is unknown, which
can be modeled as noise, bias or error uncertainty. Physics-aware machine learning
combines the strengths of DL and physics-based modeling to optimize trade-offs
between prior versus new knowledge, models vs. data, uncertainty, complexity,
and computation time, for greater accuracy and robustness. This paper addresses
the challenge of designing physics-aware synthetic data generation techniques
for training deep models for ATR. In particular, physics-based methods for data
synthesis, the limitations of current generative adversarial network (GAN)-based
methods, new ways domain knowledge may be integrated for new GAN architectures and domain adaptation of signatures from different, but related sources of
RF data, are presented. The use of a physics-aware loss term with a multi-branch
GAN (MBGAN) resulted in a 9% improvement in classification accuracy over that
attained with the use of real data alone, and a 6% improvement over that given
using data generated by a Wasserstein GAN with gradient penalty. The implications for DL-based ATR in Dynamic Data-Driven Application Systems (DDDAS)
due to fully-adaptive transmissions are discussed.

Keywords: Automatic target recognition · cognitive radar · adversarial networks

1 Introduction

Physics-aware machine learning (PhML) is an emerging field within ML that strives
to integrate physics-based models with data-driven DL to reap the benefits of both
approaches. Physics-based models represent the high level of domain knowledge gained
from a study of the electromagnetic backscatter from surfaces and objects of the years.
It can also capture phenomenological factors integral to the sensing scenario as well

© The Author(s), under exclusive license to Springer Nature Switzerland AG 2024
E. Blasch et al. (Eds.): DDDAS 2022, LNCS 13984, pp. 114–122, 2024.
https://doi.org/10.1007/978-3-031-52670-1_11

as known sensor properties. However, physics-based models are less adept at capturing the nuances of environment-specific, sensor-specific, or subject-specific properties, which lie at the heart of Data-Driven Dynamic Adaptive Systems (DDDAS). Here, deep learning can provide tremendous insight through data-driven learning.

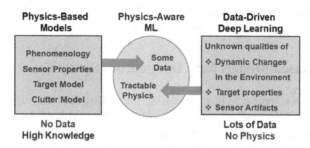

Fig. 1. The physics-aware ML tradeoff.

DL relies on the availability of massive amounts of data to learn a model for the underlying relationships reflected in the data. Unfortunately, in sensing problems, it is not common to have a lot of data. The limitations in training sample support ultimately also limit the accuracy and efficacy of DL in sensing. Moreover, no model is perfect - while more complex models could surely be developed to improve accuracy, the dynamic nature of the sensing environment ensures that there will always be some part of the signal that is unknown. This is where leveraging data-driven DL can provide a powerful tool when used in tandem with physics-based models. The resulting hybrid approach, PhML (Fig. 1) combines the strengths of DL and physics-based modeling to optimize trade-offs between prior versus new knowledge, models vs. data, uncertainty, complexity, and computation time, for greater accuracy and robustness. In this way, PhML also represents an integral concept to DDDAS, which uses applications modeling, mathematical algorithms and measurement systems to work with dynamic systems. ATR is a dynamic system challenged by variations of the sensor, target and environment in both space and time. This paper develops DDDAS-based ATR [1–3] methods empowered by PhML to improve the ability of deep models to learn from data via guidance of physics-based models.

Much of current literature involving physics-aware machine learning has focused on the solution of ordinary differential equations (ODEs) [4, 5], data-driven discovery of physical laws [6, 7], uncertainty quantification [8] and data generation - both to synthesize data for validation on simulated data in cases where acquiring real data is not feasible and for physics-guided initialization to pre-train deep models. The question of whether GAN-generated samples conform to physical constraints has recently been raised in the context of turbulent flow simulation, where both deterministic constraints (conservation laws) [9] and statistical constraints (energy spectrum of turbulent flows) [10] have been proposed for incorporation into the loss function. These constraints were shown to yield improvements in performance relative to that attained by standard GANs because the synthetic samples more faithfully emulated certain physical properties of the system, while also significantly reducing (by up to %80) the training time.

This paper addresses the design of physics-aware synthetic data generation techniques for training deep models for ATR using radar micro-Doppler signatures. In particular, the limitations of current generative adversarial network (GAN)-based methods, physics-based methods for data synthesis, new ways domain knowledge may be integrated for new GAN architectures and domain adaptation of signatures from different, but related sources of RF data, are presented. The implications for DL-based ATR in DDDAS due to fully-adaptive transmissions are discussed.

2 Physics-Based Models for μD Signature Synthesis

There are two main approaches for simulating micro-Doppler (μD) signatures (μDS) [11]: kinematic modeling and motion capture (MOCAP)-based animation. Both methods are based on the idea of representing the overall target backscatter as the sum of the backscatter from a finite number of point targets. MOCAP data have been effective in capturing the time-varying positions of moving humans and animals [12–15], while fixed body approximations to the mechanics of rotating parts are utilized to model vehicles, helicopters, aircraft or drones [16, 17]. Biomechanical models of human gait, such as the Boulic walking model [18], can also be used to animate the motion of point-targets. Thus, the radar return can be modeled as the sum of returns from point targets distributed throughout the target [19]. For K points,

$$\hat{s}_h[n] = \sum_{i=1}^{K} \left(A/R_{n,i}^4\right) exp\left[-j\frac{4\pi f_t}{c}R_{n,i}\right], \tag{1}$$

where f_t is the transmit frequency, c is the speed of light, and $R_{n,i}$ is the time-varying range between the radar and i^{th} point target at discrete time n. The parameter A is modeled by the radar range equation, and is a function of the radar system parameters, atmospheric and system loss factors, and radar cross-section. The simulated μDS are computed as the square modulus the short-time Fourier Transform.

Because MOCAP-based point tracking relies on actual measurements from a radar, the size of the dataset is still limited by the human effort, time and cost of data collections. To overcome this limitation, diversification [20] can be applied to generate physically meaningful physics-based variants of target behavior within a motion class. In the case of human recognition, skeleton-level diversification can be accomplished by scaling in time and size, while applying perturbations on the trajectory of each to generate statistically independent samples (Fig. 2). Data augmentation on the point-target model is important because augmentation methods used in image processing, such as rotation and flipping, result in physically impossible variations of the μDS, which degrades model training. In this way, a small amount of MOCAP data can be leveraged to generate a large number of synthetic μDS for model training.

To give an example of the benefit of this approach, this diversification technique was applied to 55 MOCAP samples, generating a total of 32,000 synthetic samples. The synthetic samples were then used to initialize a 15-layer residual DNN (DivNet-15), which was then fine-tuned with approximately 40 samples/class. Note that the depth of 15 layers was much deeper than the maximum depth of 7-layers for a convolutional neural network (CNN) trained with real data only. The 95% accuracy attained by DivNet-15 surpassed alternative approaches, including convolutional autoencoders, genetic algorithm optimized frequency warped cepstral coefficients [21], transfer learning from VGGnet, a CNN or autoencoder (AE) trained on real data, and support vector machine classifier using 50 features extracted from μD.

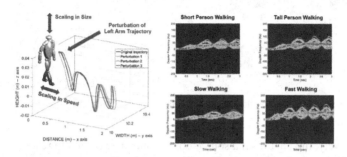

Fig. 2. Diversification of MOCAP-based synthetic micro-Doppler [20].

Thus, physics-aware initialization with knowledge transfer from model-based simulations is a powerful technique for overcoming the problem of limited training data and can also improve generalization performance by exploiting the simulation of scenarios for which real data acquisition may impractical.

3 Adversarial Learning-Based μD Signature Synthesis

While model-based training data synthesis has been quite effective in replication of target signatures, it does not account for other sources of noise and interference, such as sensor artifacts and ground clutter. Because interference sources may be device specific or environment-specific, data-driven methods for data synthesis such as adversarial learning are well-suited to account for such factors. Adversarial learning can be exploited in several different ways to learn and transfer knowledge in offline model training, as illustrated in Fig. 3; for example,

- To improve realism of synthetic data generated from physics-based models;
- To adapt data from a different source to resemble data from the target domain; and
- To directly synthesize both target and clutter components of measured RF data.

The main benefit of using adversarial learning to improve the realism of synthetic images generated from physics-based models is that it preserves the micro-Doppler signature properties that are bound by the physical constraints of the human body and kinematics while using the adversarial neural network to learn features in the data unrelated to the target model, e.g. sensor artifacts and clutter. The goal for improving realism [22] is to generate training images that better capture the characteristics of each class, and thus improve the resulting test accuracy. However, as the goal of the refiner is merely to improve its similarity to real data, a one-to-one correspondence is maitained between synthetic and refined samples. In other words, however much data we have at the outset, as generated by physics-based models, is the same amount of data that we have after the refinement process - no additional data is synthesized.

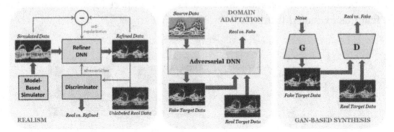

Fig. 3. Ways of exploiting adversarial learning for training data generation.

Alternatively, the data from a source domain may be adapted to resemble real data acquired in the target domain [23]; then, the adapted data is used for network initialization. In this approach, the source domain can be real data acquired using a different RF sensor with different transmit parameters (frequency, bandwidth, pulse repetition interval), while the target domain is that which is to be classified. For example, consider the case where the target domain is RF data acquired with a 77 GHz frequency modulated continuous wave (FMCW) automotive radar, but there is insufficient data to adequately train a DNN for classification. Perhaps data from some other sensor, however, is available: this could be data from a publicly released dataset, or data from a different RF sensor. Suppose we have ample real data from two other RF sensors - a 10 GHz ultra-wide band impulse radar and a 24 GHz FMCW radar. Although the data from these three RF sensors will be similar for the same activity, there are sufficient differences in the μD that direct transfer learning suffers from great performance degradation. While the classification accuracy of 77 GHz data with training data from the same sensor can be as high as 91%, the accuracy attained when trained on 24 GHz and 10 GHz data is just 27% and 20% [24], respectively. This represents over 65% poorer accuracy. On the other hand, when adversarial domain adaptation is applied to first transform the 10 GHz and 24 GHz data to resemble that of the target 77 GHz data, classification accuracies that surpass that of training with just real target data can be achieved [25].

Effective data synthesis requires being able to *represent both target as well as clutter components*. While model-based methods effectively capture target kinematics, they do not capture environmental factors. Conversely, while GANs can capture sensor artifacts and clutter, they face challenges in accurately representing target kinematics. The efficacy of different GANs to accurately synthesize radar μD signatures varies. Auxiliary conditional GANs (ACGANs) generate crisper μD signatures than conditional variational autoencoders (CVAEs) [26]. For radar μD applications, however, fidelity cannot be evaluated by considering image quality alone. A critical challenge is the possibility of generating misleading synthetic signatures. The μD signature characteristics are constrained not only by the physics of electromagnetic scattering, but also by human kinematics. The skeleton physically constrains the possible variations of the spectrogram corresponding to a given class. But, GANs have no knowledge of these constraints. It is thus possible for GANs to generate synthetic samples that may appear visually similar but are in fact incompatible with possible human/animal motion.

Examples of erroneous data generation are given in Fig. 4(a) for an Auxiliary-Conditional GAN (ACGAN), which visually show some of the physically impossible or out-of-class samples synthesized by GANs. In fact, *classification accuracy greatly increases* when such fallacious synthetic samples are identified and discarded from the training data. For example, when an ACGAN was used to synthesize 40,000 samples, 9000 samples were identified as kinematic outliers and discarded from the training dataset, but in doing so the classification accuracy increased by 10% [26].

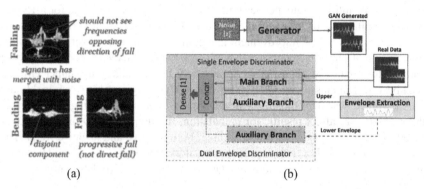

Fig. 4. (a) Errors observed ACGAN-synthesized μDS; (b) MBGAN

4 Physics-Aware GAN Design

Physics-aware GAN design aims at preventing the generation of kinematically flawed data that is inconsistent with possible target behavior by integrating physics-based models and domain knowledge into data-driven DL. This can be done through modifying the GAN architecture, the way the GAN is trained, and by developing more meaningful loss functions. Current GAN architectures typically use distance measures rooted in statistics, such as cross-entropy or Earth mover's distance, to derive a loss metric reflective of the degree of discrepancy between true and predicted samples. Statistical distance metrics, however, do not reflect discrepancies in the underlying kinematics of actual target motion versus predicted motion. However, target kinematics are physically constrained. One way of informing a GAN of such constraints is through the design of a physics-based loss term that is added to the standard statistical loss term:

$$Loss = (D(x) - D(G(z))) + \gamma \left(\|\nabla_{\hat{x}} D(\hat{x})\|^2 \right) + Loss_{phyiscs}, \tag{2}$$

where the first term is the critic loss, the second term is the gradient penalty, x is the real data instance, z is noise, γ is a hyperparameter, and $D(\cdot)$ and $G(\cdot)$ represent the discriminator and generator functions, respectively. In this way, a penalty will now be incurred if the resulting signature exhibits deviant physical properties. In the case of micro-Doppler, one property of special significance is the envelopes of the signature, as they form a physical upper bound of the radial velocity incurred during motion. Thus, physics-based loss metrics reflective of consistency in envelope between synthetic and real samples are proposed for µDS classification.

Envelope information is provided to the GAN in two different ways: 1) through the addition of one or two auxiliary branches in the discriminator, which take as input the upper and lower envelope of the µDS, as shown in Fig. 4(b) of the resulting Multi-Branch GAN (MBGAN); and 2) physics-based loss metrics such as the Dynamic Time Warping (DTW) distance. In time series analysis, DTW is an algorithm for measuring the similarity between two temporal sequences that may vary in time or speed [27]. Utilizing a convolutional autoencoder trained on GAN-synthesized data for a 15-class activity recognition problem, we found that use of physics-aware loss in the MBGAN architecture resulting a 9% improvement in classification accuracy over the use of real data alone, and 6% improvement over that of a Wasserstein GAN with gradient penalty. For ambulatory data use of just a single auxiliary branch taking the upper envelope worked best. But in classification of 100 words of sign language [28], the dual branch MBGAN worked best. This illustrates the degree to which understanding the kinematics of the problem can affect results: for ambulation, the dominant µD are positive and reflected in the upper envelope. But in signing, both positive and negative frequencies hold great significance. For both datasets (ambulatory and signing), we found that direct synthesis of µDS outperformed domain adaptation from other RF data sources.

5   Conclusion

This paper provided a broad overview of physics-aware machine learning, including use of synthetic datasets to train deep models, with focus on aspects relevant to classification of RF µDS. The design of physics-aware GANs that surpass alternatives to data synthesis, including physics-based models, standard GANs, and domain adaptation was presented. This work has significant implications for cognitive radar design, where the adaptive transmissions result in physics-based mismatch between training and test datasets. Extension of physics-aware techniques to domain adaptation can help mitigate potential degradation in accuracy incurred due to RF frequency mismatch.

Acknowledgements. This work was supported AFOSR Award #FA9550-22-1-0384.

References

1. Blasch, E., Seetharaman, G., Darema, F.: Dynamic data driven applications systems (DDDAS) modeling for automatic target recognition. In: Proceedings SPIE, vol. 8744 (2013)
2. Ahmadibeni, A., Jones, B., Smith, D., Shirkhodaie, A.: Dynamic transfer learning from physics-based simulated SAR imagery for automatic target recognition. In: Darema, F., Blasch, E., Ravela, S., Aved, A. (eds.) DDDAS 2020. LNCS, vol. 12312, pp. 152–159. Springer, Cham (2020). https://doi.org/10.1007/978-3-030-61725-7_19

3. Metaxas, D., Kanaujia, A., Li, Z.: Dynamic tracking of facial expressions using adaptive, overlapping subspaces. In: Shi, Y., van Albada, G.D., Dongarra, J., Sloot, P.M.A. (eds.) ICCS 2007. LNCS, vol. 4487, pp. 1114–1121. Springer, Heidelberg (2007). https://doi.org/10.1007/978-3-540-72584-8_146

4. Chen, R.T.Q., Rubanova, Y., Bettencourt, J., Duvenaud, D.: Neural ordinary differential equations. In: Proceedings of the 32nd International Conference on NIPS, Red hook, NY, USA, pp. 6572–6583 (2018)

5. Raissi, M., Perdikaris, P., Karniadakis, G.: Physics-informed neural networks: a deep learning framework for solving inverse forward and inverse problems involving nonlinear partial differential equations. J. Comput. Phys. **378**, 686–707 (2019)

6. Chen, Z., Liu, Y., Sun, H.: Deep learning of physical laws from scarce data. arXiv, abs/2005.03448 (2020)

7. De Oliveria, L., Paganini, M., Nachman, B.: Learning particle physics by example: location-aware generative adversarial networks for physics synthesis. Comput. Softw. Big Sci. **1**(1), (2017). https://link.springer.com/article/10.1007/s41781-017-0004-6

8. Yang, Y., Perdikaris, P.: Adversarial uncertainty quantification in physics-informed neural networks. J. Comput. Phys. **394**, 136–152 (2019)

9. Yang, Z., Wu, J., Xiao, H.: Enforcing imprecise constraints on generative adversarial networks for emulating systems. Commun. Comput. Phys. **30**, 635–665 (2021)

10. Wu, J., Kashinath, K., Albert, A., Chirila, D.B., Prabhat, Xiao, H.: Enforcing statistical constraints in generative adversarial networks for modeling chaotic dynamical systems. J. Comput. Phys. **406**, 109–209 (2020)

11. Ram, S.S., Gurbuz, S.Z., Chen, V.C.: Modeling and simulation of human motions for micro-Doppler signatures. In: Amin, M. (ed.) Radar for In-Door Monitoring: Detection, Classification, and Assessment, CRC Press (2017)

12. Blasch, E., Majumder, U., Minardi, M.: Radar signals dismount tracking for urban operations. In: Proceedings of SPIE, vol. 6235, May (2006)

13. Majumder, U. Minardi, M., Blasch, E., Gorham, L., Naidu, K., Lewis, T., et al.: Radar signals dismount data production. In: Proceedings of SPIE, vol. 6237 (2006)

14. Ram, S.S., Ling, H.: Simulation of human micro-Dopplers using computer animation data. In: Proceedings IEEE Radar Conference (2008)

15. Erol, B., Gurbuz, S.Z.: A kinect-based human micro-doppler simulator. IEEE Aerosp. Electron. Syst. Mag. **30**(5), 6–17 (2015)

16. Passafiume, M., Rojhani, N., Collodi, G., Cidronali, A.: Modeling small UAV micro-Doppler signature using millimeter-wave FMCW radar. Electronics **10**(6), 747 (2021)

17. Moore, M., Robertson, D.A., Rahman, S.: Simulating UAV micro-Doppler using dynamic point clouds. In: Proceedings IEEE Radar Conference, pp. 1–6 (2022)

18. Boulic, R., Magnenat-Thalmann, N., Thalmann, D.: A global human walking model with real-time kinematic personification. Vis. Comput. **6**, 344–358 (2005)

19. Van Dorp, P., Groen, F.C.A.: Human walking estimation with radar. IEE Proc. Radar Sonar Navigation **150**(5), 356–365 (2003)

20. Seyfioglu, M.S., Erol, B., Gurbuz, S.Z., Amin, M.: DNN transfer learning from diversified micro-Doppler for motion classification. IEEE TAES **55**(5), 2164–2180 (2019)

21. Erol, B., Amin, M.B., Gurbuz, S.Z.: Automatic data-driven frequency-warped cepstral feature design for micro-Doppler classification. IEEE Trans. Aerosp. Electron. Syst. **54**(4), 1724–1738 (2018)

22. Shrivastava A., et al.: Learning from simulated and un-supervised images through adversarial training. In: IEEE Proceedings of the CVPR, pp. 2242–2251 (2017)

23. Wang, M., Deng, W.: Deep visual domain adaptation: a survey. Neurocomputing **312**, 135–153 (2018)

24. Gurbuz S, et al.: Cross-frequency training with adversarial learning for radar micro-Doppler signature classification. In: Proceedings of the SPIE, vol. 11408, pp. 1–11 (2020)
25. Gurbuz, S.Z., Rahman, M.M., Kurtoglu, E., et al.: Multi-frequency RF sensor fusion for word-level fluent ASL recognition. IEEE Sens. J. **22**, 11373–11381 (2021)
26. Erol, B., Gurbuz, S.Z., Amin, M.G.: Motion classification using kinematically sifted ACGAN-synthesized radar micro-Doppler signatures. IEEE Trans. Aerosp. Electron. Syst. **56**(4), 3197–3213 (2020)
27. Berndt, D.J., Clifford, J.: Using dynamic time warping to find patterns in time series. In: KDD Workshop, Seattle, WA (1994)
28. Kurtoğlu, E., Gurbuz, A.C., Malaia, E.A., Griffin, D., Crawford, C., Gurbuz, S.Z.: ASL trigger recognition in mixed activity/signing sequences for RF sensor-based user inter-faces. IEEE Trans. Hum.-Mach. Syst. **52**(4), 699–712 (2022)

DDDAS for Optimized Design and Management of 5G and Beyond 5G (6G) Networks

Nurcin Celik[1]([✉]), Frederica Darema[2], Temitope Runsewe[1], Walid Saad[3], and Abdurrahman Yavuz[1]

[1] Department of Electrical and Computer Engineering, Virginia Tech, Arlington, VA 22203, USA
ceik@miami.edu
[2] InfoSymbiotic Systems Society, Boston, USA
[3] Department of Industrial and Systems Engineering, University of Miami, Miami, FL 33146, USA

Abstract. The technologies vested by the introduction of fifth generation (5G) networks as well as the emerging 6G systems present opportunities for enhanced communication and computational capabilities that will advance many large-scale critical applications in the critical domains of manufacturing, extended reality, power generation and distribution, water, agriculture, transportation, healthcare, and defense and security, among many others. However, for these enhanced communication networks to take full effect, these networks, including wireless infrastructure, end-devices, edge/cloud servers, base stations, core network and satellite-based elements, should be equipped with real-time decision support capabilities, cognizant of multilevel and multimodal time-varying conditions, to enable self-sustainment of the networks and communications infrastructures, for optimal management and adaptive resource allocation with minimum possible intervention from operators. To meet the highly dynamic and extreme performance requirements of these heterogeneous multi-component, multilayer communication infrastructures on latency, data rate, reliability, and other user-defined metrics, these support methods will need to leverage the accuracy of full-scale models for multi-objective optimization, adaptive management, and control of time-varying and complex operations. This paper discusses how algorithmic, methodological, and instrumentation capabilities learned from Dynamic Data Driven Applications Systems (DDDAS)-based methodologies can be applied to enable optimized and resilient design and operational management of the complex and highly dynamic 5G/6G communication infrastructures. Such smart DDDAS capabilities are unswervingly proven for more than two decades on adaptive real-time control of various systems requiring the high accuracy of full-scale modeling for multi-objective real-time decision making with efficient computational resource utilization.

Keywords: DDDAS · 5G · 6G · Internet of Everything (IoE) · Wireless Communications

E. Blasch et al. (Eds.): DDDAS 2022, LNCS 13984, pp. 123–132, 2024.
https://doi.org/10.1007/978-3-031-52670-1_12

1 Introduction

The fifth generation (5G) systems have emerged in the past few years in order to support a plethora of multimedia and Internet of Things (IoT) applications ranging from high-speed mobile TV to smart agriculture and smart city services. Primary innovations of the fifth generation networks (5G) were along the lines of three key services: Enhanced mobile broadband (eMBB) to support high-speed wireless access, massive machine-type communications (mMTC) to support services that are characterized by a large number of connected devices, transmitting mostly non-delay-sensitive data (e.g., smart grid, smart buildings, smart cities, etc.) [1], and ultra-reliable and low latency communications (URLLC) to support a diverse set of IoT applications. However, these existing 5G technologies will still fall short in meeting the stringent demands of emerging applications such as extended reality (XR) and the metaverse, haptics, brain computer interactions, connected autonomy, driverless ground vehicles and unmanned aerial vehicles [2] as well as large-scale critical infrastructures such as manufacturing systems, power generation and distribution, water and sewer systems, agriculture, healthcare, and defense and security [3]. In particular, these applications will impose *extreme* quality-of-service (QoS) and quality-of-experience (QoE) requirements that may require new services that span across eMBB, URLLC, and mMTC. For example, as shown in [4], XR services require a new breed of high rate, highly reliable low latency communication (HRLLC) services that combine the rate needs of eMBB with the stringent latency and reliability needs of URLLC. To respond to this rapid proliferation of these new applications, the Beyond 5G (B5G) – the sixth generation (6G) of wireless systems [5] efforts have started to shape up across academia, industry, and the government.

6G promises to deliver substantial gains in terms of rate (1 Tbps as opposed to 100 Gbps for 5G), latency (sub-millisecond as opposed to few milliseconds for 5G), reliability (seven nines as opposed to five nines for 5G), and energy (zero-energy goal) while also operating as a *self-sustaining* system that can cater to the dynamics of its environment by merging artificial intelligence (AI) with data-driven protocols. The promises of 6G cannot be fully realized unless a range of multilevel and multimodal technologies are programmed together into the communications infrastructure. Among these multilevel and multimodal technologies, some key innovations support high-frequency (millimeter wave, sub-terahertz, and terahertz) communication for heterogeneous services, integration of sensing and communications, coordination of multi-connectivity, and AI-native protocols. In addition to the terrestrial wireless and fiber optic systems that will be integrated in 5G and 6G, non-terrestrial satellites constellations and drones will also support optimized connectivity of devices according to the specified QoS, and where satellites will be utilized to solve "the last mile" problem (e.g., in rural areas where cellular infrastructure is physically impossible or prohibitively expensive) [6]. The disaggregated, virtualized, self-driven, application-specific, and software-oriented network architecture in other related technologies such as radio access networks (RAN) services, and open RAN (O-RAN enable *lower latency, agility, deployment flexibility, real-time responsiveness, and reduced operating costs*. O-RAN will potentially facilitate the deployment and services provided by 5G and 6G.

Architecting these different multilevel and multimodal technologies together into a communications infrastructure that can deliver the full promises of 6G is challenging.

As discussed, in order to take full effect, enhanced 5G and 6G infrastructure nodes including end-devices, edge/cloud servers, base stations, and core network elements, should be equipped with a real-time data-driven decision support system enabling its self-sustainment for optimal management and adaptive resource management with minimum possible intervention from operators. This system should also be able to meet the stringent QoS needs in terms of rate, latency, and reliability, while operating in a sustainable, energy-efficient ways. Resilience against system anomalies and environmental dynamics further requires tomorrow's wireless cellular systems to embody intelligent analytics, data and resource management and different forms of coupling and decoupling mechanisms depending on the priority and cost of the tasks assigned. Because deployment of a separate communication network for each application scenario is not practical, *network slicing* has been proposed to dynamically allocate the flexible network resources to logical network slices according to customized service requirements on demand while meeting its highly dynamic and extreme performance requirements on latency, data rate, reliability, and other user-defined metrics. While data analytics methods (i.e., machine learning) are commonly used to address various problems within a single system, the mere application of these methods fall short of solving problems at the systems-of-systems level (i.e., context-aware globally optimal control of wireless communication networks) due to the high complexity and heterogeneity, as well as the dynamically, time-varying inter-operational aspects involved.

To this end, wireless communication networks can immensely benefit from Dynamic Data Driven Applications Systems (DDDAS)-based methods, which can be used to optimize design features in network and communications infrastructures, such as to dynamically allocate the network resources to a given network slice, or support energy efficiencies in 5G and 6G network infrastructures as highlighted in IEEE/INGR reports [Darema in 7, 8, 9]. DDDAS based methods can be used to adaptively manage the associated heterogeneity, hierarchical, decentralized control and real-time monitoring of anomalies and dynamics, and adaptive multi-objective optimization needs for such environments, and address the "systems-analytics" needed, rather than simply "data analytics". On the one hand, large-scale multi-objective optimization problems naturally arise in a plethora of 5G and 6G resource management and network slicing systems, and, on the other hand, DDDAS has been proven as an effective framework to solve such large-scale problems in the context of other domains such as the power grid. As such it is natural to ask for example: *What can DDDAS provide for supporting energy-efficient and sustainable wireless cellular systems, including 5G and 6G?*

The main contribution of this paper to expose how algorithmic, methodological, and instrumentation capabilities learned from DDDAS-based methodologies can be applied to enable i) optimized and resilient design, ii) self-sustaining resource management, iii) operational management, and delivery service quality of the complex and highly dynamic 5G/6G communication infrastructures for *optimized connectivity of devices, low latency, reliability, and agility in* extreme URLLC and HRLLC *and deployment flexibility, real-time responsiveness, and reduced operating costs in* eMBB and mMTC. Such smart DDDAS capabilities are unswervingly proven for more than two decades on adaptive real-time control of various systems, where multi-objective optimization and real-time decision-support with the accuracy of full-scale models is provided through

DDDAS-based methods. Inherent in DDDAS-based methods is the integration of high-end and mid-range computing with the real-time, Edge Computing. Thus, in the context of 5G and 6G systems, DDDAS naturally encompasses edge computing (EC) architectures – a hallmark of tomorrow's cellular systems – to further bring down the latency requirements and enable distributed computing of 5G/6G problems at the network edge; and, alleviating the need for central communications with a distant cloud over the core network. In this work, we present some preliminary results on an example of the DDDAS framework for adaptive and optimized 5G and 6G services case-examples, to support a microgrid testbed over a 5G/6G communication network.

2   Background of 5G and 6G Technologies

While communication networks have undergone significant improvements, there is still the prospect of increasing amounts of data traffic that exceeds the growth rates of new wireless communication technologies. This, together with other complexities of 5G and 6G networks calls for advanced methods for supporting such systems, and, to this end, DDDAS can provide the needed powerful and systematic approaches to further advance these emerging communication technologies. DDDAS refers to the concept of dynamically integrating computation and instrumentation into a feedback control loop in which collected real-time data can be dynamically incorporated into executing models of an application. Furthermore, in reverse, the instrumentation process can be controlled by the executing model [10]. DDDAS is highly effective for a wide range of application areas due to its ability to efficiently analyze and manage dynamic conditions in complex systems. Real-time monitoring and control of dynamic systems using the DDDAS paradigm and ensuing frameworks has shown promising results, in application areas such as semiconductor manufacturing facilities [11], self-aware drones [12], transportation systems [13], resilient cyber-space [14] and smart (power)grids [15–17].

5G and 6G wireless cellular systems generate vast amounts of data about the network conditions that are generated through instrumentation of the network-infrastructure, making them a rich resource for network operators to exploit to achieve more efficient design and operational management. For 6G, leveraging data-driven, self-sustaining protocol that is capable of catering to dynamic and multiple functions to autonomously reconfigure the network itself is a major requirement [5]. Thus, despite the growing complexity of networks, the enormous data volumes generated through the instrumentation of the network-infrastructure become an important means of alleviating such complexity. In this regard, we envision that DDDAS-based frameworks play a significant role in the wireless communications field, since the core of DDDAS is dynamic data assimilation and taking proactive and more informed measures for optimum system resiliency and efficiency. In 5G/6G networks, resource management processes such as network slicing will play a fundamental role. In essence, resource management in a wireless cellular system is a large-scale multi-objective optimization problem that seeks to allocate network resources, either slices (for network slicing) or resource blocks (RBs) which are the smallest time-frequency units for physical resource allocation [5], as a way to balance QoS tradeoffs across metrics such as energy, rate, latency, and reliability, among others. Naturally, the resource management problem will have to deal with the dynamics of the environment, such as the changes in the application services (e.g., mobility of users, change of traffic patterns, etc.) as well as the heterogeneity of the services. A major challenge here is to be able to provide a real-time decision support system that is able to handle data, and that can introduce minimal overhead to the system while solving such large-scale wireless-centric problems. This is where DDDAS can play an instrumental role.

While DDDAS can be used to support the diverse set of services we discussed in Sect. 1, we present two use cases where *i*) DDDAS can be used to maintain QoS at the desired levels under dynamically changing networks and *ii*) a use case shown in Fig. 1 that uses the DDDAS framework for supporting the deployment of a micro-grid testbed over a 5G/6G communication network.

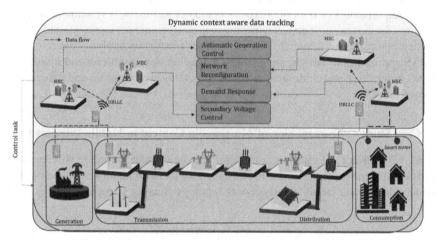

Fig. 1. DDDAS for 5G networks handling microgrid operations

3 DDDAS for 5G Network Design and Optimization

The DDDAS framework can support methods required to design and manage the operation for resilience in the complex, heterogeneous and multimodal 5G&6G communications infrastructures. Opportunities where DDDAS-based methods can be used for supporting 5G and 6G systems, are in addressing two key challenges presented as examples here and addressed in the subsequent case studies discussed in Sects. 3.1 and 3.2: 1) Efficient management of energy in wireless networks is very important, due to the much higher power of the higher frequency signals, and the need to support new applications with improved QoS. This becomes even more critical for 5G and more so 6G systems that must leverage high-frequency bands, while supporting energy needs of communication messages and signals. Prior work that focuses on energy often has overlooked the QoS needs of the users, particularly when it comes to services that require HRLLC-type requirements – a key feature for 6G. Communication efficiencies are also key, and DDDAS-based methods presented in Sect. 3.1 on transmission of multimedia content to mobile wireless users over 5G millimeter wave networks, can improve the QoS. 2) Self-sustaining real-time decision making in a dynamic manner is central for solving resource management problems in 5G and 6G systems. Remarkably, most existing approaches, such as optimization-based and reinforcement learning-based solution, still require a major overhead in terms of latency and complexity. And, even more importantly, these approaches are often restricted to static environments, and they cannot scale to the faster, non-stationary dynamics envisioned for the emerging 5G applications and more so for tomorrow's 6G applications. Such latency and stationarity limitations render existing resource management techniques not suitable for latency and reliability-critical services, thus motivating the need for a novel DDDAS-based framework, as per the example case study presented in Sect. 3.2.

Overall, in order to meet the needs of energy efficiencies in wireless networks and the QoS requirements of both the network operators and users, a systematic approach that allows for the real-time decision making in these dynamic environments and a multi-objective optimization is crucial. Thus, this paper focuses on applying the referenced DDDAS approaches to design, and manage these complex, multilevel and multimodal 5G&6G networking and communication infrastructures in a cognizant and dynamically optimized way.

3.1   Case Study I: Transmission of Multimedia Content to Mobile Wireless Users

In [18], we consider the transmission of multimedia content to mobile wireless users over 5G millimeter wave networks. In particular, we focus on rapidly changing network dynamics due to user mobility and network topology changes as depicted in Fig. 2.

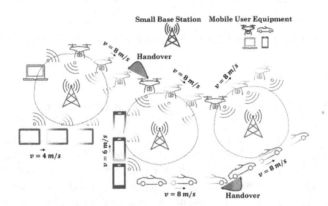

Fig. 2. Dynamic matching problem in 5G millimeter wave networks

This will, in turn, demonstrate an early result for our ability to design intelligent, distributed algorithms (based on multi-agent game theory) that can maintain QoS under dynamically changing 5G network; a result that can be further expanded with the use of more advanced DDDAS approaches in the future. In this prior work, using game theory, we show how a 5G system can particularly leverage high-speed millimeter wave links to opportunistically pre-store content at the mobile user, thus allowing it to smoothly transition between two base stations, as it moves in the network. This, in turn, leads to efficient resource management across base stations. A key result can be found in the work by co-author Saad [Fig. 7, 18], updated and shown here, as Fig. 3; in this figure is shown how the proposed system can maintain a smooth network QoS for a mobile user that is transitioning at high speed from being associated with the network slices of a given base station to those of another one.

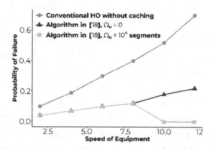

Fig. 3. Speed of mobile equipment vs. failure probability, reproduced from [18]

In particular, a comparison is made how an algorithm that is aware of the dynamics can enable coordination so as to maintain a low handover failure (HOF) probability, i.e., a low probability that the high-speed user fails to associate with the needed network slices/resources. This figure particularly shows the average HOF probability versus the speed of the users, and it compares the dynamic-aware approach with a conventional scheme that does not make use of the dynamics (and of millimeter wave links) to store information at the device. These prior results demonstrate that the proposed algorithm can significantly reduce the HOF probability by leveraging the information on the dynamics (i.e., trajectory) and the network's topology. [Fig. 7, 18] also shows that the algorithm is considerably robust against HOF. For instance, the HOF probability decreases for speeds beyond 8 m/s, since higher speed allows the user to move for longer distances before the opportunistically stored data is used. The results showcase that being aware of the dynamics of the environment leads to better decisions, cognizant and optimized, and which will help the network maintain higher QoS levels. A key next step would be to introduce a larger-scale DDDAS framework that can be used to extend this work across other types of resource management problems beyond HOF management and data pre-storage.

3.2 Case Study II: Collaborative Microgrids

The impact of dynamically adjusting the wireless network configuration is shown in a case of DDDAS-based coordination of three interconnected microgrids. These micro-grids are equipped with numerous sensors whose readings hold importance that are subject to change, as the dynamic environment of the microgrids dictates. The importance of the sensory data is determined by real-time data filtering and detection techniques.

More detailed explanation on the application of DDDAS framework within the concept of microgrids *per se* can be found in [11, 15, 17]. In this study, we did not only change the fidelity of the data (i.e., number of samples collected and integrated into the decision-making process from a specific sensor) but also the network configuration by dynamically assigning transmission of the high priority sensory data to the slices that have superior capabilities. Figure 4a shows the dynamic assignment of the sensors to the network slices according to the importance ranking of the sensors.

A key simulation experiment is conducted to assess the effectiveness of such a dynamic assignment approach on the total operational cost of the system. Figure 4b shows the difference between static and dynamic assignment of the sensors to the different network slices. In the former approach, the sensors' assignment is determined by analyzing the historical data. In contrast, in the latter approach, the assignments are made based on the real-time ranking of the sensors. In parallel, Fig. 4c shows the total energy consumption incurred by communication and computation in a microgrid cluster using both approaches. These results constitute only a proof-of-concept that showcases how DDDAS can potentially provide a real-world, real-time decision support for 5G system network slicing. Many future extensions are planned with results expected in the near future, including more realistic modeling of the wireless system and metrics (e.g., accounting for precise formulations of latency and reliability, incorporation of high-frequency bands and integrated sensing and communication for 6G, etc.), incorporation of additional services, and the integration of larger-scale dynamics.

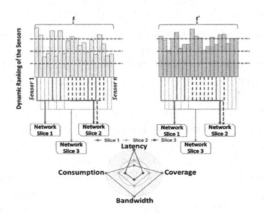

Fig. 4a. Dynamic sensor assignments to the network slices

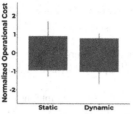

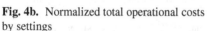

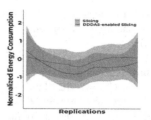

Fig. 4b. Normalized total operational costs by settings

Fig. 4c. Normalized total energy consumption

4 Conclusion

The paper discusses and demonstrates in example cases how 5G networks can benefit from DDDAS-based methods and frameworks akin to those presented in this paper. Emerging and communication systems (i.e., 5 G/B 5G (6G)) are appealing, due to their desired and advanced properties, the dramatically increasing amount of data traffic in today's IoT applications transcends the capacity growth rates of new wireless communication technologies. Furthermore, ensuring efficiencies in the design and the operational management of the 5G and 6G infrastructures themselves, is in tandem and synergistically related to the applications that will exploit the 5G and 6G capabilities. Thus, the deployment and management of the communication networks require further attention to achieve conflicting objectives of spectrum efficiency, cost/energy efficiency and QoS in terms of low latency and high reliability. Addressing this need, the DDDAS framework is very promising since it offers bi-directional information flow between decision-making mechanism and the real-time data collected from the system to harness the large amount of data systematically.

Acknowledgements. This study was supported by the Air Force Office of Scientific Research (AFOSR) Award No: FA9550-19-1-0383.

References

1. Soldani, D., Guo, Y.J., Barani, B., Mogensen, P., Chih-Lin, I., Das, S.K.: 5G for ultra-reliable low-latency communications. IEEE Network **32**(2), 6–7 (2018)
2. Khan, L.U., Yaqoob, I., Imran, M., Han, Z., Hong, C.S.: 6G wireless systems: a vision, architectural elements, and future directions. IEEE Access **8**, 147029–147044 (2020)
3. Sengupta, S.: IoE: An innovative technology for future enhancement. In: Computer Vision and Internet of Things, pp. 19–28. Chapman and Hall/CRC (2022)
4. Chaccour, C., Naderi Soorki, M., Saad, W., Bennis, M., Popovski, P.: Can terahertz provide high-rate reliable low latency communications for wireless VR?. IEEE Internet Things J. **9**, 9712–9729 (2022)
5. Saad, W., Bennis, M., Chen, M.: A vision of 6G wireless systems: applications, trends, technologies, and open research problems. IEEE Network **34**, 134–142 (2020)

6. Tikhvinskiy, V., Koval, V.: Prospects of 5g satellite networks development. In: Moving Broadband Mobile Communications Forward-Intelligent Technologies for 5G and Beyond (2020)
7. IEEE International Network Generations Roadmap (INGR): Energy efficiency. Retrieved 11 July 2022. https://futurenetworks.ieee.org/images/files/pdf/INGR-2022-Edition/IEEE_I NGR_EE_Chapter_2022-Edition-FINAL.pdf
8. IEEE International Network Generations Roadmap (INGR): Artificial intelligence and machine learning. Retrieved 26 July 2022. https://futurenetworks.ieee.org/images/files/pdf/ INGR-2022-Edition/IEEE_INGR_AIML_Chapter_2022-Edition-FINAL.pdf
9. IEEE International Network Generations Roadmap (INGR): Applications & services. Retrieved 26 July 2022. https://futurenetworks.ieee.org/images/files/pdf/INGR-2022-Edition/ IEEE_INGR_AppsSvcs_Chapter-2022-Edition-FINAL.pdf
10. Blasch, E.P., Darema, F., Bernstein, D.: Introduction to the dynamic data driven applications systems (DDDAS) paradigm. In: Blasch, E.P., Darema, F., Ravela, S., Aved, A.J. (eds.) Handbook of Dynamic Data Driven Applications Systems, 2nd edition, vol. I, pp. 1–32, Springer, Cham (2022). https://doi.org/10.1007/978-3-030-74568-4_1
11. Celik, N., Lee, S., Vasudevan, K., Son, Y.J.: DDDAS-based multi-fidelity simulation framework for supply chain systems. IIE Trans. 42(5), 325–341 (2010)
12. Allaire, D., Kordonowy, D., Lecerf, M., Mainini, L., Willcox, K.: Multifidelity DDDAS methods with application to a self-aware aerospace vehicle. Procedia Comput. Sci. 29, 1182–1192 (2014)
13. Fujimoto, R., et al.: Dynamic data driven application simulation of surface transportation systems. In: Alexandrov, V.N., Albada, G.D., Sloot, P.M.A., Dongarra, J. (eds.) ICCS 2006. LNCS, vol. 3993, pp. 425–432. Springer, Heidelberg (2006). https://doi.org/10.1007/117585 32_57
14. Hariri, S., Al-Nashif, Y., Valerdi, R., Prowell, S., Blasch, E.: DDDAS-based resilient cyberspace. In: Presentation Proceedings of AFOSR DDDAS PI Meeting, October 2, 2
15. Damgacioglu, H., Bastani, M., Celik, N.: A dynamic data-driven optimization framework for demand side management in microgrids. In: Blasch, E., Ravela, S., Aved, A. (eds.) Handbook of Dynamic Data Driven Applications Systems, pp. 489–504. Springer International Publishing, Cham (2018). https://doi.org/10.1007/978-3-319-95504-9_21
16. Yavuz, A., Runsewe, T., Celik, N., Chaccour, C., Saad, W., Darema F.: DDDAS @ 5G and beyond 5G networks for resilient communications infrastructures and microgrid clusters. In: Blasch, E., Darema, F., Ravela, S., Aved, A. (eds.) Handbook of DDDAS (Vol. III), Springer, Heidelberg (2022)
17. Thanos, A.E., Bastani, M., Celik, N., Chen, C.H.: Dynamic data driven adaptive simulation framework for automated control in microgrids. IEEE Trans. Smart Grid 8(1), 209–218 (2015)
18. Semiari, O., Saad, W., Bennis, M., Maham, B.: Caching meets millimeter wave communications for enhanced mobility management in 5G networks". IEEE Trans. Wire-less Commun. 17(2), 779–793 (2018)

Plenary Presentations - Systems Support Methods

DDDAS-Based Learning for Edge Computing at 5G and Beyond 5G

Temitope Runsewe[1], Abdurrahman Yavuz[1], Nurcin Celik[1(✉)], and Walid Saad[2]

[1] Department of Industrial and Systems Engineering, University of Miami, Miami,
FL 33146, USA
celik@miami.edu
[2] Department of Electrical and Computer Engineering, Virginia Tech, Arlington,
VA 22203, USA

Abstract. The emerging and foreseen advancements in 5G and Beyond 5G (B5G) networking infrastructures enhance both communications capabilities and the flexibility of executing computational tasks in a distributed manner, including those encountered in edge computing (EC). Both, 5G & B5G and EC environments present complexities and dynamicity in their multi-level and multimodal infrastructures. DDDAS-based design and adaptive and optimized management of the respective 5G, B5G, and EC infrastructures are needed, to tackle the stochasticity inherent in these complex and dynamic systems, and to provide quality solutions for respective requirements. In fact, both emerging communication and computational technologies and infrastructure systems can benefit from their symbiotic relationship in optimizing their corresponding adaptive and optimized management. EC enabled by 5G (and future B5G) allows efficient distributed execution of computational tasks. DDDAS-based methods can support the adaptivity in bandwidth and energy efficiencies in 5G and B5G communications. On the other hand, EC has become a very attractive feature for critical infrastructure such as energy grids as it allows for secure and efficient real-time data processing. In order to fully exploit the advantages of EC, the communication network should be able to tackle the changing requirements related to the task management within edge servers. Thus, leveraging the DDDAS paradigm, we jointly optimize the scheduling and offloading of computational tasks in an EC-enabled microgrid considering both physical constraints of microgrid and network requirements. The results showcase the superiority of the proposed DDDAS-based approaches in terms of network utilization and operational efficiencies achieved, with microgrids as case example.

Keywords: Task Scheduling · Task Offloading · EC enabled Microgrid

1 Introduction

In the era of the fifth generation (5G) and beyond networks (6G being the immediate beyond 5G), edge computing (EC) has emerged as a revolutionary technology that offers computational capabilities at the edge [1–3]. EC could allow for low-latency, high-bandwidth communication infrastructure that will advance numerous applications and

E. Blasch et al. (Eds.): DDDAS 2022, LNCS 13984, pp. 135–143, 2024.
https://doi.org/10.1007/978-3-031-52670-1_13

services relating to the Internet of Things (IoT) (e.g., smart cities, autonomous vehicles, smart grids) [1]. Over 10 billion IoT devices were connected in 2021, and that number is expected to increase to 25 billion by 2030. An IoT-based microgrid system is at the forefront of this advancement as it provides intelligence, visibility, control, and communication to optimize energy operations [4]. Due to the need for low latency computing in microgrid applications, EC is a promising approach to enable effective microgrid operations because it can offload computation-intensive tasks to EC servers over wireless cellular networks [5].

DDDAS-based methods for managing microgrids use edge-based computing to enable multidirectional flow of energy and data with real-time monitoring and reaction to supply and demand changes. As a result, such approaches unlock the capability of constantly monitoring the power network [6], and to make predictions of onset of failures and allow emergency preparations and recovery from faults automatically. By enabling real-time changes in energy provisioning, the ability to account for future demands and make informed decisions, sets the DDDAS-based methods apart from traditional power grids management methods. The success of smart systems to optimize infrastructure functions and services depends on the efficient handling of data. Thus, integration of all data into a single, integrated system which combines edge as a seamless part of cloud computing is paramount to ensuring optimal performance [7]. Due to the dynamic and highly stochastic nature of the power grid, control of the operations to achieve operational efficiency requires the incorporation of real-time data into decision-making processes[8]. Although EC emerges as a promising option of in-situ (at the instrumentation side) to manage massive amounts of data mitigating the need to move such data, there are several challenges to overcome (e.g., high energy consumption, task scheduling/offloading, and resource allocation).

There has been an increase in EC related research [1, 9–14]. The authors in [9] presented an optimization problem to minimize the energy consumption of a Mobile Edge Computing (MEC) offloading system. The work in [10] also presented an optimization problem to minimize system consumption considering delay and energy for EC system. In [11], the authors investigated the partial computational offloading for a single mobile device within EC networks. In a large-scale antenna heterogeneous EC network, the work in [12] jointly optimized the unloading ratio, calculation frequency and offloading time to minimize the energy consumption of the system. In [13], data partitioning is used to improve the energy efficiency of massive multiple input multiple output EC. Although there have been various research efforts addressing EC related issues, there is still limited research available on EC enabled microgrid. The authors in [14] proposed a massive multiple input multiple output technology to improve energy efficiency and transmission rate while meeting the requirements of low latency in the smart grid system. An optimization problem is also formulated in [1] to minimize the energy consumption of microgrid-enabled EC networks considering computational and latency constraints. Although the above studies have addressed various issues in EC and EC enabled microgrid, a strong coordination between the two considering their physical constraints is yet to be addressed (e.g., power flow constraints, network speed and bandwidth, and operational constraints). This study presents a context aware dynamic data driven applications systems (DDDAS) based framework that offers the potential to provide adaptive assessments of the EC enabled microgrid based on dynamic data.

The DDDAS paradigm can harness data in intelligent ways to provide new capabilities, including decision support systems, efficient data collection, resource management, and data mining [15]. In DDDAS, the model of a system is augmented ("learns") using targeted system-instrumentation data (dynamic data inputs), and in return the system models control the instrumentation of the system. Such methods ("DDDAS-based") have shown to be key for enabling autonomic operations. DDDAS has been applied with immense success in a variety of engineering and scientific applications where modeling and data collection are utilized. Examples include uncertainty quantification [16, 17], power systems [18–20], computing and security [21]. The main contribution of the present paper is the development of a framework for EC-enabled microgrids that is focused on optimizing the scheduling and offloading of computational tasks while being cognizant of both the physical constraints of the microgrid and network requirements of the EC servers. Using the DDDAS framework, the proposed framework maintains the systems' feasibility under varying circumstances based on the real-time data/measurements observed from both the network and the microgrid. The task scheduling module prioritizes the computation of critical tasks whereas the microgrid controller prioritizes the critical loads. Using a bidirectional feedback loop between modules, we find the (near) optimal schedule for the tasks with the least possible compromise.

The proposed framework (the DDDAS feedback-control loop) presented in Fig. 1 consists of three major components: (a) a database with information about the state of the microgrid (total demand, market price and weather conditions), (b) an EC enabled microgrid model and (c) a task scheduling and offloading optimization model.

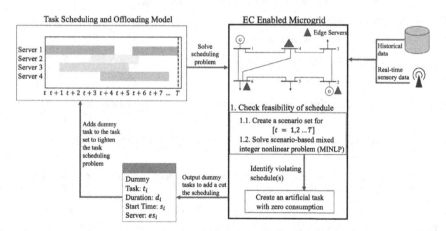

Fig. 1. Proposed DDDAS framework for EC enabled microgrid.

The rest of this paper is organized as follows. In Sect. 2, the modules that are employed to build DDDAS-based context-aware scheduling and offloading of the EC tasks are explained. In Sect. 3, a numerical analysis conducted using a modified version of IEEE-30 network (a representative test case which is a simple approximation of the American

Electric Power system)[22]. Finally, In Sect. 4, the initial findings and future directions of addressing 5G/B5G within DDDAS framework have been discussed.

2 Proposed Joint Optimization Approach

The joint optimization algorithm built to facilitate the scheduling of EC tasks iterates towards (near) optimal solution by solving dynamically generated subproblems. First, the master problem is formulated as a mixed integer nonlinear problem (MINLP). Due to its combinatorial properties, the problem is classified as an NP-hard problem [23]. Hence, finding the global optimal solution is extremely challenging and time consuming. Therefore, in order to obtain a low complexity solution, we decompose the formulated problem into two subproblems: 1) tasks scheduling and offloading (TSO) and 2) micro-grid operational planning (MOP); these two subproblems are discussed in Sects. 2.1 and 2.2 below. Table 1 shows the notations for parameters used in both task scheduling and EC enabled microgrid optimization models.

Table 1. Parameters of optimization models

Parameters:		
$\mathcal{S}$	Set of edge computing servers	
$\mathcal{N}, \mathcal{L}, \Omega, \mathcal{T}$	Set of buses, lines, scenarios, and time periods	
$\mathcal{R}$	Set of buses with renewable energy sources	
$\mathcal{D}$	Set of buses with droop-controlled generation units	
$\mathcal{E}$	Set of buses with edge servers	
$\mathcal{K}$	Set of tasks to be scheduled	
$\mathcal{D}_{is}$	Duration of task i at edge server s	$\forall i \in \mathcal{K}, \forall s \in \mathcal{S}$
$\mathcal{P}_i^C$	Preconditions for task i	$\forall i \in \mathcal{K}$
$\mathcal{D}_{its}^P, \mathcal{D}_{its}^Q$	Active/reactive power demand for bus i at time t in scenario s	$\forall i \in \mathcal{N}, \forall s \in \Omega, \forall t \in \mathcal{T}$
$\mathcal{E}_{ikt}^P, \mathcal{E}_{ikt}^Q$	Active/reactive power demand of executing task k in edge server at bus i at time t	$\forall i \in \mathcal{E}, \forall k \in \mathcal{K}, \forall t \in \mathcal{T}$
$\mathcal{R}_{its}^P, \mathcal{R}_{its}^Q$	Active/reactive power output of renewable energy source at bus i at time t in scenario s	$\forall i \in \mathcal{R}, \forall s \in \Omega, \forall t \in \mathcal{T}$
$\|Y_{ij}\|, \theta_{ij}$	Magnitude and angle for line between buses i and j	$\forall (i, j) \in \mathcal{L}$
$\mathcal{S}_{ik}^T, \mathcal{D}_{ik}$	Start time, duration of task k in the edge server i	$\forall i \in \mathcal{E}, \forall k \in \mathcal{K}$
PRI_i^T, PRI_j^L	Priority of task i and load j	$\forall i \in \mathcal{K}, \forall j \in \mathcal{N}$

Table 2 below shows the notations for the decision variables used in both task scheduling and EC enabled microgrid optimization models.

Table 2. Decision variables used to build optimization models

Decision Variables:		
$\int \sqcup_{is}$	Starting time of the of task i at edge server s	$\forall i \in \mathcal{K}, \forall s \in \mathcal{S}$
$\S_{is}$	1, if task i is assigned to edge server s; 0, otherwise	$\forall i \in \mathcal{K}, \forall s \in \mathcal{S}$
$\sqsubseteq_{its}, \delta_{its}$	Voltage magnitude and angle of bus i at time t in scenario s	$\forall i \in \mathcal{N}, \forall s \in \Omega,$ $\forall t \in \mathcal{T}$
ω_{ts}	System frequency at time t in scenario s	$\forall t \in \mathcal{T}, \forall s \in \Omega$
$\rfloor_{it}$	1, if the load i is not shed at time t; 0, otherwise	$\forall i \in \mathcal{N}, \forall t \in \mathcal{T}$
$\int_{ikt}$	1, if edge server at bus i starts executing task k at time t; 0, otherwise	$\forall i \in \mathcal{E}, \forall t \in \mathcal{T}$ $\forall k \in \mathcal{K}$
∇_{ikt}	Auxiliary binary variable indicating that edge server at bus i can execute task k at time t	$\forall i \in \mathcal{E}, \forall t \in \mathcal{T}$ $\forall k \in \mathcal{K}$
$\updownarrow_{\sqrt{}}^{i}, \setminus \amalg_{i}$	Droop coefficients of DG unit at bus i	$\forall i \in \mathcal{DR}$
$\sqrt[g]{}_{its}, \amalg_{its}^{g}$	Active/reactive power generation at bus i at time t in scenario s	$\forall i \in \mathcal{DR},$ $\forall s \in \Omega, \forall t \in \mathcal{T}$
$\sqrt[d]{}_{its}, \amalg_{its}^{d}$	Active/reactive power demand at bus i at time t in scenario s	$\forall i \in \mathcal{N}, \forall s \in \Omega,$ $\forall t \in \mathcal{T}$

2.1 Task Scheduling and Offloading (TSO)

Microgrid task assignment can prove challenging as its components must work cooperatively and at times without supervision. For a microgrid to be self-sufficient, the microgrid controller must assign to the EC servers several tasks, including (i) monitoring breaker position, (ii) monitoring microgrid bus parameters (e.g., voltage and frequency, which are primarily monitored in islanded mode), and (iii) communicating node status information. The computational tasks sent to the EC servers arrive randomly with varying needs of computational power and energy. To ensure these tasks are executed with the set priority, the EC and microgrid must be coordinated effectively without ignoring their constraints. The algorithm starts with a solution for a relaxed problem in which we find a schedule without considering the physical constraints of the microgrid, and we send the schedule to the microgrid operational planning (MOP) module developed within our DDDAS framework. This module controls microgrid operations while being cognizant of the environmental dynamics via systematic incorporation of real-time into the optimization process. Here, we modify the optimization model in order to give the necessary feedback to the task scheduling of edge servers deployed within the microgrid as shown in Fig. 2. When a high priority task cannot be completed within a set delay by the edge servers, this task is offloaded to the cloud servers. The TSO is modeled in Python using convex hull reformulation within generalized disjunctive programming library and solved with branch and cut algorithm [24].

2.2 Microgrid Operational Planning

For our MOP model formulation, an islanded microgrid [25] was used while being cognizant of both the physical constraints and other demands needed to be met by the microgrid. The microgrid is also equipped with demand side management capability that allows microgrid to shed the loads when it is needed. This problem is formulated as a MINLP and solved by using Bonmin solver in Pyomo [26]. A major aspect of the MOP model is its ability to dynamically maintain supply and demand balance without compromising high priority loads or computational tasks. Since the EC enabled microgrid obtains real time data from the task scheduling and offloading model, the MOP is able to ascertain the feasibility of executing a set of tasks under the constraints of computational capability and maximum tolerable delay. As a result of this information from the MOP, the TSO model is able to reschedule the task within the limits of the constraints, as shown in Fig. 2.

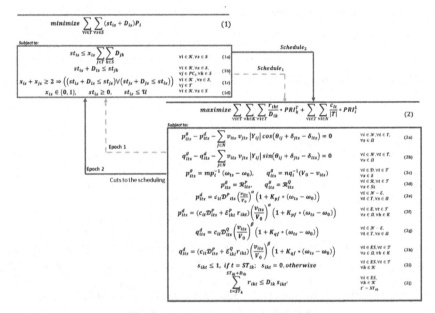

Fig. 2. DDDAS-based learning model

3 Simulation Results and Analysis

We illustrate the effectiveness of the proposed approach with a case study in which we have seven tasks to be scheduled. These tasks are assumed to have different prerequisites, for instance, the order of processing Task 2 must be Server 1, Server 2 and Server 3, respectively. Task scheduling module first solves the problem (1) (see Fig. 2) without any consideration of energy availability of the dynamic environment of the edge servers and sends the relaxed solution (see Fig. 3a) to the MOP module.

The schedule sent by the TSO module is treated as a parameter within the MOP module. However, in order to construct informative feedback for TSO module, whether a given task will be computed at a scheduled period is introduced as a decision variable in Problem 2 (see Fig. 2). As a result, the MOP module returns the periods and servers where computing the tasks will jeopardize network's supply and demand balance. Then, the output is used to create additional constraints that will be added to the TSO module. "Task_D" and "Task_D2" in Fig. 3b are created in our case to avoid scheduling tasks for the time intervals of (15,35) and (4,16). After adding the constraint to the Problem 1, the resulting schedule is given in Fig. 3b. The algorithm terminates if a given schedule is feasible for both modules. With the feedback mechanism in place, we dynamically reduce the feasible solution space of the scheduling and offloading problem lightening the burden of (near) real-time decision-making.

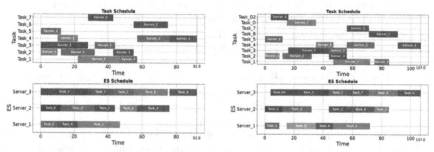

Fig. 3a. Schedule at Epoch 1   **Fig. 3b.** Schedule at Epoch 2

4 Conclusion

In this paper, we have presented a first step towards introducing 5G EC as an uncharted application for DDDAS (that is, using the DDDAS framework for context aware EC network management while considering multiple objectives posed by both decision-making mechanisms of the microgrid and EC networks). We have proposed a resource-aware approach utilizing DDDAS framework for EC servers that are deployed within a microgrid. We have showed that the flexibility and stability of each system can be enhanced with the bidirectional information flow between applications while accounting for the effects of environmental uncertainty. Although the focus of this study was to dynamically adjust optimization models for collaborative prioritization of critical tasks and loads, the relationship between applications can be extended to cover different objectives such as minimizing the cost of operations or the delay of computation.

Acknowledgements. This study was supported by the Air Force Office of Scientific Research Award No: FA9550-19-1-0383. Authors would like to extend their sincere thanks to Dr. Frederica Darema for her invaluable and constructive feedback, and guidance during the development of this research work.

References

1. Munir, M.S., Abedin, S.F., Tran, N.H., Hong, C.S.: When edge computing meets microgrid: a deep reinforcement learning approach. IEEE Internet Things J. **6**(5), 7360–7374 (2019)
2. Munir, M.S., Abedin, S.F., Tran, N.H., Han, Z., Hong, C.S.: A multi-agent system toward the green edge computing with microgrid. In: 2019 IEEE Global Communications Conference (GLOBECOM), pp. 1–7 (2019)
3. Saad, W., Bennis, M., Chen, M.: A vision of 6G wireless systems: applications, trends, technologies, and open research problems. IEEE Network **34**(3), 134–142 (2019)
4. Li, C.S., Darema, F., Kantere, V., Chang, V.: Orchestrating the cognitive internet of things. In: IoTBD, pp. 96–101 (2016)
5. Pu, T., et al.: Power flow adjustment for smart microgrid based on edge computing and multi-agent deep reinforcement learning. J. Cloud Comput. **10**(1), 1–13 (2021)
6. Hong, G., Hanjing, C.: An edge computing architecture and application oriented to distributed microgrid. In: 2021 IEEE Intl Conf on Parallel & Distributed Processing with Applications, Big Data and Cloud Computing, Sustainable Computing and Communications, Social Computing and Networking, pp. 611–617. IEEE (2021)
7. Gustafson, K.: How edge-to-cloud computing powers smart grids and smart cities | HPE. Accessed 14 July 2022
8. Bastani, M., Damgacioglu, H., Celik, N.: A δ-constraint multi-objective optimization framework for operation planning of smart grids. Sustain. Cities Soc. **38**, 21–30 (2018)
9. Zhang, K., Mao, Y.M.: Energy-efficient offloading for mobile edge computing in 5G heterogeneous networks. IEEE Access **4** (2016)
10. Sun, Y., Hao, Z., Zhang, Y.: An Efficient offloading scheme for MEC system considering delay and energy consumption. J. Phys. Conf. Ser. **960**(1), 012002 (2018). IOP Publishing
11. Ali, Z., Abbas, Z.H., Abbas, G., Numani, A., Bilal, M.: Smart computational offloading for mobile edge computing in next-generation Internet of Things networks. Comput. Netw. **198**, 108356 (2021)
12. Hao, Y., Ni, Q., Li, H., Hou, S.: Energy-efficient multi-user mobile-edge computation offloading in massive MIMO enabled HetNets. In: IEEE International Conference on Communications, Shanghai, pp. 1–6 (2019)
13. Malik, R., Vu, M.: Energy-efficient offloading in delay-constrained massive MIMO enabled edge network using data partitioning. IEEE Trans. Wireless Commun. **19**(10), 6977–6991 (2020)
14. Qin, N., Li, B., Li, D., Jing, X., Du, C., Wan, C.: Resource allocation method based on mobile edge computing in smart grid. IOP Conf. Ser. Earth Environ. Sci. **634**(1), 012054 (2021). IOP Publishing
15. Blasch, E., Bernstein, D., Rangaswamy, M.: Introduction to dynamic data driven applications systems. In: Handbook of Dynamic Data Driven Applications Systems, 2nd edn. Springer, Cham (2022)
16. Linares, R., Vittaldev, V., Godinez, H.C.: Dynamic data-driven uncertainty quantification via polynomial chaos for space situational awareness. In: Handbook of Dynamic Data Driven Applications Systems, pp. 75–93. Springer, Cham (2018). https://doi.org/10.1007/978-3-319-95504-9_4
17. Celik, N., Son, Y.J.: Sequential Monte Carlo-based fidelity selection in dynamic-data-driven adaptive multi-scale simulations. Int. J. Prod. Res. **50**(3), 843–865 (2012)
18. Hunter, M., Biswas, A., Chilukuri, B., Guin, A., Fujimoto, R., Rodgers, M.: Energy-aware dynamic data-driven distributed traffic simulation for energy and emissions reduction. In: Handbook of Dynamic Data Driven Applications Systems, pp. 475–495. Springer, Cham (2022). https://doi.org/10.1007/978-3-030-74568-4_20

19. Damgacioglu, H., Bastani, M., Celik, N.: A dynamic data-driven optimization framework for demand side management in microgrids. In: Handbook of Dynamic Data Driven Applications Systems, pp. 489–504. Springer, Cham (2018). https://doi.org/10.1007/978-3-319-95504-9_21

20. Thanos, A.E., Shi, X., Sáenz, J.P., Celik, N.: A DDDAMS framework for real-time load dispatching in power networks. In: Winter Simulations Conference, pp. 1893–1904 (2013)

21. Blasch, E., Aved, A., Bhattacharyya, S.S.: Dynamic data driven application systems (DDDAS) for multimedia content analysis. In: Handbook of Dynamic Data Driven Applications Systems, pp. 631–651. Springer, Cham (2018). https://doi.org/10.1007/978-3-319-95504-9_28

22. Alsac, O., Stott, B.: Optimal load flow with steady state security. IEEE Trans. Power Appar. Syst. **93**(3), 745–751 (1974)

23. Bienstock, D., Verma, A.: Strong NP-hardness of AC power flows feasibility. Oper. Res. Lett. **47**(6), 494–501 (2019)

24. Chen, Q., et al.: GDP: an ecosystem for logic based modeling and optimization development. Optimiz. Eng. **23**(1), 607–642 (2022)

25. Jain, M., Gupta, S., Masand, D., Agnihotri, G., Jain, S.: Real-time implementation of islanded microgrid for remote areas. J. Control Sci. Eng. (2016)

26. Hart, W.E., et al.: Pyomo-Optimization Modeling iin Python, vol. 67. Springer, Cham (2017)

Monitoring and Secure Communications for Small Modular Reactors

Maria Pantopoulou[1,2(✉)], Stella Pantopoulou[1,2], Madeleine Roberts[3], Derek Kultgen[1], Lefteri Tsoukalas[2], and Alexander Heifetz[1]

[1] Nuclear Science and Engineering Division, Argonne National Laboratory, Argonne, IL 60439, USA
{mpantopo,spantopo}@purdue.edu, {dkultgen,aheifetz}@anl.gov
[2] School of Nuclear Engineering, Purdue University, West Lafayette, IN 47906, USA
tsoukala@purdue.edu
[3] Department of Physics, University of Chicago, Chicago, IL 60637, USA
mbroberts@uchicago.edu

Abstract. Autonomous, safe and reliable operations of Small Modular Reactors (SMR), and advanced reactors (AR) in general, emerge as distinct features of innovation flowing into the nuclear energy space. Digitalization brings to the fore of an array of promising benefits including, but not limited to, increased safety, higher overall efficiency of operations, longer SMR operating cycles and lower operation and maintenance (O&M) costs. On-line continuous surveillance of sensor readings can identify incipient problems, and act prognostically before process anomalies or even failures emerge. In principle, machine learning (ML) algorithms can anticipate key performance variables through self-made process models, based on sensor inputs or other self-made models of reactor processes, components and systems. However, any data obtained from sensors or through various ML models need to be securely transmitted under all possible conditions, including those of cyber-attacks. Quantum information processing offers promising solutions to these threats by establishing secure communications, due to unique properties of entanglement and superposition in quantum physics. More specifically, quantum key distribution (QKD) algorithms can be used to generate and transmit keys between the reactor and a remote user. In one of popular QKD communication protocols, BB84, the symmetric keys are paired with an advanced encryption standard (AES) protocol protecting the information. In this work, we use ML algorithms for time series forecasting of sensors installed in a liquid sodium experimental facility and examine through computer simulations the potential of secure real-time communication of monitoring information using the BB84 protocol.

Keywords: Advanced Reactors · Sensor Monitoring · Quantum Communication

1 Introduction

Advanced high temperature fluid small modular reactors (SMR), such as sodium fast reactors (SFR) [1, 2] and molten salt cooled reactors (MSCR) [3] are among the promising options for the future of nuclear energy [4]. These types of reactors operate at

© The Author(s), under exclusive license to Springer Nature Switzerland AG 2024
E. Blasch et al. (Eds.): DDDAS 2022, LNCS 13984, pp. 144–151, 2024.
https://doi.org/10.1007/978-3-031-52670-1_14

ambient pressure and at temperatures of several hundred $^\circ$C to achieve efficient thermal to electric energy conversion and require longer time periods for refueling compared to existing light water reactors. To achieve low operation and maintenance (O&M) costs, the SMR's are expected to operate in autonomous remote mode with minimal human staffing. Monitoring of sensor readings can be achieved with the use of machine learning (ML) algorithms, which perform forecasting of time series of sensors. Thermocouples are the most common sensors in high temperature thermal hydraulic components of the SMR. Continuous monitoring of thermocouple readings is crucial to detect any signs of failure in these sensors, so they can be replaced in time [5–7]. Additionally, data obtained from the monitoring of thermocouple sensors need to be securely transmitted to remote user/operator. Achieving autonomous operation of SMR thus also requires the development of advanced methods for secure communications of reactor monitoring data.

The main challenge for prediction with ML models is, in most cases, the complexity of the dependence in the input variables. Recurrent neural networks (RNN) are a category of networks capable of processing time series mostly due to their internal memory structure. However, these types of algorithms can experience the problem of the vanishing gradient, where a network cannot learn the data sequences properly [8]. This issue can be solved with the use of long short-term memory (LSTM) networks, which are a special case of RNN [9–13]. These networks consist of cells, which in turn consist of three gates; an input gate, a forget gate, and an output gate. These gates have the ability to learn the relevant information during the training process. Specifically, the forget gate makes a decision regarding keeping or ignoring information. The input gate decides on the addition of information from the current step. The output gate determines the following hidden state.

Security threats, such as cyberattacks pose additional issues and make the need for secure communications more and more imminent. Sensitive data are often encrypted and then transmitted through channels along with keys which help decode the useful information. However, the methods of classical computing used for this purpose render this process particularly vulnerable to threats. Quantum communications, which are based on quantum computing obey the laws of quantum physics, and thus offer an opportunity for secure data transmission [14]. This is because of the particles used (qubits), which change their state whenever they are measured, due to the property of superposition they acquire [15, 16]. Quantum key distribution (QKD) is a process used for generating and transmitting keys between two parties using qubits [17, 18]. There has been recent interest in using QKD for secure communications with power systems [19], underwater environments [20], and smart grids [21]. A commonly used protocol for QKD is the BB84 protocol, developed in the 1980s by Bennett and Brassard [22].

Typically, QKD algorithms are paired with some encryption protocol, such as a symmetric advanced encryption standard (AES) protocol, to encrypt the message through a conventional classical channel. AES was developed by National Institute of Standards and Technology (NIST) [23], and is a symmetric encryption algorithm that requires both users to share the same key (for encryption and decryption). Although it is harder to implement compared to other encryption algorithms, it is much stronger. The keys can be either 128, 192 or 256 bits long. This work proposes for the first time to integrate secure QKD-based communication with ML-based SMR monitoring. More specifically, it explores the use of LSTM networks for time series forecasting of sensors installed in a liquid sodium experimental facility and examines the potential of secure real-time communication of monitoring information using the BB84 protocol through computer simulations. In this paper, data is obtained from the Mechanisms Engineering Test Loop (METL) facility, which is an intermediate-scale liquid metal experimental facility at Argonne National Laboratory [24]. Figure 1 shows a photograph of METL (left) and a schematic of the envisioned scheme of the secure remote monitoring. The BB84 protocol is implemented to generate a key which is then used to encrypt the predicted values of the thermocouple time series with an AES protocol. The remote user can then decrypt the received message using the same key.

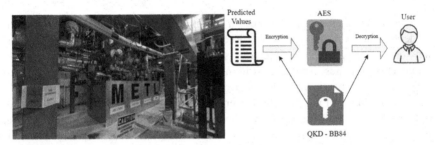

Fig. 1. The envisioned secure communication system for remote monitoring of METL facility (photograph on the left).

2　Model Development and Results

2.1　Liquid Sodium Vessel Monitoring Using LSTM Models

In this work, we examined the ability of two types of LSTM neural networks to predict values of temperature time series obtained with K-type Process Control Thermocouples (PCTCs) in METL. METL has four experimental test vessels (two of 18 inches and two of 28 inches) and more than one thousand sensors. A schematic of the test vessel connections is shown in Fig. 2. One of the monitored variables is sodium vapor temperature above the liquid in the vessel. The dataset that was used in the study consists of almost 225500 readings, in units of °C, of PCTCs measurements of the vapor temperature. The time series were measured with 1 s resolution over the course of three days using the control interface installed at METL. For the temperature ranges in this study, the PCTC measurement uncertainty was about ± 1.5 °C (0.75%).

In order to evaluate the performance of LSTM networks, the dataset was split into three segments: 70% for training, 10% for validation and 20% for testing. The segment chosen for testing purpose included the highest temperatures in the time series. Time series forecasting was performed, in two separate cases, with a Vanilla LSTM and with a stacked LSTM with two LSTM layers. We tried to minimize the complexity of the structure of both networks by just adding a dropout layer after every LSTM layer with dropout rate equal to 20%, to avoid the overfitting phenomenon and improve the performance of the model. Both Vanilla LSTM and stacked LSTM networks were implemented using Python TensorFlow in an Intel® Core™ i5-10210U CPU @ 1.60GHz with 8 GB RAM. The number of hidden layers used for the implementation was 20, and both models were trained for 40 epochs. The optimizing algorithm was Adam, and the learning rate was reduced from 0.001 to 0.0001 to achieve better results.

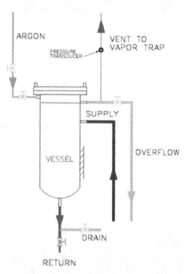

Fig. 2. Schematics of test vessel connections in METL facility [24]. Time series of PCTC monitoring vapor were analyzed in this study.

We present predicted time series in blue color and original data in red color in Fig. 3 and Fig. 4 for the vapor PCTC for Vanilla LSTM and stacked LSTM networks respectively after 40 epochs of training the networks. As shown in the graphs, both networks' performance is high as the predicted values are almost identical with the original data. Vanilla LSTM is shown to perform slightly better than stacked LSTM in the part where a higher fluctuation in the vapor temperature appears. In order to estimate the accuracy of the results that the two developed models provide, root mean square error (RMSE) was used. RMSE can be calculated using the following formula

$$\text{RMSE} = \sqrt{\frac{\sum_{i=1}^{N} \|y_n - \hat{y}_n\|^2}{N}} \tag{1}$$

where y_n is the original data, $\hat{y}_n$ corresponds to the predicted data, and N is the total number of measurements in the testing subset. The RMSE values calculated for the Vanilla LSTM and the stacked LSTM networks are much smaller than the measurement uncertainty of the PCTC which is 1.5 °C. More specifically, RMSE values are equal to 0.021 and 0.03, respectively. Finally, the execution time needed for the prediction of the testing data was calculated for the specific operating system and was equal to 2.524 s.

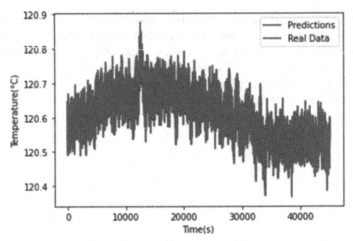

Fig. 3. Original (red) and predicted (blue) thermocouple time series using Vanilla LSTM for the vapor PCTC.

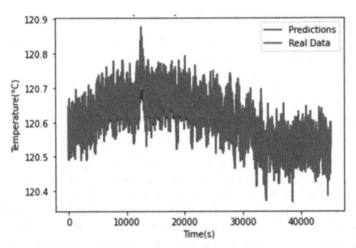

Fig. 4. Original (red) and predicted (blue) thermocouple time series using stacked LSTM for the vapor PCTC.

2.2   Secure Communications Using BB84 and AES Protocols

The simulation of a key transmission through a fiber-optic channel with the BB84 protocol was performed with SeQUeNCe [25], which is an open-source software package for quantum communications. In SeQUeNCe, the user can create the classical and quantum channels, and then specify parameters such as pathlength and channel losses, which affect the time needed to generate and share keys. This process is followed by the procedure of error correction using the Cascade protocol [26], which is often used along with the QKD algorithm. Ultimately, the purpose is to achieve transmitting a key with 0% error rate.

For the specific study, attenuation of 0.22 dB/km in wavelength of 1550 nm was used in SeQUeNCe to model channel attenuation, and signal depolarization of 3% was also considered. We performed one iteration to calculate the time that is needed for the generation, distribution, and correction of a 256-bit key over 10 km distance, which will then be used with an AES protocol to encrypt the temperature value that needs to be securely transmitted every time. The operating system used for the simulation was an Intel® Core™ i5-10210U CPU @ 1.60 GHz with 8 GB RAM, and the execution time was 6.78 s. The reported time includes the generation, transmission, and correction of the 256-bit key. Nuclear system monitoring sensors have a range of time delays, from milliseconds (pressure transmitters) to seconds (thermocouples) to a minute for hydrogen sensing for sodium leak detection in SFR. Thus, a general estimate of message transmission frequency to and from SMR could be on the order of one message per second. Therefore, the time of protocol duration for QKD is, qualitatively, compatible with real time communication requirements with SMR.

The next step is to use an AES protocol to perform encryption and decryption of the transmitted message. For this purpose, an AES-256 protocol was developed using the PyCryptodome Python package. Every time a predicted temperature value needs to be sent to the remote user, it is encrypted using the 256-bit key generated by the BB84 protocol. The remote user, who possesses the same key, decrypts the ciphertext and is able to read the original message. The AES-256 protocol was simulated in the same operating system as BB84, and the total execution time needed for the encryption and decryption of the message was almost 13 ms. The execution times needed for each part of the communication process are presented in Table 1.

Table 1. Execution time needed for each process of the communication.

Process for one sample value	Execution Time (s)
QKD and Error Correction (256-bit key)	6.78
Encryption with AES-256	17.56×10^{-4}
Decryption with AES-256	10.92×10^{-3}

3 Conclusions

Autonomous, safe and reliable operations of advanced reactors emerge as distinct features of innovation flowing into the nuclear energy space. In this work, we used ML for time series forecasting of sensors installed in a liquid sodium experimental facility and examined the potential of secure real-time communication of monitoring information using the BB84 protocol through simulations. Specifically, we used a Vanilla LSTM model, as well as a stacked LSTM model for the prediction of a PCTC time series. Results showed that both ML models can predict time series values with RMSE significantly lower than the thermocouple measurement uncertainty. The execution time needed for the prediction of one sample was calculated to be equal to 56 ms. For the purpose of secure transmission of the predicted data, the BB84 protocol was used to generate and share a 256-bit key over 10 km distance. This was paired with the AES-256 protocol, to encrypt the data before transmission and decrypt in the end. The total time needed for the communication process was calculated to be equal to 6.79 s per sample. Preliminary results indicate feasibility of the approach described in this study to enabling remote autonomous operation of SMR.

Acknowledgment. This work was supported in part by the U.S. Department of Energy, Advanced Research Projects Agency-Energy (ARPA-E) Generating Electricity Managed by Intelligent Nuclear Assets (GEMINA) program under Contract DE-AC02-06CH11357, and in another part by a donation to AI Systems Lab (AISL) by GS Gives and the Office of Naval Research under Grant No. N00014-18-1-2278. Experimental data was obtained from the Mechanisms Engineering Test Loop (METL) liquid sodium facility at Argonne National Laboratory.

References

1. Aoto, K., et al.: A summary of sodium-cooled fast reactor development. Prog. Nucl. Energy **77**, 247–265 (2014)
2. Kim, J.B., Jeong, J.Y., Lee, T.H., Kim, S., Euh, D.J., Joo, H.K.: On the safety and performance demonstration tests of prototype Gen-IV sodium-cooled fast reactor and validation and verification of computational codes. Nucl. Eng. Technol. **48**, 1083–1095 (2016)
3. Blandford, E., et al.: Kairos power thermal hydraulics research and development. Nucl. Eng. Des. **364**, 110636 (2020)
4. Ho, M., Obbard, E., Burr, P.A., Yeoh, G.: A review on the development of nuclear power reactors. Energy Procedia **160**, 459–466 (2019)
5. Mandal, S., Santhi, B., Sridhar, S., Vinola, K., Swaminathan, P.: Nuclear power plant thermocouple sensor-fault detection and classification using deep learning and generalized likelihood ratio test. IEEE Trans. Nucl. Sci. **64**, 1526–1534 (2017)
6. Mandal, S., Santhi, B., Sridhar, S., Vinola, K., Swaminathan, P.: A novel approach for fault detection and classification of the thermocouple sensor in nuclear power plant using singular value decomposition and symbolic dynamic filter. Ann. Nucl. Energy **103**, 440–453 (2021)
7. Mandal, S., Santhi, B., Sridhar, S., Vinola, K., Swaminathan, P.: Minor fault detection of thermocouple sensor in nuclear power plants using time series and analysis. Ann. Nucl. Energy **134**, 383–389 (2019)
8. Pascanu, R., Mikolov, T., Bengio, Y.: On the difficulty of training recurrent neural networks. In: Proceedings of the 30th International Conference on Machine Learning, Proceedings of Machine Learning Research, vol. 28, no. 31, pp. 310–1318 (2013)
9. Hochreiter, S., Schmidhuber, J.: Long short-term memory. Neural Comput. **9**(8), 1735–1780 (1997)
10. Ankel, V., Pantopoulou, S., Weathered, M., Lisowski, D., Cilliers, A., Heifetz, A.: One-Step Ahead Prediction of Thermal Mixing Tee Sensors with Long Short-Term Memory (LSTM) Neural Networks. Argonne National Laboratory (No. ANL/NSE-20/37) (2020)
11. Pantopoulou, S., et al.: Monitoring of temperature measurements for different flow regimes in water and Galinstan with long short-term memory networks and transfer learning of sensors. Computation **10**, 108 (2022)
12. Wang, P., Zhang, J., Wan, J., Wu, S.: A fault diagnosis method for small pressurized water reactors based on long short-term memory networks. Energy **239**, 122298 (2022)

13. Miki, D., Demachi, K.: Bearing fault diagnosis using weakly supervised long short-term memory. Nucl. Sci. Technol. **57**, 1091–1100 (2020)
14. Heifetz, A., et al.: Perspectives on secure communications with advanced reactors: ultrasonic and millimeter waves classical and quantum communications. In: ANS Annual Meeting Embedded Conference 12th Nuclear Plant Instrumentation, Control and Human-Machine Interface Technologies (NPIC-HMIT2021) (2021)
15. National Academies of Sciences, Engineering, and Medicine. Quantum computing: progress and prospects (2019)
16. Heifetz, A., Agarwal, A., Cardoso, G.C., Gopal, V., Kumar, P., Shahriar, M.S.: Super-efficient absorption filter for quantum memory using atomic ensembles in a vapor. Opt. Commun. **232**(1–6), 289–293 (2004)
17. Nurhadi, A.I., Syambas, N.R.: Quantum key distribution (QKD) protocols: a survey. In: 2018 4th International Conference on Wireless and Telematics (ICWT), 12 July 2018. IEEE (2018)
18. Liu, R., Rozenman, G.G., Kundu, N.K., Chandra, D., De, D.: Towards the industrialisation of quantum key distribution in communication networks: a short survey. IET Quant. Commun. (2022)
19. Zhao, B., et al.: Performance analysis of quantum key distribution technology for power business. Appl. Sci. **10**(8), 2906 (2020)
20. Raouf, A.H.F., Safari, M., Uysal, M.: Performance analysis of quantum key distribution in underwater turbulence channels. JOSA B **37**(2), 564–573 (2020)
21. Alshowkan, M., Evans, P.G., Starke, M., Earl, D., Peters, N.A.: Authentication of smart grid communications using quantum key distribution. Sci. Rep. **12**(1), 1–13 (2022)
22. Shor, P.W., Preskill, J.: Simple proof of security of the BB84 quantum key distribution protocol. Phys. Rev. Lett. **85**, 441 (2000)
23. Dworkin, M.J., et al.: Advanced encryption standard (AES) (2001)
24. Kultgen, D., Grandy, C., Kent, E., Weatherd, M., Andujar, D., Reavis, A.: Mechanisms Engineering Test Loop – Phase I Status Report, Argonne National Laboratory, ANL-ART-148 (2018)
25. Wu, X., et al.: SeQUeNCe: A customizable discrete-event simulator of quantum networks. Quant. Sci. Technol. **6**, 4 (2021)
26. Martinez-Mateo, J., Pacher, C., Peev, M., Ciurana, A., Martin, V.: Demystifying the information reconciliation protocol cascade. arXiv preprint arXiv:1407.3257 (2014)

Data Augmentation of High-Rate Dynamic Testing via a Physics-Informed GAN Approach

Celso T. do Cabo, Mark Todisco, and Zhu Mao[✉]

Worcester Polytechnic Institute, Worcester, MA 01609, USA
zmao2@wpi.edu

Abstract. High-rate impact tests are essential in the prediction of component/system behavior subject to unplanned mechanical impacts which may lead to several damages. However, significant challenges exist when identifying damages in such events, considering its complexity, these diagnostic challenges inspire the employment of data driven approaches as feasible solution for such problems. However, most deep machine learning techniques require big amount of data to support an effective training to reach an accurate result, while the data collected from each test is extremely limited yet performing multiple tests to collect data is oftentimes unrealistically expensive. Therefore, data augmentation is very important to enhance the learning quality. Generative Adversarial Network (GAN) is a deep learning algorithm able to generate synthetic data under a recorded testing environment. However, a GAN uses random input as seeds to generate adversarial models, and with sufficient training, it may produce synthetic data with good quality. This paper proposes a hybrid approach which employs the output from an oversimplified FE model as the seed to drive a GAN generator, such that a drastic amount of computation is saved, and the GAN training will converge faster and more accurately than using just the random noise seeds. Variational Autoencoder (VAE) is combined with the approach to reduce the data dimension and the extracted features are classified via a Support Vector Machine (SVM). Results show that using the proposed physics-informed approach will improve the accuracy of the damage classifier and reduce the classification uncertainty, compared to using the original small dataset without augmentation.

Keywords: Deep Learning · Generative Adversarial Networks · Variational Autoencoder · Structural Health Monitoring · High-Rate Dynamics · Data Augmentation · Support Vector Machine

1 Introduction

Structural health monitoring (SHM) is often used to obtain the state awareness of a variety of systems. However, it commonly involves the application of sensors over a long period of time and record gradual changes in the structure. Meanwhile, high-rate dynamic phenomena usually happen in a small duration of time, commonly leading to catastrophic damages of such systems. The high-rate dynamical systems can be defined as systems that are subject to loads or impacts that happens in a short duration of time,

E. Blasch et al. (Eds.): DDDAS 2022, LNCS 13984, pp. 152–162, 2024.
https://doi.org/10.1007/978-3-031-52670-1_15

from 30 μs to 100 ms and that the amplitude of the impact can generate an acceleration response up to 100 g [1–3]. Those conditions will lead to a system where there is a high uncertainty on the loading condition, in addition, its response will be non-stationary, and its dynamics will be difficult to model. Some examples of high-rate dynamic systems are occurrence of blasts in civil structures or collisions in automotive applications and anomalies in aerospace applications. Considering the complexity of the systems subject to such high-rate impacts, damage detection or localization of such applications poses to be challenging.

Oftentimes, for cases whereas it is difficult to classify the data solely based on physical characteristics or traditional signal processing techniques, machine learning approaches poses to be applicable, considering its capabilities for pattern recognition and classification for complex cases. For instance, studies such as [4] where the condition of operation for drilling oil and gas was observed, and later processed using dimension reduction algorithms and machine learning algorithms were performed. The application of deep learning techniques for prognosis and remaining useful life of rotary machines were also done by [5, 6]. Regarding electronic equipment, [7] analyzed the consequences of mechanical shocks in PCs motherboards. In the domain of high-rate dynamics, [8, 9] tried to classify different damage levels on a board using dimensioning reduction from autoencoders in time-series dataset, and later classify it using support vector machine (SVM).

However, since high-rate impact tests are often destructive, it is unfeasible to repeat the test several times to collect the required quantity of data to train a reliable and robust classifier. An alternative for the deficiency of the data for this situation would be to employ data augmentation, adding noise to the original signal sometimes can be used for such purposes, however, for time-series datasets, it would not increase the variability of the data and would not be a feasible solution. Instead, generative adversarial networks (GANs), initially introduced by [10] and later modified into the conditional generative adversarial networks (cGANs) by [11] for multiclass classification, can synthetically generate new datasets that may be used for data augmentation. Those algorithms were initially developed to generate images and improve the accuracy of convolutional neural networks (CNNs). However, it was already employed for the synthetic generation of time-series by [12].

Although GANs and cGANs show promising results to generate new datasets, they are often driven by random noise, which can lead to a high computational resource consumption. In addition, their hyperparameter tunning can be challenging in the process of creating suitable datasets. Other variants of GANs were proposed in the past, such as [13], that uses a finite element (FE) model to generate different types of damage for a GAN development.

This paper proposes using a simple FE model output as the seed for a cGAN to generate a synthetic dataset to improve the robustness of the classifier of a high-rate dynamic impact test result. In addition, in order to reduce the computational expense, variational autoencoder (VAE) is employed. This technique, introduced by [14] uses a neural network to reduce the dimension of the original dataset avoiding losing as much information from the system as possible. In the next chapters, it is explained in detail how the test is performed, in addition to an explanation of each algorithm applied to process the data. Later the results for the classification are presented and discussed. Finally, a conclusion is drawn from the methodology developed and suggestions for future work are given.

2 Theory Background

2.1 Dataset and Preprocessing

The dataset was based on a series of impacts tests generated by a drop tower in a three-board electronic assembly, provided by [15]. Figure 1 shows a model of the board used in the test and the locations of the sensors used to record the data. For the test it was employed three strain gauges, identified in Fig. 1 as SG. In addition, an internal accelerometer is added in the middle board. Finally, the acceleration data obtained in the fixture of the testing subject is recorded by a reference accelerometer that can be seen in Fig. 2.

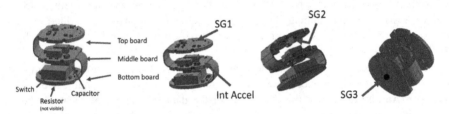

Fig. 1. Experimental three board electronic assembly with an RC circuit [15].

The testing data consists of time series recorded at a 1 MHz sampling rate during 5.5 ms duration. The data is divided into six different classes, from the first test, where the test subject is still healthy, to five different levels of damage, where the last level one of the capacitors loses its functionality, so the electronic board can no longer be used.

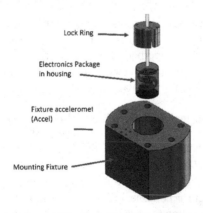

Fig. 2. Fixture setup of the three-board electronic assembly [15].

Since the data for each class generated consist into a long time series sample as shown in Fig. 3 for the strain gauge #1 measurements. The preprocessing of the data is then performed by using a moving window to down sample the original datapoints and thus generate a number of small segments from the original set, allowing a single test of 5.5 ms to generate 50 segments of 1.9 ms each.

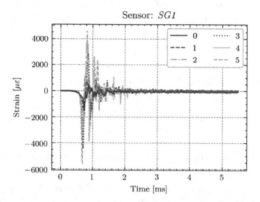

Fig. 3. Strain gauge one measurement for all tests [16].

From the time series, a spectrogram with the frequency content over time was extracted and used as input for the algorithms described afterwards. Since there are five sensors placed on the structure, and each time series recorded was converted into a corresponding spectrogram, the input for the algorithm explained next will have a 5-spectrogram input for each set as shown in Fig. 4.

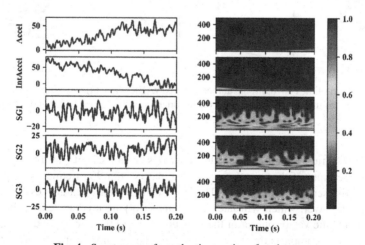

Fig. 4. Spectrogram from the time series of each sensor

2.2 Convolutional Variational Autoencoders (CVAE)

The variational autoencoders (VAE) are neural networks capable of condense the data into a smaller number of variables. Often used for dimension reduction of nonlinear inputs, it consists in two neural networks. One of them will have the same number of features or points of the available data as input and a smaller dimension as output, while the other will be a mirrored neural network that will reconstruct the original signal. In this paper, a convolutional variational autoencoder is adopted, which will convert the spectrogram obtained from a time-series data into a small latent dimension dataset, and more details of the algorithm may be found in [9].

Originally, the VAE will compact the dataset into a selected number of dimensions and use the statistical information of the dataset to generate new points into the dimensions, using a Gaussian distribution as described in Eq. (1):

$$Z = \mu + \sigma \times \varepsilon \tag{1}$$

where, μ is the mean and σ is the standard deviation of the data in each one of the dimensions. ε is a normal distributed random number, and Z is the encoded dataset. As shown in [16], the use of a high number of dimensions for the classification gives a reasonable accuracy even with a small dataset, but the encoded data loses its interpretability. On the other hand, for low dimensions, the classification will have a poorer performance, but the interpretability of the data is easier. Therefore, a total of three latent dimensions are selected in the encoder, and it is aimed to increase the accuracy of the classifier by augmenting the dataset. A schematic of the structure used for the CVAE algorithm can be seen in Fig. 5.

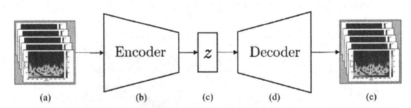

Fig. 5. Flowchart of the convolutional autoencoder (CVAE), where (a) is the input spectrograms, (b) the encoder, (c) the latent variables, (d) the decoder and (e) is the reconstructed spectrograms [9].

The spectrogram obtained from the timeseries from each of the five sensors are adopted as the input of the encoder, and a 3D matrix with 962 frequency steps, 1924 timesteps and 5 sensors is generated. The input will then have features extracted from a convolutional layer, which will later have a max pooling layer to reduce the size of the dataset, and the vector output of the max pooling will then be applied into a neural network that will compress the data into the three latent dimensions. Later, the neural network will calculate the mean (μ) and standard deviation (σ) of each datapoint for each latent dimension. Those values will be added into the normal distribution described by Eq. (1) to generate the final latent space dataset Z.

To validate the encoder, the decoder of the algorithm, shown in Fig. 5 (d), will use the latent space dataset Z to reconstruct the original input. For that, the input of the decoder will be fed into a neural network, which will regenerate a vector with the information of the original set. Later this process will have an un-pooling layer and a convolution transpose that will recreate a similar 3D matrix as output for the original spectrogram used as input. Finally, both the original and recreated spectrograms will be compared to check how much information of the model was lost. Therefore, the loss function for the CVAE is defined as the average between the reconstruction loss and how well the model is described by a normal distribution. The reconstruction loss is shown in Eq. (2).

$$L_R = \sum_{i=1}^{N} (x_i - \hat{x}_i)^2 \tag{2}$$

where, x_i is the original datapoint being encoded and $\hat{x}_i$ is the reconstructed datapoint after decoding. In Eq. (3) it is presented the loss function term that penalizes if the encoded data is not well represented by a Gaussian distribution.

$$L_L = \frac{1}{2} \sum_{i=1}^{N} \left[1 + \log(\sigma_i)^2 - \mu_i^2 - \sigma_i^2 \right] \tag{3}$$

where, σ_i and μ_i are the standard deviation and mean of the latent dimension for a datapoint i in the latent dimension. Finally, the loss function for the autoencoder will be calculated based on Eq. (4).

$$\mathcal{L}_{Encoder} = \frac{L_R + L_L}{N} \tag{4}$$

2.3 FEM Based Conditional Generative Adversarial Networks (cGAN)

Conditional generative adversarial networks (cGANs) are neural networks used to generate synthetic data. For this purpose, two neural networks, the generator and discriminator are trained simultaneously, one generating signals from a random input, while the second tries to identify if the data generated is real or fake. The difference between the original GAN and the cGAN is that the second has another input feature the class of the original dataset, thus, allowing the output to be multiclass. A new modification in the cGAN is proposed that instead of using a random noise as input of the generator, an oversimplified FEM model will generate similar spectrograms from the real data, and then the generator will update this dataset to become closer to the real data. The flowchart of the FEM based cGAN can be seen in Fig. 6.

To reduce even further the computational cost, the dataset is firstly encoded by the CVAE and then input into the cGAN, and this will reduce the original dimension of the dataset, allowing it to easily generate new datasets for a larger number of datapoints. In addition, one of the main issues when dealing with cGANs is what is called mode collapsing, where the generator encounters a region which the discriminator will always accept as a real dataset, then the generator will create all datapoints in the same position. To counter this issue the loss function of the cGAN was altered. The original binary cross entropy loss function for the cGAN can be seen in Eq. (5).

$$L = -w_n \left[y_n \cdot \log x_n + (1 - y_n) \cdot \log(x_n - 1) \right] \tag{5}$$

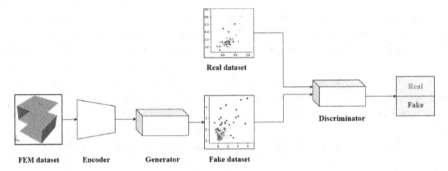

Fig. 6. Finite element based conditional generative adversarial network (cGAN).

where, L is the loss function for a point n, w_n is the weight of the loss, y_n is the real value of the class of the datapoint and x_n the prediction of its class. It was added another term in Eq. (5) to avoid that the points would be focused on a single region of the plot. That term was based on the Gaussian mixture model available on Eq. (6) and will penalize the generated point if it is close to the mean of the original dataset.

$$G_M = \frac{(x_n - \mu_k)^2}{\sigma_k^2} \tag{6}$$

where, G_M is the Gaussian mixture model, x_n is the current datapoint location, μ_k the mean of the dataset available on the batch k and σ_k^2 its standard deviation.

2.4 Support Vector Machine (SVM)

For the classification of the datasets, initially it was chosen to use a classic machine learning classifier. Since support vector machine (SVM) is a classifier that can work with multidimensional problems as well as it is a fast classifier that can works with limited amount of data. The SVM classifier basically works using kernel functions to separate the datasets into each respective class. This kernel function can be a simple linear function, to a polynomial, sigmoid or radial basis function (RBF). For this problem, the radial basis function was selected, considering the complexity of the dataset, whereas a linear function would not classify well the dataset and among the other nonlinear kernels the RBF had the best accuracy [16]. The RBF function can be seen in Eq. (7).

$$K(X_1, X_2) = e^{-\gamma \|X_1 - X_2\|^2} \tag{7}$$

where, $K(X_1, X_2)$ is the kernel function and $\|X_1 - X_2\|^2$ is the Euclidean distance between the datapoints. γ is the weight for each datapoint in the kernel function. The classifier will then generate regions for each class by minimizing the error function, described in Eq. (8).

$$t(w, \xi) = \frac{1}{2}\|w\|^2 + \frac{C}{N}\sum_{i=1}^{N} \xi_i \tag{8}$$

where, ξ_i is the error at the point i, w is the weight from the kernel function and C is a regulation factor that will prevent the model to become complex and overfit the data.

3   Validation of Damage Classifications

3.1   Convolutional Variational Autoencoder (CVAE)

The spectrograms calculated from the time-series vectors extracted from the five sensors were used as input for the convolutional variational autoencoder (CVAE) in a group. Thus, the input consisted of 5 spectrograms of the windowed dataset. Then, those spectrograms were compressed into a defined number of latent dimensions and since the variational autoencoder is employed, it is possible to extract the information as normal distributions.

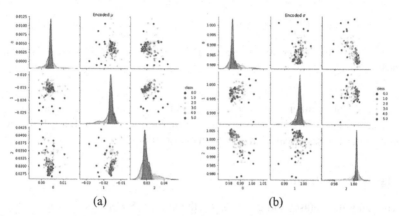

(a)   (b)

Fig. 7. Encoded mean (a) and standard deviation; (b) for all classes using three latent dimensions

In Fig. 7 it is presented the mean (a) and standard deviation (b) for each datapoint of the set. Each small plot is a plot of different latent dimensions, and the histogram in the diagonal represents the distribution of the data within a latent dimension, each of the colors also represent each class of the classification problem, where 0 is the healthy scenario and 5 the highest amount of damage that caused failure to the system. Although using a small number of latent dimensions reduces the accuracy of the encoding process and the amount of information from the system lost is higher than for a greater number of dimensions, it was decided to use a small latent space due to the classification process. Since the original dataset is small, for the classification, a higher number of points is needed to avoid overfitting.

3.2   FEM Based Conditional Generative Adversarial Network

After encoding the dataset into a smaller dimension, the mean and standard deviation are extracted for each time-series. For instance, compacting the time-series to 3 latent dimensions was employed. Since it is a small number of dimensions, allowing fast computing and good interpretability of the data. The comparison between the original dataset and the synthetic data generated by the cGAN can be seen in Fig. 8.

As it can be seen from Fig. 8, the synthetic generated data has a similar behavior and distribution as the original dataset. Since the generated dataset from cGAN is based

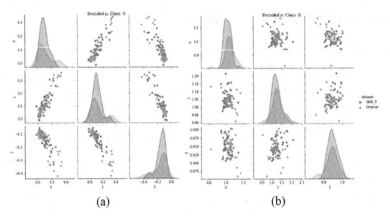

Fig. 8. Comparison between the GAN generated data and the original dataset for (a) mean; and (b) standard deviation of the first class.

on a FE model, the number of data create is illimited, allowing now the classifier to use as many datapoints as needed for training and the original dataset for validation. The results for other classes are also similar to the ones in Fig. 8.

3.3 Classification Results

After encoding the dataset and generate synthetic data, SVM was used to classify the level of damage from the board. SVM is employed since it possesses the capability to classify small datasets. In addition, a radial based kernel (RBF) is applied considering the complexity of the data. As metrics to evaluate the performance of the classification, it was used accuracy, which is available in Eq. (9) [17]:

$$A = \frac{TP + TN}{TP + TN + FP + FN} \tag{9}$$

where, A is the accuracy of the model, TP is the number of true positive values, TN is the number of true negative values, FP the false positive and FN the false negative classifications. Figure 9 shows the accuracy of the classification, when using only the generated data, a mix of the generated and real data or only the real data for training. For this case, the original dataset had 300 datapoints for all classes, it was generated via cGAN another 1,800 datapoints equally distributed in the six classes.

Figure 10 presents the confusion matrices for the mix of generated data with real data and using only the real dataset. As can be seen, although the accuracy for both models are under 50%, the model using only the original dataset will mostly predict all datapoints as only one of the damage levels. This behavior is also present in the mixed dataset. However, with significant less predictions in a single class and a more distributed prediction for the other classes as well, which corresponds to the increase in 10% of the accuracy from the original model.

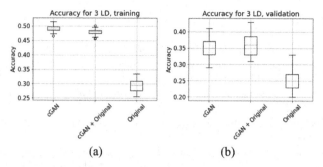

Fig. 9. Accuracy for classification using SVM, (a) training set, (b) validation set

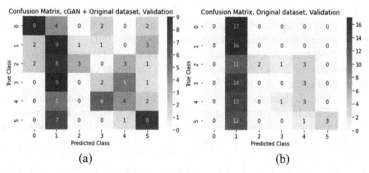

Fig. 10. Confusion matrix for validation of using (a) cGAN and original dataset and (b) only the original dataset.

4   Conclusions

High-rate dynamic impacts are essential for prognostics and the prevention of catastrophic failure of many different systems. Considering the complexity of the data obtained by such systems and the difficulty to perform several tests to generate data, this paper proposes a generative adversarial approach to augment the limited data. In particular, cGAN using FE model prediction as the seed is proved to be a reliable way to generate synthetic data while keeping a small computational cost. It also shows that even for SVM that can perform classification well with small datasets, augmenting the data will improve the robustness of the classifier.

Acknowledgements. This study is based upon work supported by the Air Force Office of Scientific Research under award number FA95501810491. Any opinions, finding, and conclusions or recommendations expressed in this material are those of the authors and do not necessarily reflect the views of the United States Air Force. The authors would also like to thank Dr. Jacob Dodson at the Air Force Research Laboratory for providing the high-rate data in this study.

References

1. Dodson, J., Joyce, B., Hong, J., Laflamme, S., Wolfson, J.: Microsecond state monitoring of nonlinear time-varying dynamic systems. In: Smart Materials, Adaptive Structures and Intelligent Systems (2017)
2. Dodson, J., et al.: High-rate structural health monitoring and prognostics: an overview. Data Sci. Eng. **9**, 213–217 (2022)
3. Hong, J., Laflamme, S., Dodson, J., Joyce, B.: Introduction to state estimation of high-rate system dynamics. Sensors **18**(1), 217 (2018)
4. Ignova, M., Matheus, J., Amaya, D., Richards, E.: Recognizing abnormal shock signatures during drilling with help of machine learning. In: SPE Middle East Oil and Gas Show and Conference (2019)
5. Ma, M., Mao, Z.: Deep-convolution-based LSTM network for remaining useful life prediction. IEEE Trans. Indust. Inf. **17**(3), 1658–1667 (2020)
6. Ma, M., Mao, Z.: Deep wavelet sequence-based gated recurrent units for the prognosis of rotating machinery. Struct. Health Monit. **20**(4), 1794–1804 (2021)
7. Pitarresi, J., Roggeman, B., Chaparala, S., Geng, P.: Mechanical shock testing and modeling of PC motherboards. In: 2004 Proceedings of the 54th Electronic Components and Technology Conference (IEEE Cat. No. 04CH37546) (2004)
8. Todisco, M., Mao, Z.: Damage quantification of high-rate impacts using hybrid deep learning models. In: ASME International Mechanical Engineering Congress and Exposition (2021)
9. Todisco, M., Mao, Z.: High-rate damage classification and lifecycle prediction via deep learning. In: Madarshahian, R., Hemez, F. (eds.) Data Science in Engineering, Volume 9. CPSEMS, pp. 225–232. Springer, Cham (2022). https://doi.org/10.1007/978-3-030-76004-5_25
10. Goodfellow, I., et al.: Generative adversarial nets. Adv. Neural Inf. Process. Syst. **27** (2014)
11. Mirza, M., Osindero, S.: Conditional generative adversarial nets. arXiv preprint arXiv:1411.1784 (2014)
12. Smith, K.E., Smith, A.O.: Conditional GAN for timeseries generation. arXiv preprint arXiv:2006.16477 (2020)
13. Gao, Y., Liu, X., Xiang, J.: FEM simulation-based generative adversarial networks to detect bearing faults. IEEE Trans. Indust. Inf. **16**(7), 4961–4971 (2020)
14. Kingma, D.P., Welling, M.: Auto-encoding variational bayes. arXiv preprint arXiv:1312.6114 (2013)
15. Beliveau, A., Hong, J., Dodson, J. Davies, M.: Dataset-3-high-rate-in-situ-damage-of-electronic-packages (2020). https://github.com/High-Rate-SHM-Working-Group/Dataset-3-High-Rate-In-Situ-Damage-of-Electronics-Packages
16. Todisco, M.: Structural Damage Classification of High-Rate Dynamic Systems via Hybrid Deep Learning (Publication Number 28714990) [M.S.Eng., University of Massachusetts Lowell]. (2021)
17. Bishop, C.M., Nasrabadi, N.M.: Pattern Recognition and Machine Learning, vol. 4. Springer (2006)

Unsupervised Wave Physics-Informed Representation Learning for Guided Wavefield Reconstruction

Joel B. Harley[1]([✉]), Benjamin Haeffele[2], and Harsha Vardhan Tetali[1]

[1] University of Florida, Gainesville, FL 32603, USA
joel.harley@ufl.edu
[2] Mathematical Institute for Data Science, Johns Hopkins University, Baltimore, MD 21218, USA

Abstract. Ultrasonic guided waves enable us to monitor large regions of a structure at one time. Characterizing damage through reflection-based and tomography-based analysis or by extracting information from wavefields measured across the structure is a complex dynamic-data driven applications system (DDDAS). As part of the measurement system, guided waves are often measured with in situ piezo-electric sensors or wavefield imaging systems, such as a scanning laser doppler vibrometer. Adding sensors onto a structure is costly in terms of components, wiring, and processing and adds to the complexity of the DDDAS while sampling points with a laser doppler vibrometer requires substantial time since each spatial location is often averaged to minimize perturbations introduced by dynamic data. To reduce this burden, several approaches have been proposed to reconstruct full wavefields from a small amount of data. Many of these techniques are based on compressive sensing theory, which assumes the data is sparse in some domain. Among the existing methods, sparse wavenumber analysis achieves excellent reconstruction accuracy with a small amount of data (often 50 to 100 measurements) but assumes a simple geometry (e.g., a large plate) and assumes knowledge of the transmitter location. This is insufficient in many practical scenarios since most structures have many sources of reflection. Many other compressive sensing methods reconstruct wavefields from Fourier bases. These methods are geometry agnostic but require much more data (often more than 1000 measurements). This paper demonstrates a new DDDAS approach based on unsupervised wave physics-informed representation learning. Our method enables learning full wavefield representations of guided wave datasets. Unlike most compressive sensing methodologies that utilize sparsity in some domain, the approach we developed in our lab is based on injecting wave physics into a low rank minimization algorithm. Unlike many other learning algorithms, including deep learning methods, our approach has global convergence guarantees and the low rank minimizer enables us to predict wavefield behavior in unmeasured regions of the structure. The algorithm can also enforce the wave equation across space, time, or both dimensions simultaneously. Injecting physics also provides the algorithm tolerance to data perturbations. We demonstrate the performance of our algorithm with experimental wavefield data from a 1m by 1m region of an aluminum plate with a half-thickness notch in its center.

E. Blasch et al. (Eds.): DDDAS 2022, LNCS 13984, pp. 163–172, 2024.
https://doi.org/10.1007/978-3-031-52670-1_16

Keywords: DDDAS · wave-informed machine learning · signal processing · acoustics

1 Introduction

1.1 Overview

Structural health monitoring is a Dynamic Data Driven Applications System (DDDAS) in which the characteristics of a structure or material dynamically vary over time due to gradual growth of damage as well as other environmental and operational variations. All these changes affect our sensor systems. Ultrasonic guided waves are a common sensing modality in structural health monitoring that allow us to characterize and identify damage in large regions of structures [1, 2]. Guided waves can be measured with in situ piezoelectric sensors or wavefield imaging systems, such as scanning laser Doppler vibrometers [3]. Even with a full wavefield imaging system, characterizing and understanding guided wave propagation can be challenging. This is due to their frequency-dependent propagation and complex interactions with structural components [4]. As a result, significant efforts have been dedicated to advanced algorithms and techniques for analyzing and characterizing guided wave data. Specifically, this paper aims to obtain compact representations of guided wave data that characterize the propagation environment with minimal assumptions. We only assume the wave equation is satisfied and based on the theory presented in [5]. Minimal assumptions are necessary in many scenarios since we often have minimal knowledge about the environment and no knowledge about the damage and its effects on the guided waves propagation.

Guided wave data has been analyzed through reflection-based [6–8], tomography-based [9], and wavefield-based analysis methods [3, 10]. Several wavefield analysis methods learn or extract special representations of the wavefields. These representations may have a physical basis, such as the modal dispersion curves [11] of guided waves, which characterize based on the material properties (e.g., velocities, densities, and thickness) and on known physics. Among these methods, sparse wavenumber analysis can extract dispersion curves with limited data [12, 13]. Most of these methods utilize a large pool of representations (often represented by a matrix) based on an analytical solution to the problem at hand (as in [14]). Yet, these fail for two reasons. First, the space of possible solutions is often infinite. In the case of the wave equation, for example, we may have one or more real-valued wavenumbers and/or frequencies that must be identified. Estimating these values is often not trivial. More importantly, there are many known representations that solve a differential equation [13, 15] but may not be compact for the specific problem, thereby becoming an ineffective representation. For example, the Fourier representation is always a solution to the wave equation [16, 17]. Yet, in the presence of discontinuities in time or space, it is not a compact representation as an infinite number of Fourier components are required. Hence, structural health monitoring systems cannot effectively represent cracks or delaminations with a Fourier basis.

There are past also efforts to integrate physics into other data-driven models. In addition, data assimilation methods used in time-series forecasting often inject physics into their frameworks [18] and act in conjunction with powerful, but often constrained, models to predict the future state of a system. This is related to, but separate, from the approach in this paper, where we aim to learn the underlying characteristics/representations of the data or environment. These representations could be used to highlight variations in such a time series, enabling us to integrate our approach with data assimilation strategies. Full waveform inversion (FWI) [19] similarly computes the characteristics of the environment based on a chosen physical model. However, the physical models used in FWI are highly constrained (i.e., often requiring many built-in assumptions) and usually coupled with expensive finite difference methods.

In contrast with each of these methods that rely on highly constrained models, our only constraint is that the data must satisfy the wave equation. While there is similar prior work on enforcing wave-physics into a dictionary learning framework with sparsity constraints in [20], this approach lacks the *global optimality guarantees, convergence guarantees,* and *algorithmic interpretability* that are present in this paper [5, 21]. Hence, we present a method to characterize material properties as well as decompose data that has strong algorithmic guarantees and assumes only fundamental physical knowledge. In addition, our physics-based decomposition can isolate and show how different wave modes propagate and change in the wavefield. We refer to our approach as wave-informed regression. We demonstrate this DDDAS approach with simulated wavefield data from a 1m by 1m region of an aluminum plate. The guided waves are generated by a 50 kHz frequency pulse.

2 Wave-Informed Regression Methodology

Wave-informed regression is based on learning a linear collection of modes that best represent wavefield data. These modes are learned by solving an optimization problem with three components: a mean squared error loss, a wave-informed loss, and mode number loss. We describe each of these components in the following subsections.

2.1 Mean Squared Error Loss

We represent a linear collection of modes to be learned as the columns of a matrix $\mathbf{D}$. The sum of these modes reconstructs the wavefield. Hence, our first loss term minimizes the mean squared error between our reconstruction and wavefield data such that

$$c_{\text{MSE}} = \|\mathbf{x} - \mathbf{D}1\|_{\mathbf{F}}^2$$

where the expression $\| \cdot \|_{\mathbf{F}}^2$ represents the squared Frobenius norm, defined by the squared sum of all the matrix or vector elements. The full wavefield data $\mathbf{x}$ is a vectorized form of a wavefield image $\mathbf{X}$ at a single frequency ω

$$\mathbf{x} = \text{vec}(\mathbf{X}), \mathbf{X}_{ij} = X(\omega, x_i, y_j).$$

For simplicity of notation, we assume the optimization is always performed at a single frequency ω. Each column of $\mathbf{D}$ represents a different vectorized spatial wave mode. The vector 1 represents a vector of all ones. Hence $\mathbf{D}1$ is the sum of columns in $\mathbf{D}$. Note that without additional information, optimizing this loss is highly underdetermined – there exists an infinite number of possible solutions. Figure 1(a) illustrates an example of **X** around a center frequency of 50 kHz. We also illustrate this data in the wavenumber domain in Fig. 1(b), with the axes representing the magnitudes of the horizontal and vertical wavenumbers. We can observe two modes of propagation at two distinct wavenumbers in these figures. However, it is difficult to observe that there is a spatial region with wavenumbers that are 10% higher than others. Hence, there are a total of four spatially dependent modes in this data.

2.2 Wave-Informed Loss

To obtain a meaningful modal representation, we add a loss function that represents the wave, or Helmholtz equation. The wave equation loss function is defined by

$$c_{\text{wave}} = \|\mathbf{L}_{x,y}\mathbf{D} - \mathbf{K}\mathbf{D}\|_{\text{F}}^2$$

where **K** is a diagonal matrix of squared wavenumbers k_m^2. The matrix $\mathbf{L}_{x,y}$ represents an operator for the approximate second derivative in the x-direction added to the approximate second derivative in the y-direction. Note that each column of **D** is a vectorized image. The second derivative has multiple numerical approximations. In this paper, we use the second-order central difference approximation [22]. Note that this approximation is good for low frequencies (or wavenumbers) and poor for high frequencies (or wavenumbers). From a numerical perspective, a better second derivative operator can be obtained by stacking the discretized continuous eigen-functions corresponding to the continuous second derivative and computing a new Laplacian $\mathbf{L}_{x,y}$ using continuous eigen-values and eigen-functions.

2.3 Mode Number Loss

Adding the two previously discussed loss functions there are still many possible optimal **D** matrices, including Fourier-like matrices. Therefore, we consider the true solution to be the one that minimizes the number of modes by including a third cost:

$$c_{\text{size}} = \|\mathbf{D}\|_{\text{F}}^2.$$

Intuitively, when there are many modes, the Frobenius norm will be large. When there is a small number of modes, the Frobenius norm will be small. Therefore, minimizing this cost will simultaneously minimize the number of modes learned. We further refer to [23], which shows that penalizing the Frobenius norm is equivalent to penalizing the nuclear norm of the matrix (enforcing a sparsity in the number of bases). Hence, we choose c_{size} as the squared Frobenius norm, as in [5] since it helps in obtaining an algorithm that solves the optimization problem to global optimality. Without this term, there would be many possible solutions to the optimization.

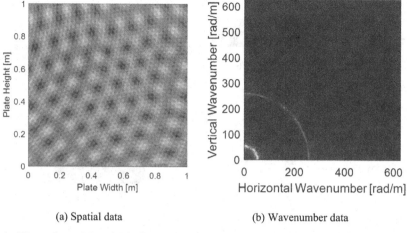

(a) Spatial data   (b) Wavenumber data

Fig. 1. Illustration of (*a*) single frequency spatial wavefield data and (b) the equivalent two-dimensional wavenumber domain. In (b), we observe two strong rings, representing the zeroth-order symmetric and zeroth-order asymmetric modes across the simulated plate.

2.4 Wave-Informed Regression

When we add our costs together and incorporate regularization constants, we obtain

$$\min_{\mathbf{D},\mathbf{K},M} \|\mathbf{x} - \mathbf{D}1\|_{\mathrm{F}}^2 + \lambda\left(\|\mathbf{D}\|_{\mathrm{F}}^2 + \frac{1}{\gamma^2}\|\mathbf{LD} - \mathbf{DK}\|_{\mathrm{F}}^2\right)$$

where M is the number of modes in the data. This formulation includes two regularization terms. The λ term represents a tradeoff between our mean squared error and our two regularizers while the γ^2 term represents the tradeoff between minimizing the number of modes and satisfying the wave equation.

2.5 Wave-Informed Algorithm

While we have established an optimization that defines wave-informed regression, creating an algorithm for solving this optimization is not trivial. This is because the optimization is not convex (i.e., there is more than one local minimum) and the variable to be optimized $\mathbf{D}$ is very high dimensional. As a result, there are no standard optimization algorithms that can be applied to this problem. An algorithm must be custom designed to solve wave-informed regression. We briefly outline a simplified version of the algorithm that solves this optimization, detailed here [5, 20, 21]. In our version of the algorithm, we assume a fixed number of modes from the very start. The algorithm identifies each value of k_i^2 in an iterative manner. After obtaining each k_i^2, it then updates $\mathbf{D}$. Learning k_i^2 is equivalent to learning an optimal filter in the wavenumber domain (see [5, 20, 21] for the exact connections to signal processing) and is closely related to standard estimation problems (specifically, spectral estimation [24]), only in our case we estimate both the modes and the wavenumbers. Learning $\mathbf{D}$ then combines and shapes these filters together to optimally reconstruct the data. Figure 2 outlines the algorithm in the form of a flowchart.

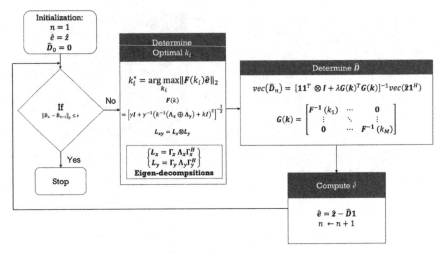

Fig. 2. A simplified wave informed regression algorithm

3   Simulation Setup

We test our algorithm with a guided Lamb wave simulation. In the simulation, ultrasonic guided waves travel in multiple directions, originating from outside the frame. There is one region where the guided waves travel with a different wavenumber, which possibly represents delamination [25], corrosion [26], or complex structural elements [27]. The spatial frame is 1 m by 1 m with a grid spacing of 5 mm. We simulate two modes, the zeroth-order symmetric mode and the zeroth-order asymmetric mode. Two more modes are produced by our small region (shown in Fig. 3(a)), which increases the wavenumbers by 10%. The guided wave data is simulated with a sampling rate of 1 MHz and a Gaussian transmission of 50 kHz center frequency and 25 kHz bandwidth. For wave-informed

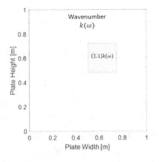

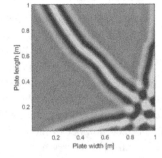

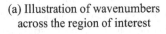

(a) Illustration of wavenumbers across the region of interest

(b) Spatial wavefield data at one point in time

Fig. 3. Illustration of (a) the two different wavenumber regions present in our simulation setup and (b) a single-time snapshot of the S0 mode in our guided waves. The waves originate from three separate sources.

regression, we choose $\lambda = 1$ and $\gamma^2 = \pi^2/(N_x N_y)$, where $N_x N_y$ is the total number of elements in x. This defines γ as the width of one sample in the wavenumber domain. We choose to obtain $M = 3$ modes. Figure 3(b) illustrates an snapshot of these waves at a single point in time. We know there are four modes in the data (A0 and S0 in the large and small regions). However, the S0 mode in the large region is difficult to distinguish from the S0 mode in the small region due to how similar the wavenumbers are.

4 Results

Figure 4 illustrates the first three modes extracted by wave-informed regression, in space and wavenumber. The second column illustrates that each mode corresponds to a particular wavenumber radius. The first column shows that the second mode is missing

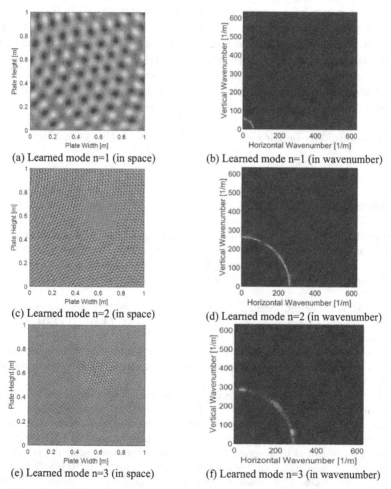

(a) Learned mode n=1 (in space)

(b) Learned mode n=1 (in wavenumber)

(c) Learned mode n=2 (in space)

(d) Learned mode n=2 (in wavenumber)

(e) Learned mode n=3 (in space)

(f) Learned mode n=3 (in wavenumber)

Fig. 4. Illustrations of the three modes (represented in the spatial and wavenumber domains) learned by wave-informed regression.

the small region with different wavenumbers. This region is observable in the third mode. The error for the wavenumbers estimated for each mode is 1.7 m^{-1}, 1 m^{-1}, and 2.3 m^{-1}, respectively. Overall, we demonstrate the ability to learn representations that spatially separate different modes or wavenumbers. We illustrate the results for two different points in time. The two points in time show the two different modes. The low wavenumber S0 mode is shown on the top and the high wavenumber A0 mode is shown on the bottom. As before, we see that the second mode has a spatial "hole" where the wavenumber changes, and the third mode fills that gap. Hence, we successfully extract the time-domain behavior of individual spatially varying modes.

5 Conclusions

This paper has illustrated preliminary results for wave-informed regression, a representation learning framework for combining wave physics within a DDDAS framework. The algorithm extracts the wavenumber of each mode in data as well as its spatially varying components. The algorithm is unsupervised and therefore requires no prior training data. However, in the future, this approach can be applied to both supervised and unsupervised learning strategies. The benefits of these physics-informed learning techniques enable machine learning in structural health monitoring to be both more reliable as well as more interpretable and the algorithms are also not data hungry.

Acknowledgments. This work is partially supported by NSF EECS-1839704 and NSF CISE-1747783.

References

1. Cawley, P.: Practical guided wave inspection and applications to structural health monitoring. In: Proceedings of the Australasian Congress on Applied Mechanics, p. 10 (2007)
2. Moll, J., et al.: Open Guided Waves: online platform for ultrasonic guided wave measurements. Struct. Health Monit. **18**(5–6), 1903–1914 (2019). https://doi.org/10.1177/1475921718817169
3. Staszewski, W.J., Lee, B.C., Mallet, L., Scarpa, F.: Structural health monitoring using scanning laser vibrometry: I. Lamb wave sensing. Smart Mater. Struct. **13**(2), 251 (2004). https://doi.org/10.1088/0964-1726/13/2/002
4. Dawson, A.J., Michaels, J.E., Michaels, T.E.: Isolation of ultrasonic scattering by wavefield baseline subtraction. Mech. Syst. Signal Process. **70–71**, 891–903 (2016). https://doi.org/10.1016/J.YMSSP.2015.09.008
5. Tetali, H.V., Harley, J.B., Haeffele, B.D.: Wave-informed matrix factorization with global optimality guarantees. arXiv preprint https://arxiv.org/abs/2107.09144v1 (2021)
6. Harley, J.B., Moura, J.M.F.: Data-driven matched field processing for Lamb wave structural health monitoring. J. Acoust. Soc. Am. **135**, 3 (2014). https://doi.org/10.1121/1.4863651
7. Harley, J.B.: Statistical lamb wave localization based on extreme value theory. Proc. Rev. Prog. Quant. Nondestruct. Eval. **1949**, 090004 (2018). https://doi.org/10.1063/1.5031567
8. Perelli, A., De Marchi, L., Marzani, A., Speciale, N.: Frequency warped cross-wavelet multiresolution analysis of guided waves for impact localization. Signal Process. **96**(PART A), 51–62 (2014). https://doi.org/10.1016/J.SIGPRO.2013.05.008

9. Rao, J., Ratassepp, M., Fan, Z.: Guided wave tomography based on full waveform inversion. IEEE Trans. Ultrason. Ferroelectr. Freq. Control **63**, 5 (2016). https://doi.org/10.1109/TUFFC.2016.2536144

10. Sheremet, A., Qin, Y., Kennedy, J.P., Zhou, Y., Maurer, A.P.: Wave turbulence and energy cascade in the hippocampus. Front. Syst. Neurosci. **12** (2019). https://doi.org/10.3389/fnsys.2018.00062

11. Harley, J.B., Schmidt, A.C., Moura, J.M.F.: Accurate sparse recovery of guided wave characteristics for structural health monitoring. In: Proceedings of the IEEE International Ultrasonics Symposium, pp. 158–161 (2012). https://doi.org/10.1109/ULTSYM.2012.0039

12. Sabeti, S., Harley, J.B.: Spatio-temporal undersampling: recovering ultrasonic guided wavefields from incomplete data with compressive sensing. Mech. Syst. Signal Process. **140**, 106694 (2020). https://doi.org/10.1016/j.ymssp.2020.106694

13. Alguri, K.S.S., Melville, J., Harley, J.B.: Baseline-free guided wave damage detection with surrogate data and dictionary learning. J. Acoust. Soc. Am. **143**(6), 3807–3818 (2018). https://doi.org/10.1121/1.5042240

14. Lai, Z., Alzugaray, I., Chli, M., Chatzi, E.: Full-field structural monitoring using event cameras and physics-informed sparse identification. Mech. Syst. Signal Process. **145**, 106905 (2020). https://doi.org/10.1016/J.YMSSP.2020.106905

15. Mesnil, O., Ruzzene, M.: Sparse wavefield reconstruction and source detection using compressed sensing. Ultrasonics **67**, 94–104 (2016). https://doi.org/10.1016/J.ULTRAS.2015.12.014

16. di Ianni, T., de Marchi, L., Perelli, A., Marzani, A.: Compressive sensing of full wave field data for structural health monitoring applications. IEEE Trans. Ultrason. Ferroelectr. Freq. Control **62**(7), 1373–1383 (2015). https://doi.org/10.1109/TUFFC.2014.006925

17. Sawant, S., Banerjee, S., Tallur, S.: Performance evaluation of compressive sensing based lost data recovery using OMP for damage index estimation in ultrasonic SHM. Ultrasonics **115**, 106439 (2021). https://doi.org/10.1016/J.ULTRAS.2021.106439

18. Ravela, S.: Two extensions of data assimilation by field alignment. In: Shi, Y., et al. (eds.) Computational Science – ICCS 2007, pp. 1147–1154. Springer, Heidelberg (2007). https://doi.org/10.1007/978-3-540-72584-8_150

19. Huang, R., Zhang, Z., Wu, Z., Wei, Z., Mei, J., Wang, P.: Full-waveform inversion for full-wavefield imaging: decades in the making. Lead. Edge **40**(5), 324–334 (2021). https://doi.org/10.1190/TLE40050324.1

20. Tetali, H.V., Alguri, K.S., Harley, J.B.: Wave physics informed dictionary learning in one dimension. In: Proceedings of the IEEE International Workshop on Machine Learning for Signal Processing (MLSP), vol. 2019, pp. 1–6 (2019). https://doi.org/10.1109/MLSP.2019.8918835

21. Tetali, H.V., Harley, J.B., Haeffele, B.D.: Wave physics-informed matrix factorizations. IEEE Trans. Signal Process. **72**, 535–548 (2024). https://doi.org/10.1109/TSP.2023.3348948

22. Yang, D., Tong, P., Deng, X.: A central difference method with low numerical dispersion for solving the scalar wave equation. Geophys. Prospect. **60**(5), 885–905 (2012). https://doi.org/10.1111/J.1365-2478.2011.01033.X

23. Recht, B., Fazel, M., Parrilo, P.A.: Guaranteed minimum-rank solutions of linear matrix equations via nuclear norm minimization. SIAM Rev. **52**(3), 471–501 (2010). https://doi.org/10.1137/070697835

24. Marple, S.L.: Tutorial overview of modern spectral estimation. ICASSP, IEEE Int. Conf. Acoust. Speech Signal Process. Proc. **4**, 2152–2157 (1989). https://doi.org/10.1109/ICASSP.1989.266889

25. Gao, F., Hua, J., Wang, L., Zeng, L., Lin, J.: Local wavenumber method for delamination characterization in composites with sparse representation of Lamb waves. IEEE Trans. Ultrason.

Ferroelectr. Freq. Control **68**(4), 1305–1313 (2021). https://doi.org/10.1109/TUFFC.2020. 3022880

26. Tian, Z., Xiao, W., Ma, Z., Yu, L.: Dispersion curve regression – assisted wideband local wavenumber analysis for characterizing three-dimensional (3D) profile of hidden corrosion damage. Mech. Syst. Signal Process. **150**, 107347 (2021). https://doi.org/10.1016/J.YMSSP. 2020.107347

27. Moll, J., Wandowski, T., Malinowski, P., Radzienski, M., Opoka, S., Ostachowicz, W.: Experimental analysis and prediction of antisymmetric wave motion in a tapered anisotropic waveguide. J. Acoust. Soc. Am. **138**(1), 299 (2015). https://doi.org/10.1121/1.4922823

Passive Radio Frequency-Based 3D Indoor Positioning System via Ensemble Learning

Liangqi Yuan[1], Houlin Chen[2], Robert Ewing[3], and Jia Li[1]

[1] Oakland University, Rochester, MI 48309, USA
{liangqiyuan,li4}@oakland.edu
[2] University of Toronto, Toronto, ON M5S 1A1, Canada
houlin.chen@mail.utoronto.ca
[3] Air Force Research Laboratory, WPAFB, Riverside, OH 45433, USA
robert.ewing.2@us.af.mil

Abstract. Passive radio frequency (PRF)-based indoor positioning systems (IPS) have attracted researchers' attention due to their low price, easy and customizable configuration, and non-invasive design. This paper proposes a PRF-based three-dimensional (3D) indoor positioning system (PIPS), which is able to use signals of opportunity (SoOP) for positioning and also capture a scenario signature. PIPS passively monitors SoOPs containing scenario signatures through a single receiver. Moreover, PIPS leverages the Dynamic Data Driven Applications System (DDDAS) framework to devise and customize the sampling frequency, enabling the system to use the most impacted frequency band as the rated frequency band. Various regression methods within three ensemble learning strategies are used to train and predict the receiver position. The PRF spectrum of 60 positions is collected in the experimental scenario, and three criteria are applied to evaluate the performance of PIPS. Experimental results show that the proposed PIPS possesses the advantages of high accuracy, configurability, and robustness.

Keywords: Indoor positioning system · Passive radio frequency · Signal of opportunity · Ensemble learning · Machine learning

1 Introduction

Signals of opportunity (SoOP) for implementing indoor positioning system (IPS) has shown progress in recent years [1,2]. SoOP refers to some non-task signals that are used to achieve specified tasks, such as Wi-Fi, cellular network, broadcasting, and other communication signals for positioning tasks. These communication signals have different frequencies according to different functions. For example, the frequency of broadcast signals is tens to hundreds of MHz, and the frequency of Wi-Fi can reach 5 GHz. Each SoOP has different performances for different tasks, which will be affected by the local base stations, experiment scenarios, and task

Supported by Air Force Office of Scientific Research.

settings. SoOP aim to facilitate high-precision positioning in GPS-shielded environments while avoiding the need for additional signal sources. However, how to use a single receiver for positioning in an environment where the signal source is unknown is still an open problem. Therefore, a passive radio frequency (PRF) system is proposed to integrate these communication signals due to the design of a customizable frequency band. Finding the frequency band most impacted for positioning is the most significant prior. In addition, PRF can capture scenario signatures, including liquids, metal objects, house structures, etc., which has been proven to further improve the performance of a positioning system.

Dynamic Data Driven Applications System (DDDAS) frameworks have already shown their application prospects, such as in the fields of environmental science, biosensing, autonomous driving, etc. The application of the DDDAS framework to these domains varies, depending on the input variables and output decisions of the system. Table 1 shows some examples of instantaneous and long-term DDDAS. Currently, most of DDDAS are emphasized to instantaneous DDDAS, which require us to react immediately to dynamic data input. For example, hurricane forecasting is an instantaneous DDDAS, and if it doesn't react in time, there will be some serious consequences. But long-term DDDAS also has its benefits, and there are no serious consequences for not responding immediately, such as an energy analysis DDDAS is used to save consumption. The advantage of long-term DDDAS is dynamic data input, which can effectively reduce consumption, improve accuracy, and enhance robustness.

Table 1. Instantaneous DDDAS vs. Long-Term DDDAS.

Instantaneous	Long-Term
Weather forecasting [3]	Energy analysis [9]
Atmospheric contaminants [4]	Materials Analysis [10]
Wildfires detection [5]	Identification of biomarkers in DNA methylation [11]
Autonomous driving [6]	Multimedia content analysis [12]
Fly-by-feel aerospace vehicle [7]	Image processing [13]
Biohealth outbreak [8]	Our proposed positioning system

Due to the uncertainty of SoOPs and scenario signature, IPSs need to conform to the paradigm of DDDAS [14,15]. For the PRF positioning system, the selection of frequency band is a dynamic issue, which is determined according to the scenario signature. Therefore, the computational feedback in DDDAS is required to reconfigure the sensor for frequency band selection. Selecting some frequency bands from the full frequency band can effectively save sampling time, computing resources, and increase the robustness, etc. [16]. Moreover, the customizable frequency band can be used in a variety of different tasks, such as human monitoring, navigation, house structure detection, etc. [17–19]. Therefore, the PRF-based systems under the DDDAS framework need to dynamically optimize the frequency parameter according to its usage scenarios and task settings to obtain higher adaptability, accuracy, and robustness.

Ensemble learning is used as the strategy for the positioning regression task due to its ability to integrate the strengths of multiple algorithms [20,21]. Ensemble learning includes three strategies, namely boosting, bagging, and stacking [22], depending on whether the base estimator is parallel or serial [23]. The boosting strategy is a serial strategy where the posterior estimator learns the wrong samples of the prior estimator, which reduces the bias of the model. However, this strategy overemphasizes the wrong samples and thus may lead to larger variance and weaker generalization ability of the model. Both bagging and stacking strategies are parallel structures, which can reduce the variance and enhances the generalization ability. Compared to the bagging strategy, which uses averaging as the final estimator, stacking uses a regressor as the final estimator. Compared to linear or weighted averaging, the model can further reduce model bias by analyzing the decisions of the base estimators.

This paper proposes a PRF-based 3D IPS, named PIPS, for the positioning regression task. Within the DDDAS framework, the performance of the PIPS system is enhanced by adaptive frequency band selection, which continues the most impacted frequency band found in the previous work [16]. PRF spectrum data was collected at 60 gridded positions in the scenario. The spectrum data set for positioning is trained in three ensemble learning strategies. Root mean square error (RMSE) is used to evaluate the accuracy of PIPS, coefficient of determination R^2 is used to evaluate the reliability, and 95% confidence error (CE) is used to evaluate the optimality. Experiments demonstrate that the proposed PIPS exhibits its potential for accurate object locating tasks.

This paper is presented as follows. In Sect. 2, the details and sensor settings of the proposed PIPS are illustrated. The experimental setup and results are shown in Sect. 3. Section 4 gives some discussions on the advantages of PIPS under the DDDAS framework prior to the conclusion and future work demonstrated in Sect. 5.

2 Frequency-Adaptive PIPS

PIPS achieves sensing by passively accepting the PRF spectrum in the scenario. Software-defined radio (SDR) is used to control the PRF sensor for data collection, including the frequency band $\mathbb{B}$, step size Δ, sampling rate R_s, etc. Reasonable selection of the parameters of the PRF sensor in PIPS is crucial. The diagram of frequency band selection by PIPS under the framework of DDDAS is shown in Fig. 1. The DDDAS framework is used to reconfigure the parameters of the PRF sensor, which is achieved through SDR. The parameters of the PRF sensor, especially the center frequency, are dynamically reconfigured to adapt to the signatures of different scenarios.

With the support of initial parameters $\mathbb{B}$, Δ, and R_s, the data set collected by the PRF sensor $D \in \mathbb{R}^{n \times m}$ and its corresponding position label set $C \in \mathbb{R}^{n \times 3}$. D is the PRF spectrum, that is, the average powers collected over the frequency band. Although it is feasible to use the average power corresponding to the full frequency band as the feature vector for positioning, it will greatly increase the

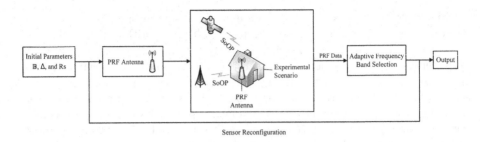

Fig. 1. DDDAS framework reconfigures the parameters $\mathbb{B}$, Δ, and R_s of the PRF sensor in PIPS.

sampling time. Therefore, it is necessary to optimize the initial parameters $\mathbb{B}$, Δ, and R_s under the DDDAS framework. The proposed PIPS system can be defined as the following function $f : C \rightarrow D$,

$$d = f(c; \mathbb{B}, \Delta, R_s), \tag{1}$$

where d and c are a pair of samples in D and C, which also represent a pair of corresponding PRF spectrum and coordinate. Equation 1 shows the collection of PRF data at the corresponding coordinates given the parameters. By training on the ensemble learning model on the collected data and making predictions, the estimated coordinates can be obtained:

$$\hat{c} = f^{-1}(d; \mathbb{B}, \Delta, R_s). \tag{2}$$

The PRF spectral data collected by the PRF sensor in the experimental scenario contains the SoOP and the signature of the experimental scenario. The PRF sensor in PIPS is reconfigured after the adaptive band selection algorithm is used to find the most impacted band for the positioning task. After the k-th optimization, the collected data set under the optimized parameter $\mathbb{B}_k$, Δ_k, and R_{s_k} is defined as $D_k \in \mathbb{R}^{n \times m_k}$. Dynamic reconfiguration may be performed once or multiple times, depending on the properties of the SoOP in the scenario, including received signal strength (RSS), center frequency, integration of multiple signal sources, etc. The dynamic needs of the configuration are mainly changing between different scenarios, implemented tasks, and SoOPs. When the SoOP remains unchanged, its dynamic configuration is only needed once to find the optimized parameters, which can reduce the waste of computing resources while achieving system applicability. The $\mathbb{B}_1(\text{MHz}) \in \{91.2, 93.6, 96.0, 98.4, 100.8\}$ found in previous work are used in the experiments for preliminary validation of the proposed PIPS. The PRF sensor used in our experiment is RTL-SDR RTL2832U because of its cheap and easy-to-configure characters, as shown in Fig. 2.

3   Experiment and Results

This section is organized as follows. In the experimental scenario, spectrum data are collected at 60 positions for the frequency band $\mathbb{B}_1$ that has the most impact

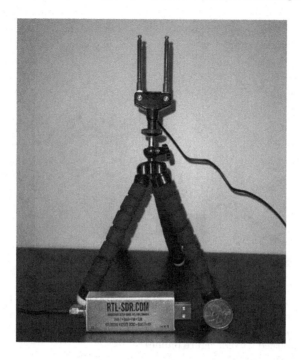

Fig. 2. RTL-SDR RTL2832U is used as PRF sensor to collect PRF spectrum.

on the positioning. Using single regressors as a baseline, three ensemble learning strategies of boosting, bagging, and stacking are compared. Three criteria are used as evaluation methods. This section focuses on the setup of experimental scenarios and the comparison and evaluation of strategies and models.

3.1 Experimental Setup

Data collection is done in an indoor home scenario, as shown in Fig. 3. In order to avoid the impact of sampling distance on performance and also to better compare with other state-of-the-art technologies, one meter is selected as the sampling distance in the three directions of length, width, and height. According to past experience, some sources that may have an impact on the PRF spectrum are marked in Fig. 3, such as a host computer, operator, TV, Wi-Fi router, printer, etc. The experimental scenario with a length of 6.15 m, a width of 4.30 m, and a height of 2.42 m is used as a preliminary verification of the PRF positioning. We collected 100 samples at each position, and a total of 6000 samples were divided into training and test data sets in a ratio of 0.7 and 0.3. Using scikit-learn, the model was built on TensorFlow and trained with a Nvidia GeForce RTX 3080 GPU.

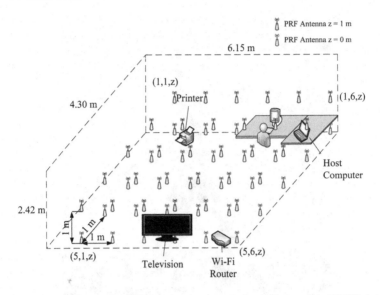

Fig. 3. Illustration of an indoor living room scenario is used to collect PRF data at 60 positions. The red and blue antennas are represented as 0 and 1 m from the bottom of the antenna to the ground, respectively. Other potentially disturbing objects and human are also marked. (Color figure online)

3.2   Results and Evaluation

To better demonstrate the effectiveness of the collected PRF spectrum data for positioning, principal component analysis (PCA) is used to reduce the dimensionality of the PRF spectrum data and visualization. The raw PRF spectrum is 5D since the most impacted frequency band used in data collection are five frequencies, while PCA reduces it to 3D for visualization. Using PCA is just for visualization, while the raw data set is used to train the ensemble learning model for the positioning task. Figure 4 shows PRF spectrum data dimensionally reduced by PCA. Data at different positions and heights can form a cluster in the PCA space, which can prove that there are differences in data at different positions, which is also the fundamental reason for positioning.

For the proposed model, it is required to compare with the baseline in terms of performance and complexity. For ensemble learning models, some single regressors are used as the baseline, including Support Vector Regression (SVR), K Nearest Neighbors Regression (KNR), Gaussian Process Regression (GPR), Decision Trees Regression (DTR), and Multi-layer Perceptron (MLP). The performance is compared by three evaluations: Root mean square error (RMSE), coefficient of determination R^2, and 95% CE. RMSE is targeted at applications that require lower average errors but less stringent positioning systems, such as warehouse patrol robots. RMSE of the test data set can be expressed as

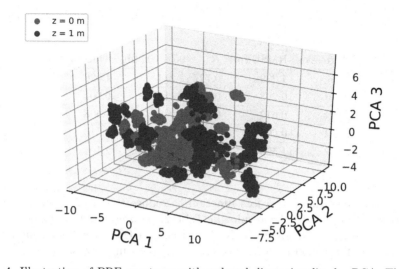

Fig. 4. Illustration of PRF spectrum with reduced dimensionality by PCA. The red and blue dots indicate that spectrum data was collected at 0 and 1 m from the ground, respectively. (Color figure online)

$$\text{RMSE} = \sqrt{\frac{\|C^* - \hat{C}^*\|^2}{n^*}}, \tag{3}$$

where $C^* \in \mathbb{R}^{n^* \times 3}$ is the label of test data set, $\hat{C}^*$ is the estimated label obtained by ensemble model. 95% CE is the corresponding error when the cumulative distribution function of RMSE reaches 95%, which can be expressed as

$$95\%\text{CE} = F_{\text{RMSE}}^{-1}(0.95), \tag{4}$$

where F is the cumulative distribution function of RMSE, the 95% CE is aimed at systems that are more critical to accuracy, such as firefighting robots. It requires a higher confidence level to limit the robot's error to a strict value. The time complexity is considered to be equivalent to model fitting time. Coefficient of determination and time complexity are not our main concerns. Since the proposed PIPS is an application system, it is necessary to pay more attention to the customer-oriented performance of the application. Each model was trained with its default parameters for initial comparison. The performance and complexity of some regressors are shown in Table 2.

It can be seen from Table 2 that KNR has the best performance, which will be used as the baseline for PIPS to compare with the ensemble learning strategy. Different models under three ensemble learning strategies are used to train on our positioning data set. For serial boosting strategies, there are three main extensions, including Adaptive Boosting Regression (ABR), Gradient Boosting Regression (GBR), and Histogram-based GBR (HGBR). ABR makes the posterior estimator focus more on samples that cannot be solved by the prior estimator through an adaptive weight method. Both GBR and HGBR are ensembles

Table 2. Single regressors to implement positioning tasks and serve as baselines for PIPS.

Regression	RMSE (m)	R^2	95% CE (m)	Time (s)
SVR	1.229	0.777	2.214	1.026
KNR	0.268	0.986	0.412	0.002
GPR	0.612	0.967	1.248	1.508
DTR	0.603	0.930	1.111	0.016
MLP	1.506	0.562	2.534	2.104

of regression trees that use a loss function to reduce the error of the previous estimator. According to the results in Table 2, we selected four models with different accuracies, namely SVR, KNR, GPR, and DTR, for further analysis. Table 3 shows the model performance under the boosting strategy.

Table 3. Performance of ensemble learning models under the boosting strategy.

Ensemble Strategy	Base Estimator	RMSE (m)	R^2	95% CE (m)	Time (s)
ABR	SVR	0.828	0.881	1.419	88.368
	KNR	0.324	0.985	0.095	2.859
	GPR	0.825	0.859	1.442	278.900
	DTR	0.324	0.983	0.095	2.193
GBR	DTR	0.807	0.879	1.575	1.698
HGBR	DTR	0.457	0.960	1.027	1.161

It can be seen that the ensemble learning model under the boosting strategy has no advantage in RMSE compared to a single regressor, but it greatly reduces 95% CE, especially for ABR with KNR and DTR as base estimators. This means that most of the samples have errors less than 0.095, but there are also a few samples with large errors that increase the value of RMSE. Boosting strategies are effective in reducing the mode of error. For the bagging strategy, the base estimator is also a crucial parameter. In addition to the general bagging model, Random Forest Regression (RFR) and Extremely Randomized Trees (ERT) as bagging variants and extensions of DTR are also included as part of the comparison. Table 4 shows the performance of the models under the bagging strategy.

Through the comparison of Table 2 and Table 4, it can be found that - whether it is KNR with the best accuracy or SVR with poor accuracy, the bagging strategy cannot significantly further improve its accuracy. The final prediction of the bagging strategy will be related to each base estimator, that is, it will also be affected by the base estimator with poor accuracy. The stacking strategy aggregates base estimators through the final estimator, which gives different

Table 4. Performance of ensemble learning models under the bagging strategy.

Ensemble Strategy	Base Estimator	RMSE (m)	R^2	95% CE (m)	Time (s)
Bagging	SVR	1.124	0.775	2.116	5.028
	KNR	0.265	0.989	0.423	0.372
	GPR	0.623	0.928	1.323	41.264
RFR	DTR	0.418	0.966	0.964	0.934
ERT	DTR	0.299	0.966	0.710	0.304

weights to base estimators. We use the ten previously mentioned regressors, including the ensemble learning model as the base estimator, and then test the performance of these regressors as the final estimator. The regression results under the stacking strategy are shown in Table 5.

Table 5. Performance of ensemble learning models under the stacking strategy.

Ensemble Strategy	Final Estimator	RMSE (m)	R^2	95% CE (m)	Time (s)
Stacking	SVR	0.271	0.988	0.463	97.281
	KNR	0.259	0.990	0.446	92.678
	GPR	2.115	0.273	3.924	97.241
	DTR	0.327	0.984	0.086	93.218
	MLP	0.263	0.990	0.459	95.106
	ABR	0.334	0.984	0.258	97.657
	GBR	**0.258**	**0.990**	**0.317**	**94.338**
	HGBR	0.254	0.990	0.371	95.478
	RFR	0.255	0.990	0.431	93.835
	ETR	0.259	0.990	0.334	93.808

The stacking strategy affords the use of any model as the base estimator, so the stacking strategy can also be a strategy that integrates ensemble learning models. The results show that the stacking strategy has an advantage in performance compared to the bagging strategy, which is because the final estimator can adaptively aggregate all the base estimators. However, the stacking strategy is not dominant compared to the boosting strategy. Although stacking is stronger than boosting in RMSE and R^2, the time complexity is dozens of times.

After experiments, we found that the Stacking strategy gave the best results. Compared to the baseline, the proposed ensemble learning strategy has considerable improvement on 95% CE. In particular, the stacking strategy with DTR as the final estimator can reduce the 95% CE by 92.3%. Although 95% of the samples have relatively low errors, the average RMSE is still high, which means that minority samples with a proportion of 5% or less have considerable errors. These samples may have received interference, such as the movement of the human body, the effect of metal or liquid shielding on the PRF spectrum, etc. Therefore, GBR as the final estimator is considered as the global optimal solution, which outperforms the baseline in all aspects.

4   Discussion

DDDAS is crucial for PIPS, the main purpose of the DDDAS framework is to find the optimal solution for the positioning task in the target scenario. We implement pre-sampling in the target scenario and then use SHAP to analyze the collected samples and find the optimal frequency band, step size, and sampling rate. PIPS under the DDDAS framework has three advantages. Firstly and most importantly, the sampling time is reduced by 98%. We reduced the 400 frequencies to 5 frequencies under the DDDAS framework. If we do not use the DDDAS framework to find the optimal frequency, data collection over the full frequency band will waste a huge amount of time. Secondly, the redeployment time of the sensor is also greatly reduced. The proposed PIPS system also has excellent redeployment capabilities in new scenarios, thanks to the DDDAS framework on frequency band $\mathbb{B}$, step size Δ, sampling rate R_s optimization. To achieve the accuracy and sampling resolution described above, the time resource required for redeployment is around $300 \ s/m^3$. The training time is negligible compared to the PRF data sampling time.

Thirdly, it can potentially improve accuracy and reliability. The PIPS system uses RSS in the five most sensitive frequencies, especially since this passive RF technology can capture signatures from scenarios such as metal parts in house structures or liquids. So basically, the PRF signal collected in each scenario is unique. On the one hand, there are inevitably some interferences in the full frequency band, including natural noise and artificial signals. These noise signals are random and abrupt, which is not conducive to the stability of a positioning system. On the other hand, we don't want to include any unnecessary features in the samples. In this task, we did not use deep learning but just traditional machine learning. Traditional machine learning cannot adaptively assign weights, so unnecessary and cluttered features obviously affect the accuracy of classification. Therefore, collecting data in the most sensitive frequency bands for positioning can effectively avoid these possible interferences and reduce feature complexity to improve accuracy and reliability.

5   Conclusion

This paper proposes a PIPS under the DDDAS framework to solve the 3D positioning problem. Three ensemble learning strategies and their various variants and extensions are used to train on the collected data set. The experimental results show that the proposed ensemble learning strategy has an RMSE of 0.258 m, an R^2 of 0.990, and a 95% CE of 0.317 m, which is much better than the baselines. PIPS under the DDDAS framework is considered a potential application in specific scenarios, such as robot-patrolled factories or warehouses, due to its efficient redeployment and high accuracy.

For future work, dimensionality reduction is a potential research direction. The current work is limited to the frequency selection technology. We selected the most sensitive frequency band from the 400 frequencies in the full frequency band under the DDDAS framework. However, dimensionality reduction, while similar in terms of results, has different effects. For the dimensionality reduction method, although it can reduce the complexity of the data, it cannot reduce the sampling time. The benefits of PCA lie in privacy considerations and visualization applications. In internet of things (IoT) applications, performing PCA processing locally can reduce the dimension of data so that customer privacy can be protected after uploading to the cloud. PCA is able to reduce multi-dimensional data to three-dimensional or two-dimensional to enable visualization applications.

Acknowledgements. Thanks to Dr. Erik Blasch for concept development and co-authoring the paper. This research is partially supported by the AFOSR grant FA9550-21-1-0224. The views and conclusions contained herein are those of the authors and should not be interpreted as necessarily representing the official policies or endorsements, either expressed or implied, of the Air Force Research Laboratory or the U.S. Government.

References

1. Moghtadaiee, V., Dempster, A.G.: Indoor location fingerprinting using FM radio signals. IEEE Trans. Broadcast. **60**(2), 336–346 (2014)
2. Souli, N., Makrigiorgis, R., Kolios, P., Ellinas, G.: Real-time relative positioning system implementation employing signals of opportunity, inertial, and optical flow modalities. In: 2021 International Conference on Unmanned Aircraft Systems (ICUAS), pp. 229–236. IEEE, Athens (2021)
3. Plale, B., et al.: Towards dynamically adaptive weather analysis and forecasting in LEAD. In: Sunderam, V.S., van Albada, G.D., Sloot, P.M.A., Dongarra, J.J. (eds.) ICCS 2005. LNCS, vol. 3515, pp. 624–631. Springer, Heidelberg (2005). https://doi.org/10.1007/11428848_81
4. Patra, A.K., et al.: Challenges in developing DDDAS based methodology for volcanic ash hazard analysis - effect of numerical weather prediction variability and parameter estimation. Procedia Comput. Sci. **18**, 1871–1880 (2013)
5. Michopoulos, J., Tsompanopoulou, P., Houstis, E., Joshi, A.: Agent-based simulation of data-driven fire propagation dynamics. In: Bubak, M., van Albada, G.D., Sloot, P.M.A., Dongarra, J. (eds.) ICCS 2004. LNCS, vol. 3038, pp. 732–739. Springer, Heidelberg (2004). https://doi.org/10.1007/978-3-540-24688-6_95
6. Allaire, D., et al.: An offline/online DDDAS capability for self-aware aerospace vehicles. Procedia Comput. Sci. **18**, 1959–1968 (2013)
7. Kopsaftopoulos, F., Chang, F.-K.: A dynamic data-driven stochastic state-awareness framework for the next generation of bio-inspired fly-by-feel aerospace vehicles. In: Blasch, E.P., Darema, F., Ravela, S., Aved, A.J. (eds.) Handbook of Dynamic Data Driven Applications Systems, pp. 713–738. Springer, Cham (2022). https://doi.org/10.1007/978-3-030-74568-4_31

8. Yan, H., Zhang, Z., Zou, J.: Dynamic space-time model for syndromic surveillance with particle filters and Dirichlet process. In: Blasch, E., Ravela, S., Aved, A. (eds.) Handbook of Dynamic Data Driven Applications Systems, pp. 139–152. Springer, Cham (2018). https://doi.org/10.1007/978-3-319-95504-9_7

9. Neal, S., Fujimoto, R., Hunter, M.: Energy consumption of data driven traffic simulations. In: 2016 Winter Simulation Conference (WSC), pp. 1119–1130. IEEE (2016)

10. Mulani, S.B., Roy, S., Jony, B.: Uncertainty analysis of self-healed composites with machine learning as part of DDDAS. In: Darema, F., Blasch, E., Ravela, S., Aved, A. (eds.) DDDAS 2020. LNCS, vol. 12312, pp. 113–120. Springer, Cham (2020). https://doi.org/10.1007/978-3-030-61725-7_15

11. Damgacioglu, H., Celik, E., Yuan, C., Celik, N.: Dynamic data driven application systems for identification of biomarkers in DNA methylation. In: Blasch, E.P., Darema, F., Ravela, S., Aved, A.J. (eds.) Handbook of Dynamic Data Driven Applications Systems, pp. 241–261. Springer, Cham (2022). https://doi.org/10.1007/978-3-030-74568-4_12

12. Blasch, E., Aved, A., Bhattacharyya, S.S.: Dynamic data driven application systems (DDDAS) for multimedia content analysis. In: Blasch, E., Ravela, S., Aved, A. (eds.) Handbook of Dynamic Data Driven Applications Systems, pp. 631–651. Springer, Cham (2018). https://doi.org/10.1007/978-3-319-95504-9_28

13. Li, H., et al.: Design of a dynamic data-driven system for multispectral video processing. In: Blasch, E., Ravela, S., Aved, A. (eds.) Handbook of Dynamic Data Driven Applications Systems, pp. 529–545. Springer, Cham (2018). https://doi.org/10.1007/978-3-319-95504-9_23

14. Blasch, E., Ravela, S., Aved, A. (eds.): Handbook of Dynamic Data Driven Applications Systems. Springer, Cham (2018)

15. Fujimoto, R., et al.: Dynamic data driven application systems: research challenges and opportunities. In: 2018 Winter Simulation Conference (WSC), pp. 664–678. IEEE, Sweden (2018)

16. Yuan, L., Chen, H., Ewing, R., Blasch, E., Li, J.: Three dimensional indoor positioning based on passive radio frequency signal strength distribution. Manuscript accepted by IEEE Internet Things J

17. Yuan, L., et al.: Interpretable passive multi-modal sensor fusion for human identification and activity recognition. Sensors **22**(15), 5787 (2022)

18. Mu, H., Ewing, R., Blasch, E., Li, J.: Human subject identification via passive spectrum monitoring. In: NAECON 2021-IEEE National Aerospace and Electronics Conference, pp. 317–322. IEEE (2021)

19. Mu, H., Liu, J., Ewing, R., Li, J.: Human indoor positioning via passive spectrum monitoring. In: 2021 55th Annual Conference on Information Sciences and Systems (CISS), pp. 1–6. IEEE (2021)

20. Tran, H.Q., Nguyen, T.V., Huynh, T.V., Tran, N.Q.: Improving accuracy of indoor localization system using ensemble learning. Syst. Sci. Control. Eng. **10**(1), 645–652 (2022)

21. Iorkyase, E.T., et al.: Improving RF-based partial discharge localization via machine learning ensemble method. IEEE Trans. Power Deliv. **34**(4), 1478–1489 (2019)

22. Majumder, U., Blasch, E., Garren, D.: Deep Learning for Radar and Communications Automatic Target Recognition. Artech House, Norwood (2020)

23. Kumar, A., Mayank, J.: Ensemble Learning for AI Developers. BApress, Berkeley (2020)

Plenary Presentations - Deep Learning

Deep Learning Approach for Data and Computing Efficient Situational Assessment and Awareness in Human Assistance and Disaster Response and Battlefield Damage Assessment Applications

Jie Wei[1]([✉]), Weicong Feng[1], Erik Blasch[2], Philip Morrone[3], Erika Ardiles-Cruz[3], and Alex Aved[3]

[1] City College of New York, New York, NY 10031, USA
jwei@ccny.cuny.edu
[2] Air Force Office of Scientific Research, Arlington, VA 22203, USA
[3] Air Force Research Laboratory, Rome, NY 13441, USA

Abstract. The importance of situational assessment and awareness (SAA) becomes increasingly evident for Human Assistance and Disaster Response (HADR) and military operations. During natural disasters in populated regions, proper HADR efforts can only be planned and deployed effectively when the damage levels can be resolved in a timely manner. In today's warfare, such as battlefield and critical region monitoring and surveillance, prompt and accurate battlefield damage assessments (BDA) are of crucial importance to gain control and ensure robust operating conditions in highly dangerous and contested environments. To design an effective HADR and BDA approach, this paper utilizes the Dynamic Data Driven Applications System (DDDAS) approach within the growing utilization of Deep Learning (DL). DL can leverage DDDAS for near-real-time (NRT) situations in which the original DL-trained model is updated from continuous learning through the effective labeling of SAA updates. To accomplish the NRT DL with DDDAS, an image-based pre- and post-conditional probability learning (IP2CL) is developed for HADR and BDA SAA. Equipped with the IP2CL, the matching pre- and post-disaster/action images are effectively encoded into one image that is then learned using DL approaches to determine the damage levels. Two scenarios of crucial importance for practical uses are examined: pixel-wise semantic segmentation and patch-based global damage classification. Results achieved by our methods in both scenarios demonstrate promising performances, showing that our IP2CL-based methods can effectively achieve data and computational efficiency and NRT updates, which is of utmost importance for HADR and BDA missions.

Keywords: Situational assessment and awareness · Human Assistance and Disaster Response · Deep learning · Semantic segmentation · Image classification

E. Blasch et al. (Eds.): DDDAS 2022, LNCS 13984, pp. 187–195, 2024.
https://doi.org/10.1007/978-3-031-52670-1_18

1 Introduction

The importance of situational assessment and awareness (SAA) becomes increasingly evident for Human Assistance and Disaster Response (HADR) and military operations. During natural disasters in populated regions, proper HADR efforts can only be planned and deployed effectively when the damage levels can be resolved in a timely manner. In today's warfare, such as battlefield and critical region monitoring and surveillance, prompt and accurate battlefield damage assessments (BDA) are of crucial importance to gain control and ensure robust operating conditions in highly dangerous and contested environments. In all these situations, manual evaluations by human experts for prompt and real-time damage level assessments for Target of Interest (TOI) are impossible and unscalable: The manual labeling process is extremely labor-intensive and time-consuming as each individual location in the TOI must be patiently studied to decide the correct labeling category wherein lots of expertise on the object types and damage styles are needed; this trouble is even more grave for major disasters/military actions over a wide area when this process will direly slow down the HADR or BDA efforts. The cutting-edge hardware, such as GPU and HPC computing, and software, such as Artificial Intelligence and Machine learning (AI/ML), especially Deep learning (DL) approaches should be called upon. Furthermore, to render the work practical utility, the method should be both ***data*** and ***computationally efficient*** because.

1. The available data in HADR and BDA applications is generally small that is significantly less than enough to adequately train a DL.
2. The HADR and military personnel are not expected to have access to immense computing power in their mission on site.

To design an effective data and computing efficient HADR and BDA approach of practical utilities, this paper utilizes the Dynamic Data Driven Applications System (DDDAS) approach within the growing utilization of Deep Learning (DL). DL can leverage DDDAS for near-real-time (NRT) situations in which the original DL trained model is updated from continuous learning through the effective labeling of SAA updates. To accomplish the NRT DL with DDDAS, an *Image-based Pre- and Post-Conditional probability Learning (IP2CL)* encoding scheme is developed to effectively encode TOI, e.g., buildings in HDAR or military targets in BDA, before and after the disasters in HADR or actions in BDA, respectively, to be used for damage level classifications. Within the DDDAS framework, we applied our IP2CL to 2 different types of scenarios: 1) pixel-wise semantic segmentation where the damage levels are densely classified for each pixel of the TOI; 2) patch-based damage level classification for a small image patch centered around the TOI, unlike case 1, a single damage level is assigned for the entire patch. Both scenarios are of crucial importance for HADR and BDA: the former filters out the most urgent regions from a wide span of regions; whereas the latter focuses on one crucial TOI, which may be of central importance to the mission, to make a judgmental call.

2   Methodology

We first introduce the DDDAS framework, the IP2CL encoding scheme is described in detail, and the final 2 subsections delineate our algorithms for pixel-wise and patch-based damage level assessments.

2.1   DDDAS

DDDAS is an important framework to achieve our objectives, which combines forward estimation analysis from both collected data as well as data augmentation from models. Figure 1 shows the DDDAS concept updated with DL methods. The diagrammatic chart of DDDAS is depicted in Fig. 2 [1]. Within the last decade, the resurgence of machine learning (ML) approaches using deep network learning engenders the DDDAS concept. For example, gathering a copious amount of data for training affords an understanding of the physical environment. Within the HADR approach, the natural terrain scene is deduced as a model of the physical characteristics, and using the prior information provides a scene analysis for future data augmentation. When a natural disaster unfolds, the previous models need to be updated relative to the scene change (e.g., building destruction is a new scene) and requires model updating evident in the DDDAS framework. The challenge for both DDDAS and DL is the "labeling" of the parameters from which computationally efficient approaches are needed.

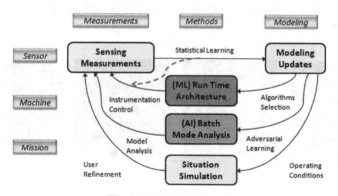

Fig. 1. DDDAS Concept.

To design an effective HADR and BDA SAA approach, this paper utilizes the DDDAS approach within the growing utilization of Deep Learning (DL). DL can leverage DDDAS for near-real-time (NRT) situations in which the original DL-trained model is updated from continuous learning through the effective labeling of SAA updates.

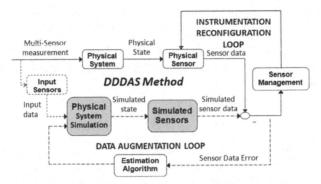

Fig. 2. DDDAS diagrammatic chart.

2.2 IP2CL

To achieve effective damage level assessment, the images, taken by satellites or airplanes, of the pre- and post-disaster/action for the TOI are normally available and well registered and matched, such as the XView2 datasets used in the Overhead Imagery Hackathon (OIH) organized by Air Force Research Laboratory, Air Force Office of Scientific Research, University of Wisconsin-Madison, and the Toyota Institute of Technology at Chicago in Sept. 2011. In Fig. 3, one sample set of pre-, and post-disaster ("*Guatemala Volcano No. 19*") and the damage designations (red: no damage, green: minor damage, yellow: total damage) of the buildings from the train set are depicted on the left 3 panels.

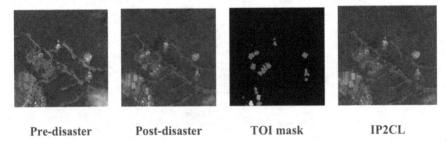

| Pre-disaster | Post-disaster | TOI mask | IP2CL |

Fig. 3. From left to right: one sample set of pre-, and post-disaster images, the ground truth of the damage mask, and the corresponding IP2CL, from the XView2 dataset.

Dr. Wei (the first author) and Dr. Blasch (the third author) led the City College of New York/AFOSR team to participate in the OIH delivering the Contrastive Learning Information Fusion Generative Adversarial Net (CLIFGAN) system [2], which consisted of the following components: 1) reduced the imagery resolution from original 1024 × 1024 to 512 × 512 to facilitate computing in laptops; 2) developed GAN to augment training data by ~30%; 3) developed a contrastive learning (CL)-based classifier to learn the representation from pre- and post-disaster images using a Siamese network; 4) used the post-disaster images with transfer learning, DeepLabV3Plus with information fusion by committee method. Our CL method (3) delivered the highest F1 score 0.674, whereas our transfer learning and information fusion method achieved an F1 score 0.664. We also reproduced former winners using the XView2 dataset, the top method, due to the resolution reduction, is 0.617. Hence the CL based method delivered the top performance in OIH, our CLIFGAN was one of the 3 winners of the OIH. To our knowledge, most winning methods over XView2 datasets used CL [3] at least as part of the strategy since there are 2 matching images to label TOI, in this case, the buildings in the disaster scene.

CL is an effective method to attach contents of similar identify while pushing away those of dissimilar identify, thus finding a broad spectrum of applications recently [4]. However, one major trouble of the CL framework in the XView2 dataset, and actually for all HADR and BDA SAA, is that: in the large scene covered by the imageries, the TOI only covers a small portion of the entire scene, e.g., more than 95% of the pixels on average in the XView2 dataset belong to the background that is of no relevance to the damage assessment at all. In consequence, it is highly questionable that the representations learned via CL are really those from the TOI, but most likely are about the background, which is of no concern to the SAA at all. Therefore, simply using the powerful CL may not achieve desirable results in HADR/BDA. One major recent breakthrough in DL is the induction of attention mechanism [5], where the valuable processing is focused more on those of more importance to the targets, which has since revolutionized Natural Language processing and computer vision with visual transformers. Unless we can find a means to direct the attention of the CL to the TOI, the efficacy of CL cannot be fully exploited, which was evidenced by the generally lackluster performance of CL in XView2 datasets.

In view of the foregoing observations, we developed the IP2CL to generate image Z from pre-image X and post-image Y, viewed as random variables with the following set of 2 conditional probability formulas:

$$P(Z_0|X_0, Y_0) = Y_0 \tag{1}$$

$$P(Z_1|X_1, Y_1) = Norm(Y_1 - X_1) \tag{2}$$

where X_c, and Y_c are the pre- and post-image with values in the range [0,1], subscript c is the indicator variable: 1 for foreground (TOI), 0 for background, Z_c is the new random variable of the same range as X and Y, $Norm(w) = \frac{w}{\max(w) - \min(w)}$ is to bring the random variable to the range [0,1]. Equations (1) and (2) essentially generate a new image Z whose background comes from the post image, the TOI is the normed version of the difference in the TOI region [6]. By bringing the difference to [0,1], the pixel-wise differences in the TOI regions are actually stressed. On the rightmost panel of Fig. 3, the IP2CL is shown, where it can be observed that the TOI regions, i.e., the buildings, are of more outstanding colors differing from the background. There are several benefits of using Z instead of the original X or Y: 1) *Data and computing efficiency*: instead of 2 color images, now we have only 1 color image to be processed, which reduced the data and computing loads. 2) *Far more choices of DL methods*: with merely 3 channels (red, green, blue) we have more choices than CL or 6-channel deep nets for data classification, both semantic segmentation, and image/object classification. 3) *Emphasis on the TOI*: due to Eq. (2) the TOI regions have larger values and variances and thus could be given more importance in classification. 4) *Simulation of human annotation*: in manual labeling by human experts, we imagine that the pixel-wise difference in TOI plays an important role to make the damage level assessment. 5) *Contextual information*: Eq. (1) states that the post images provide the most relevant contextual information to the TOI assessment as the situations after natural disasters or military actions carry more weighty information on the TOI than those before: we actually tried using the pre-disaster image or totally removing the background, both resulting in far worse classifications. The post-disaster/action information should thus be employed as dictated by Eq. (1) to provide valuable contextural information.

3   IP2CL Applications

3.1   IP2CL Applications in Semantic Segmentation

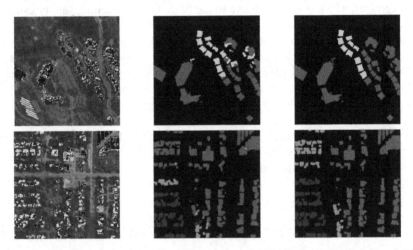

Fig. 4. Two sample results of pixel-wise segmentation using IP2CL. Left: IP2CL image, Middle: pixel-wise labels predicted by IP2CL semantic segmentation; Right: Ground truths (color code scheme: red: no damage, green: minor damage, blue: major damage, yellow: total damage). The F1 scores for the top and bottom set are 0.89 and 0.72, respectively.

For imageries covering a wide expanse as shown in Fig. 3, the semantic segmentation procedure is needed to generate dense pixel-wise damage classification. In our previous award-winning CLIFGAN [2], our CL-based method exploited both pre- and post-disaster images to achieve pixel-wise damage classification with an overall F1 score 0.674, which beat all methods CCNY/AFOSR team tried out. Equipped with the IP2CL, which can be treated as a set of normal color images with TOI stressed, by taking advantage of transfer learning: using pre-trained deep nets such as ResNet for feature generation and DeepLabV3 + for pixel-wise segmenation. In Table 1, the semantic segmentations performances delivered by the new IP2CL-based method and methods developed or reproduced by CCNY/AFOSR team in 2021 OIH are reported in Table 1.

The IP2CL captured more TOI information than post-only (66.4) and CL (67.4). In Fig. 4 2 sets of segmentation results are illustrated.

Table 1. Pixel-wise semantic segmentation performances

Algorithm	Method 1	Method 2	ccny IF method	ccny CL	IP2CL
F1 score %	50.2	61.7	66.4	67.4	**69.7**

3.2 Patch-Based Damage Classification

The pixel-wise dense damage classification reported in the preceding section is used for imageries covering large areas, in HADR/BDA applications when of the utmost interest is a specific TOI in a small region, e.g., in HADR for a specific building we would like to know if it is damaged need urgent help, or in BDA after a concentrated military action we want to know whether or not a TOI is damaged or not, accordingly different help or further military actions will be needed. In these important scenarios, the dense pixel-wise damage level classification is no longer needed; conversely, only one classification label is needed for the entire small patch where the TOI is located.

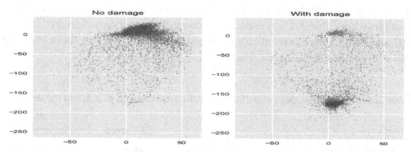

Fig. 5. Contrastive learning 2D embedding for IP2CL patches labeled "No damage" (red points) and "With damage" (green points).

In XView2 dataset, there are 5 different labels, one main reason why outstanding segmentation classification performances are elusive is due to the considerable unbalanced nature of these labels: more than 90% are background pixels, while for regions with damage, there are again ~90% are "no damage" (label 1), only <2% with "total damage", we tried various image augmentation methods, e.g., GAN, different class importance, trying to mitigate this problem, but with little success. The best F1 score by IP2CL is still slightly less than 0.70. In the patch-based scenario, our situation can be significantly improved: 1) instead of separating damage levels from "minor" up to "total", especially in military BDA applications, a crude label *"no damage"* and *"with damage"* will suffice, then the trouble caused by the extremely small number of "major" and "total" damage levels can be effectively avoided. 2) when generating the small patches of "no damage" and "with damage", we first run a statistics collection procedure by varying the patch size (128x128, 64x64, etc.) and TOI size of different types, then fix 2 thresholds (δ_1, δ_2) for "no damage" and "with damage" types. For a patch p, its class label is "no damage" or "with damage" if the maximal TOI of p has its size exceeds δ_1/δ_2, respectively, and the TOI regions belonging to the other class are erased to ensure the current patch p only has the winning class. This way we can produce patches with these 2 types of roughly similar size to facilitate effective future classification. From the XView2 datasets, after intensive offline processing, it is found that by using 64x64 patches and setting (δ_1, δ_2) to (0.12, 0.04), the training patch number for the 2 labels are 21k and 20k; and for testing set there are 7.1k vs. 6.8k for these 2 classes, hence no serious class unbalance problem anymore. Class unbalance is not a problem now.

The IP2CL patch is now a 64×64 color image/tile with binary identities: no or with damage. Now that the patches in the same class are similar, whereas those belonging to different classes are dissimilar, CL can thus be effectively applied to embed patches of the same class in a low-dimensional space of small distance, and meanwhile pushing away those of different classes. We developed a simple 4-layer CNN and 4-layer projection head and randomly pose similar and dissimilar pairs to train the Siamese net, the F1 score we achieved is 95.9 with a confusion matrix $\begin{pmatrix} 0.98 & 0.02 \\ 0.06 & 0.94 \end{pmatrix}$. As depicted in Fig. 5, the two classes are separated well in the 2-D embedding space reached by CL. As this patch-based classification is similar to the original ImageNet, transfer learning from readily available deep nets can be employed. We tried out >10 different nets from *torchvision.models* (with fine-tuning option), 6 of them achieved F1 scores above 0.90, which are listed in Table 2. CL is especially powerful as it not only delivered the top F1 score (0.959) beating the closest attention-based visual transformer by about 1% (0.950), but also with net size significantly smaller.

Table 2. Patch-based damage classification performances using IP2CL and the net size

Algorithms	F1 score	Net size (K)
Contrastive learning (4-layer CNN, 4-layer projection)	**95.9**	**8,566**
Visual transformer (ViT-b-16)	95.0	343,261
VGG19	94.2	558,326
ResNet 152	93.0	233,503
ResNet 101	92.7	170,643
Inception v3	90.5	100,810
googlenet	90.3	22,585

4 Conclusion

In this paper within the DDDAS framework, we proposed image-based pre- and post-conditional probability learning to effectively stress the changes in the target of interest, based on which outstanding pixel-wise damage and patch-based damage classification performances are consistently observed, which is of great importance to achieve data and computing efficient SAA in HADR and BDA applications.

References

1. Darema, F., Blasch, E.P., Ravela, S., Aved, A.J. (eds.): Handbook of Dynamic Data Driven Applications Systems. Springer, Cham (2018). https://doi.org/10.1007/978-3-319-95504-9
2. Wei, J., Zhu, Z., Blasch, E.P., et. al.: NIDA-CLIFGAN: natural infrastructure damage assessment through efficient classification combining contrastive learning, information fusion and generative adversarial networks. In: AI HADR 2021 Workshop, NeurIPS 2021 (Oral Presentation) (2021)
3. Zhao, X., et al.: Contrastive learning for label-efficient semantic segmentation (2021)
4. Khosla, P., et al.: Supervised Contrastive Learning, NeurIPS 2020, pp. 661–673 (2020)
5. Vaswani, A., et al.: Attention is All You Need, NeurIPS 2017, pp. 5998–6008 (2017)
6. Wei, J.: Video content classification based on 3-D eigen analysis. IEEE Trans. Image Process. **14**(5), 662–673 (2005)

SpecAL: Towards Active Learning for Semantic Segmentation of Hyperspectral Imagery

Aneesh Rangnekar$^{(\boxtimes)}$ ⓘ, Emmett Ientilucci, Christopher Kanan, and Matthew Hoffman

Rochester Institute of Technology, Rochester, NY, USA
aneesh.rangnekar@mail.rit.edu, emmett@cis.rit.edu, {kanan,mjhsma}@rit.edu

Abstract. We investigate active learning towards applied hyperspectral image analysis for semantic segmentation. Active learning stems from initially training on a limited data budget and then gradually querying for additional sets of labeled examples to enrich the overall data distribution and help neural networks increase their task performance. This approach works in favor of the remote sensing tasks, including hyperspectral imagery analysis, where labeling can be intensive and time-consuming as the sensor angle, configured parameters, and atmospheric conditions fluctuate.

In this paper, we tackle active learning for semantic segmentation using the AeroRIT dataset on three fronts - data utilization, neural network design, and formulation of the cost function (also known as acquisition factor, uncertainty estimator). Specifically, we extend the batch ensembles method to semantic segmentation for creating efficient network ensembles to estimate the network's uncertainty as the acquisition factor for querying new sets of images. Our approach reduces the data labeling requirement and achieves competitive performance on the AeroRIT dataset by using only 30% of the entire training data.

Keywords: hyperspectral · active learning · segmentation

1 Introduction

There has been significant development in designing deep neural networks for semantic segmentation across multiple computer vision applications. Most of those networks benefit from large amounts of data [1,3,9,10]. This additional data load comes with an overhead cost of pixel-level annotations that increases exponentially with the number of samples. For example, the AeroRIT semantic segmentation dataset (1973 × 3975) required around 50 h of manual labeling and multiple review rounds to ensure a good quality release.

supported by Dynamic Data Driven Applications Systems Program, Air Force Office of Scientific Research under Grant FA9550-19-1-0021 and Nvidia GPU Grant Program.

Multiple branches of machine learning deal with reducing the need for a large number of labeled examples (for example, semi-supervised learning, weakly-supervised learning, and active learning). Active learning is an approach that has gained significant traction to reduce dependency on large amounts of data for semantic segmentation ([8,12,14]). This approach focuses on tracking the most informative samples within the unlabeled data pool to add to the labeling pool via a scoring mechanism, most commonly an estimation of the network's uncertainty. This accumulation continues for multiple active learning cycles until one of the two conditions are met: 1) the network under question achieves a preset performance budget (typically, 95% of the entire data performance), or 2) the data labeling budget gets exhausted.

This paper explores the ability of neural networks to capture the information within hyperspectral signatures and function in an active learning data-training framework: we report competitive performance by utilizing significantly less labeled data, at par with fully utilizing the labeled data, which can be achieved under proper training conditions. For our analysis, we use the baseline network provided in the AeroRIT dataset [9] and make our increments to reduce the labeled data requirement by 70%.

2   Related Work

Kendall *et al.* formulated segmentation as an approximate Bayesian interpretation by applying dropout at selective layers of their architecture to formulate test-time ensembles [4]. Lakshminarayanan *et al.* showed that training multiple iterations of the same network with random initializations acts as ensembles [5], and Wen *et al.* proposed to use multiple rank-1 matrices along with a shared weight matrix to form batch ensembles as an alternative to existing methods [13]. Rangnekar *et al.* studied the effects of using these approaches for uncertainty quantification on hyperspectral imagery by training on the AeroRIT dataset and found that applying dropout during test time gives the most definitive results [7]. In our paper, we build on this approach for uncertainty quantification as the active learning acquisition factor in the learning pipeline.

3   Methodology

The objective of this paper is to get an understanding of how different components of an experiment design can function towards improving the scope of data requirements for hyperspectral semantic segmentation with limited data. We adopt the active learning approach, which works in the following manner: (1) Train on available labeled data, (2) Acquire additional labels by evaluating the network on the unlabeled data pool, (3) Add the freshly labeled data to the existing labeled data pool, and (4) Repeat (1) to (4) till convergence. Given this process, we focus on three essential adjustments: data utilization, neural network design, and acquisition factor.

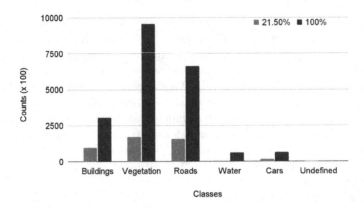

Fig. 1. The distribution of classes within the AeroRIT train set at 100% and our selected starting set with only 21.50% of the data.

3.1 Data Utilization

We use the AeroRIT dataset as the reference for our learning pipeline [9]. The train set consists of 502 image patches of 64 × 64 resolution, post ignoring the overlap and ortho-rectification artifact patches used in the original studies presented in the paper. From this train set, we fixate on a small patch of the area as our starting point, as shown in Fig. 3a. This patch consists of 108 image patches, thus giving us a starting set of 21.5% of the dataset. We treat the rest of the image patches within the train set as the unlabeled pool of available data during the learning pipeline. Figure 1 shows the statistics of each class within the respective sets, and we observe that the water class is not present in the initial labeled set. We hypothesize a well-trained network will have high uncertainty towards regions consisting of water and hence, do not consider it a concern. This would not be the case when dealing with standard gray-scale or color imagery, but the discriminative nature of hyperspectral imagery allows us to make this hypothesis.

We think it is unrealistic for any neural network to be able to achieve comparable performance to its fully labeled counterpart when using limited data. To this extent, we use data augmentation to increase the number of possible examples within the scene by applying random horizontal and vertical flips, with additive Gaussian noise in randomly selected spectral bands, random resizing, and CutOut [2]. This enables us to increase exposure to underlying data distribution and fully utilize the available data.

We increase the scope of learning further by increasing the learning schedule to account for relatively fewer data samples seen per training epoch. The network sees 502 samples per epoch on the fully labeled dataset to learn representations from scratch. Conversely, in the limited data regime, the network only sees 108 samples per epoch; hence, we do not expect it to learn at total capacity due to the nearly 1/5th reduction of samples per epoch. Hence, our first significant adjustment is to increase the number of samples seen per epoch

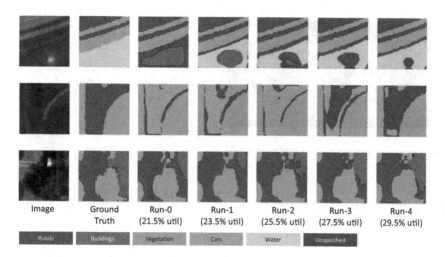

| Image | Ground Truth | Run-0 (21.5% util) | Run-1 (23.5% util) | Run-2 (25.5% util) | Run-3 (27.5% util) | Run-4 (29.5% util) |

Roads   Buildings   Vegetation   Cars   Water   Unspecified

Fig. 2. Example predictions as a progressive acquisition step in the active learning pipeline. We observe a distinct improvement across all classes as the percentage of data used grows through the active learning cycles.

during training in the limited data regime. We show in our experiments section that this dramatically improves the network's ability to learn representations from the limited set of available data.

3.2   Neural Network Design

We use the network described in [7,9] for fair comparison. The network is based on the U-Net architecture and consists of two downsampling encoder blocks, followed by a bottleneck block, and two upsampling blocks before making the final prediction [10]. Our goal is to modify the network to express uncertainty, and previous studies have shown that ensembles work well for this purpose [4,7]. The Monte-Carlo Dropout based approach is a clear choice for our task, where dropout is applied during test time for multiple network ensembles. However, dropout-based approaches have a shortcoming regarding reproducibility as the application is a function of the random probability distribution. Hence, we consider an alternative, simpler approach of Batch Ensembles.

Batch Ensembles (BE) work by sharing a tuple of trainable rank-1 matrices for every convolutional filter that is present in the neural network (Fig. 3b). The tuples act as the ensembles and, when combined with the filter weights, act as an individual ensemble members. The network in a standard manner - each mini-batch is passed through the network, wherein it is split according to the number of ensembles (for example, a network with four tuples would split a mini-batch of 32 samples into eight samples per tuple). The core idea is that the convolutional filter weight acts as the shared weight amongst all randomly initialized rank-1 tuples, which learn their own set of representations. During the evaluation

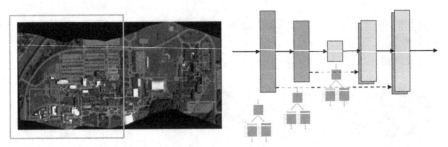

(a) The entire training set (Green) versus the starting set for active learning (Red).

(b) A graphic description of our U-Net with Batch Ensembles blocks.

Fig. 3. (a) shows the scene overview with active learning and training area. (b) shows the network used in this paper.

(inference) phase, every data point is replicated (for the above example, four times) and passed through the network to get an ensemble prediction, which is further averaged to obtain a final network prediction. In this approach, the tuple weights being fixed post learning ensure consistent predictions every time, unlike the dropout approach, while maintaining a low-cost solution to training individual ensemble instances like the Deep Ensembles approach.

Kendall *et al.* found in their study that applying dropout only to the bottleneck blocks of the encoder and decoder gave them optimum results [4]. With this motivation, we experiment with key areas to apply Batch Ensembles tuples to the convolutional filters within our network and experimentally found out the best and most consistent results were obtained by converting the convolutional filters in the encoder and bottleneck blocks to their ensemble counterparts. Our second significant adjustment is to convert a deterministic neural network into its light-weight ensemble version that can easily express uncertainty without further adjustments.

3.3 Acquisition Factor

Our goal now is to interpret the model's outputs as a function for querying a fixed budget of patches for labeling from the unlabeled pool of image patches. We simulate the process of querying for additional labels in reality by using the ground truth annotations already present for our dataset. These queried image patches are added to the labeled set for another cycle of learning. We experiment with four different approaches as acquisition factors: random (lower-bound), softmax confidence, softmax entropy, and softmax margin. We refer the readers to [11] for an in-depth explanation of these factors and experimentally find softmax entropy as the best candidate for our approach.

Table 1. Quantitative improvements over the baseline performance using a small U-Net on the AeroRIT dataset.

Training Scheme	Data Util.	Building	Vegetation	Roads	Water	Cars	mIoU
Baseline (1x data)	21.50%	61.92	93.39	74.06	0.00	31.39	51.70
5x data	21.50%	72.13	94.12	75.10	0.00	31.66	54.60
5x data (BE)	21.50%	77.00	94.50	76.99	0.00	28.85	55.47
5x data (BE)	29.50%	81.06	93.51	78.83	72.50	35.52	72.28
Oracle	100.0%	82.94	94.82	80.00	63.35	35.82	71.40

We observe that the networks quickly become confident in their predictions during the training process. Typically, the network sees enough variance in the data to understand the minor differences between classes that may have similar signatures (for example, a black car and a black roof on a building). We observe this in Fig. 2 as the network progressively makes an understanding of the similarities and differences in the signatures with more labeled data. We account for this spurious leap in the network's prediction by penalizing the confident predictions [6]. This results in an elegant win for us as a byproduct of the penalty is higher entropy, which helps express uncertainty better. Our third significant adjustment is to combine confidence penalty regularization with softmax entropy as our acquisition factor.

4   Experiments and Results

4.1   Hyperparameters

We use 50 bands in the AeroRIT dataset, ranging from 400 to 890 nm, in this paper - 31 visible and 20 infrared bands. All chips are clipped to a maximum of 2^{14}, and normalized between 0 and 1 before forward passing through the networks. All networks are initialized with Kaiming initialization, and the rank-1 matrices for batch ensembles are initialized to have a mean of 1 and a standard deviation of 0.5 in accordance with the original paper. We use an initial learning rate of $1e^{-2}$, with drops of 0.1 at 50 and 90 epochs. We train all our networks for 120 epochs with standard cross-entropy loss and use confidence penalty for all limited data training instances. We will release the code post-publication.

For the active learning scenario, we start with an initial labeled set of 108 images (21.5% of the data) and iteratively query for 10 images (2%) every active learning cycle. We do not keep a preset data budget but instead strive to obtain performance comparable with the network trained on full data (502 images).

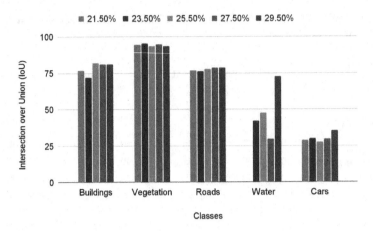

Fig. 4. Quantitative improvements of individual classes in the AeroRIT dataset as a function of active learning (data querying) cycles.

4.2 Results

Table 1 shows that the training performance dramatically benefits from increasing the learning schedule throughout the process. An increase in the number of samples per epoch (5x data) results in an improvement from a mIoU of 51.70 to 54.60, most significantly affecting the Building category during the data augmentation schemes. We also observe that shifting the model to its ensemble version (BE) further increases the mIoU by another point, yet again, mainly influencing the Building category that has two distinct white and black signatures throughout our dataset. BE also drops the IoU for the Car category by a few points, which is unexpected but is gradually over-comed through the active learning cycles (Fig. 4). Figure 4 also shows an interesting trend in the Water category, we immediately observe a leap from 0 points in the IoU to roughly 45 points, before finally improving at the final cycle to 72.5 points and beating the performance of the fully supervised network. This could indicate (and warrants analyzing) wrongly labeled instances within the training set, which the network has successfully chosen to ignore during its learning process.

We observe that using a confidence penalty helps stabilize the performance and ensure reproducibility among various random initializations of the network. We run the entire framework through a rigorous evaluation scheme by further sampling 108 grid patches across random areas in the training set, ensuring that all classes follow the data distribution in Fig. 1. Surprisingly, the networks could reach similar performances in eight of the ten trials. We repeated the same set of experiments and enhanced our analysis with test-time data augmentation via random flips but did not obtain a reasonable difference in performance.

5   Conclusion

We present an approach for learning with limited data in hyperspectral imagery by leveraging the active learning framework. We can obtain performance at par with a fully supervised network using only 30% of the data budget. In closing, our next steps are to explore the domains of self-supervised learning to have a better-initialized network, which can also incorporate pseudo information from the unlabeled data to learn better representations.

References

1. Badrinarayanan, V., Kendall, A., Cipolla, R.: Segnet: a deep convolutional encoder-decoder architecture for image segmentation. IEEE Trans. Pattern Anal. Mach. Intell. **39**(12), 2481–2495 (2017)
2. DeVries, T., Taylor, G.W.: Improved regularization of convolutional neural networks with cutout. arXiv preprint arXiv:1708.04552 (2017)
3. Kemker, R., Salvaggio, C., Kanan, C.: Algorithms for semantic segmentation of multispectral remote sensing imagery using deep learning. ISPRS J. Photogramm. Remote. Sens. **145**, 60–77 (2018)
4. Kendall, A., Badrinarayanan, V., Cipolla, R.: Bayesian segnet: model uncertainty in deep convolutional encoder-decoder architectures for scene understanding. arXiv preprint arXiv:1511.02680 (2015)
5. Lakshminarayanan, B., Pritzel, A., Blundell, C.: Simple and scalable predictive uncertainty estimation using deep ensembles. In: Advances in Neural Information Processing Systems, vol. 30 (2017)
6. Pereyra, G., Tucker, G., Chorowski, J., Kaiser, L., Hinton, G.: Regularizing neural networks by penalizing confident output distributions. arXiv preprint arXiv:1701.06548 (2017)
7. Rangnekar, A., Ientilucci, E., Kanan, C., Hoffman, M.J.: Uncertainty estimation for semantic segmentation of hyperspectral imagery. In: Darema, F., Blasch, E., Ravela, S., Aved, A. (eds.) Dynamic Data Driven Applications Systems: Third International Conference, DDDAS 2020, Boston, MA, USA, October 2-4, 2020, Proceedings, pp. 163–170. Springer International Publishing, Cham (2020). https://doi.org/10.1007/978-3-030-61725-7_20
8. Rangnekar, A., Kanan, C., Hoffman, M.: Semantic segmentation with active semi-supervised learning. arXiv preprint arXiv:2203.10730 (2022)
9. Rangnekar, A., Mokashi, N., Ientilucci, E.J., Kanan, C., Hoffman, M.J.: Aerorit: a new scene for hyperspectral image analysis. IEEE Trans. Geosci. Remote Sens. **58**(11), 8116–8124 (2020)
10. Ronneberger, O., Fischer, P., Brox, T.: U-Net: convolutional networks for biomedical image segmentation. In: Navab, N., Hornegger, J., Wells, W.M., Frangi, A.F. (eds.) Medical Image Computing and Computer-Assisted Intervention – MICCAI 2015: 18th International Conference, Munich, Germany, October 5-9, 2015, Proceedings, Part III, pp. 234–241. Springer International Publishing, Cham (2015). https://doi.org/10.1007/978-3-319-24574-4_28
11. Siddiqui, Y., Valentin, J., Nießner, M.: Viewal: Active learning with viewpoint entropy for semantic segmentation. In: Proceedings of the IEEE/CVF Conference on Computer Vision and Pattern Recognition, pp. 9433–9443 (2020)

12. Sinha, S., Ebrahimi, S., Darrell, T.: Variational adversarial active learning. In: Proceedings of the IEEE/CVF International Conference on Computer Vision, pp. 5972–5981 (2019)
13. Wen, Y., Tran, D., Ba, J.: Batchensemble: An alternative approach to efficient ensemble and lifelong learning (2020)
14. Xie, S., Feng, Z., Chen, Y., Sun, S., Ma, C., Song, M.: Deal: Difficulty-aware active learning for semantic segmentation. In: Proceedings of the Asian Conference on Computer Vision (2020)

Multimodal IR and RF Based Sensor System for Real-Time Human Target Detection, Identification, and Geolocation

Peng Cheng[1]([✉]), Xinping Lin[1], Yunqi Zhang[1], Erik Blasch[2], and Genshe Chen[1]

[1] Intelligent Fusion Technology, Inc, Germantown, MD 20874, USA
gchen@intfusiontech.com
[2] Air Force Research Laboratory, Rome, NY, USA

Abstract. The Dynamic Data Driven Applications System (DDDAS) paradigm incorporates forward estimation with inverse modeling, augmented with contextual information. For cooperative infrared (IR) and radio-frequency (RF) based automatic target detection and recognition (ATR) systems, advantages of multimodal sensing and machine learning (ML) enhance real-time object detection and geolocation from an unmanned aerial vehicle (UAV). Using an RF subsystem, including the linear frequency modulated continuous wave (LFMCW) ranging radar and the smart antenna, line-of-sight (LOS) and non-line-of-sight (NLOS) friendly objects are detected and located. The IR subsystem detects and locates all human objects in a LOS scenario providing safety alerts to humans entering hazardous locations. By applying a ML-based object detection algorithm, i.e., the YOLO detector, which was specifically trained with IR images, the subsystem could detect humans that are 100 m away. Additionally, the DDDAS-inspired multimodal IR and RF (MIRRF) system discriminates LOS friendly and non-friendly objects. The whole MIRRF sensor system meets the size, weight, power, and cost (SWaP-C) requirement of being installed on the UAVs. Results of ground testing integrated with an all-terrain robot, the MIRRF sensor system demonstrated the capability of fast detection of humans, discrimination of friendly and non-friendly objects, and continuously tracked and geo-located the objects of interest.

Keywords: LFMCW Radar · IR sensor · Human Target Detection · Geolocation · Machine learning · unmanned aerial vehicle (UAV)

1 Introduction

Unmanned aerial vehicles (UAVs) have been widely utilized in a variety of search and rescue, and surveillance applications by leveraging its mobility and operational simplicity. In some situations, a UAV's ability to recognize the actions of a human subject is desirable, then take responsive actions. Identify human targets from videos captured from a static platform is a challenging task owing to the articulated structure and range of possible poses of the human body. Human target identification is further challenged by the quality of videos which include perspective distortion, occlusion, and motion blur, or even the visibility of the target in the foggy and rainy weather conditions.

E. Blasch et al. (Eds.): DDDAS 2022, LNCS 13984, pp. 205–216, 2024.
https://doi.org/10.1007/978-3-031-52670-1_20

Using UAVs in human detection and recognition missions is a relatively new topic and most of them use traditional EO camera [1]. Some studies focused on human detection methods from aerial videos in relation to search and rescue missions [2, 3]. These studies aimed at identifying humans lying or sitting on the ground. Some notable approaches related to human identity recognition in low-resolution aerial videos are weighted voter-candidate formulation by Oreifej et al. [4] and blob matching using an adaptive reference set by Yeh et al. [5]. Monajjemi et al. [6] developed a UAV onboard gesture recognition system to identify periodic movements of waving hands from other periodic movements like walking and running in an outdoor environment.

The study presented in this paper is focused on developing a multimodal IR and RF (MIRRF) based sensor system that can be integrated with a UAV to detect and recognize human subjects and to continuously tracking the interested target in real-time. Our sensor system consists of two subsystems: 1) the RF subsystem, including the linear frequency modulated continuous wave (LFMCW) ranging radar and smart antenna, to detect and locate the line-of-sight (LOS) and non-line-of-sight (NLOS) friendly objects; 2) the IR subsystem, to detects and locates all human objects in a LOS scenario providing safety alerts to humans entering hazardous locations. Two subsystems have been integrated together and the communication between the sensor system and the host computer have been established to realize real-time human target detection and friendly human target recognition.

The key contribution of this paper is to provide a preliminary solution that a UAV will be able to use for human detection, position estimation and continuously human target tracking in real-time. Before deploying it to the UAV, in this paper, we developed the whole package of the MIRRF sensor system, including both the hardware and the software, and tested its capability of human target detection and friendly human target (blue force) recognition on the ground by mounting it onto an all-terrain robot.

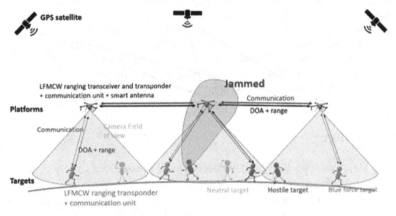

Fig. 1. Detect and localize human targets in an open field.

2   System Architecture

The proposed IR and RF mixed sensor systems for human target detection and geolocation system can be integrated with the UAV and work in many different hazardous environments. Figure 1 Reference source not found. Shows one scenario that the proposed system can be applied, i.e., detecting and localizing human targets in an open field where GPS signal is not usable. In this scenario, some UAVs have GPS access while others do not. The platforms with GPS access assist the platforms without GPS access to localize themselves. Then, all platforms perform the human target detection and localization tasks.

The whole sensor system structure for human target detection and recognition is presented in Fig. 2. All the necessary sensors that will be used for human target detection and geolocation are implemented on the platform (UAV), including the ranging radar, the IR camera, the laser range finder, the differential barometer, and the pan/tilt stage. The friendly target is equipped with the IR emitter and the RF transponder so that our MIRRF sensor system can easily recognize the friendly target from all the detected human target.

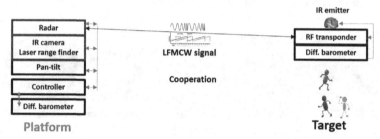

Fig. 2. Structure of multimodal IR and RF based sensor system.

2.1   RF Subsystem

The RF subsystem is constructed with a LFMCW ranging radar (the transceiver is on the platform side and the transponder is on the friendly target side) and a smart antenna (on the platform side). The LFMCW transceiver, as shown in Fig. 3(a), is constructed with a LFMCW transmitter, a LFMCW receiver with frequency/range scanning capability, and a signal processor. It is equipped with a smart antenna that is able to estimate the angle of the friendly target with respect to the platform. The smart antenna is able to reach the measurement accuracy of $0.8°$, and has the ability to suppress multipath signals reflected from ground, walls and ceilings. Figure 3(b) shows the radar transponder that will be on the friendly target side. The whole radar subsystem was tested in the indoor environment. Figure 3(c) presents the measured distance between the platform and the friendly target. The results demonstrated that our self-developed radar subsystem can consistently detect and measure the distance of the friendly target.

The following techniques are used to improve the signal to noise ratio and range detection:

1) The RF signals were sampled multiple times (such as 8 samples) and the FFT calculation was conducted for each of them. Then the results were averaged to improve the signal to noise ratio and the detection range.
2) Since the hardware gain responds is different across the base band spectrum, we need to find the local signal noise floor to use as a reference. The real signal can be compared with the local noise floor instead of the entire operation base band noise floor.
3) By running through a local average window, the proper reference level was setup locally.

Currently, the radar range cutoff is just over 27 m. If needed, we can set the parameter for a longer detection range. The distance measurement update rate is 7 times/second. The average current draw in this refresh rate is 700 mA at 6 v. The refresh rate can be much faster if we don't turn off a certain portion of the radar function to save power between each update.

The capability of our RF subsystem was also tested and verified in the outdoor open environment, as well as in the woods. Moreover, it has also been verified that the RF subsystem can consistently detect the distance of the human target who is equipped with the radar transponder even through multiple drywalls.

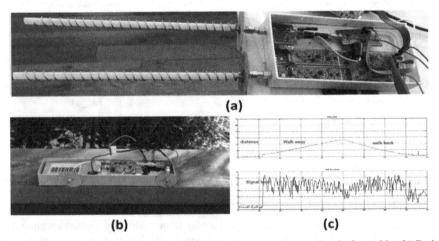

Fig. 3. LFMCW Radar with smart antenna. (a) Radar transceiver on the platform side, (b) Radar transponder on the friendly target side, and (c) distance measurement in experiment.

2.2 IR Subsystem

The IR subsystem is constructed with an IR camera, a laser rangefinder (found on the platform side), a controllable IR emitter (for the friendly target side), and a pan/tilt platform. In the IR subsystem, two different IR cameras were used for different detection ranges. Both were tested and implemented. The laser range finder, aligned with the viewing direction of the IR camera, measures a range of 100 m. The IR subsystem can consistently discriminate LOS friendly and non-friendly targets based on the IR signal emitted from the IR emitter equipped by the friendly target.

The hardware arrangement of the IR subsystem is illustrated in Fig. 4(a). The IR camera and the laser rangefinder were aligned to point at the same direction. Both devices were mounted on the pan/tilt platform and can be rotated to different directions. The laser rangefinder is used to measure the distance of the target in the center of the IR image's field of view. As presented in Fig. 4(b), from the first image, which is taken at t_1 the human target is detected from the IR image and its position in the IR image's filed of view can be calculated. Then the lateral angle position α and the vertical angle position ϕ of the target, relative to the IR camera's pointing direction, can be calculated accordingly. The calculated angle position of the target will be sent to the pan/tilt platform to move the IR subsystem so that the target can be placed at the center of the IR camera's field of view. Thus, at the time instant t_2, the distance of the target can be measured from the laser rangefinder. Figure 4(c) shows the flowchart of the working principle of the IR subsystem for detecting, tracking and measuring the distance of the interested target.

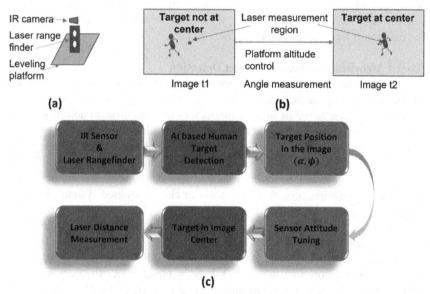

Fig. 4. IR subsystem for human target detection and tracking. (a) IR subsystem hardware setup; (b) centering on interested target for distance measurement, and (c) flowchart of the working principle for the IR subsystem.

The human target detection from the IR image was accomplished by applying the machine learning (ML) -based object detection algorithm called "YOLOv4" [7]. Multiple human targets in the field of view can be detected from each frame of the IR image. The model for the YOLO detector was specifically trained with over 1000 labeled IR images to improve the human target detection accuracy.

The details of each component in the IR subsystem, as well as the software algorithm, are described in the following sections.

IR Camera

In the IR subsystem, two different IR cameras, i.e., 9640P IR camera from ICI, and Boson 320 IR camera from Teledyne, had been implemented and tested. This is performed for different human target detection range.

The short-range Boson 320 IR camera, with a 6.3mm lens and 34° horizontal field of view (FOV), can detect a human target up to 25 m. Its body weight is only 7.5g and its size is as small as $21 \times 21 \times 11$ mm. It features an industry-leading thermal sensitivity of less than or equal to ($\leq$)20 mK and an upgraded automatic gain control (AGC) filter delivering dramatically enhanced scene contrast and sharpness in all environments. The frame rate of this camera is up to 60Hz, which is fast enough for the application of real-time human target detection. The image resolution is 320×256 pixels. The image stream is transferred from the camera to the host PC through USB cable in real-time.

The long-range 9640p is a high-quality thermal grade IR camera with the image resolution of 640×512 pixels. It uses a 50 mm Athermalized Lens with the FOV of $12.4° \times 9.3°$ and its total weight is 230 g. It can achieve a detection range of over 100 m. The maximum frame rate of this camera is 30 Hz.

ML Based Algorithm for Human Target Detection

After comparing the performance of the machine learning based object detection algorithms that can be implemented for real-time applications [7], we selected the open-source YOLO detector as the IR image analysis tool for human target detection. YOLO detector is one of the most advanced machine learning algorithms that can be applied for real-time object detection. It applies a single neural network to the full image. This network divides the image into regions, predicts bounding boxes and probabilities for each region. These bounding boxes are weighted by the predicted probabilities.

The YOLO model has several advantages over classifier-based systems. It looks at the whole image at test time, thereby its predictions are informed by the global context in the image. It also makes predictions with a single network evaluation unlike systems like R-CNN which require thousands for a single image. This makes it extremely fast, over 1000x faster than R-CNN and 100x faster than Fast R-CNN.

In order to make sure that the YOLO detector could detect the human target in different scenarios, more than 1000 pictures were acquired with the IR camera covering all these cases, as shown in Fig. 5. Moreover, we also considered the situation that only part of the human body was in the IR camera's field of view, such as the lower body, upper body, right body, and left body.

Once we have annotated the raw IR image data, the IR images and the annotation files were used as the input file for the training of the YOLO model. The YOLO model, which had already been trained with the Microsoft COCO dataset, was used as the initial model to train with the annotated IR images.

After the training of the YOLO model is completed, the performance of the trained model was evaluated with the IR images that were not used for YOLO model training. As shown in Fig. 6, the trained YOLO model can be applied to accurately detect the human targets in different scenarios, including:

Fig. 5. Examples of training image data obtained in different environments.

1) Human target detection in the indoor environment.
2) Human target detection in the outdoor environment.
3) Multiple human targets detection from the same IR image.
4) Human targe detection at different distances.
5) Human target detection with different human body gestures.

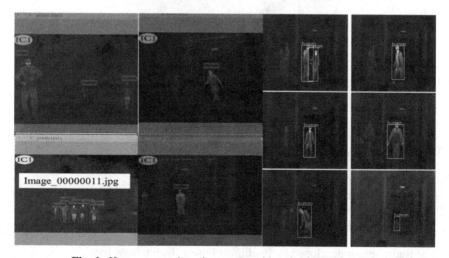

Fig. 6. Human target detection results with trained YOLO model.

It has been verified that the human target detection accuracy can reach up to 90% when the image size of the human body is greater than 40 pixels in the vertical direction. Considering the case when the Boson 320 Camera with a 6.3 mm lens was used, the corresponding human target detection distance is between 20 to 25 m. With the further increase of the human target's distance to the IR camera, the size of the human body will be even smaller. Whereas the human target can still be detected, the detection accuracy will be degraded.

Laser Rangefinder

Although the IR camera can detect the location of the human target in the field of view, it lacks the capability of measuring the distance of the detected targets. To address this issue, the laser rangefinder SF30/C from LightWare, was employed to provide the distance of the interested human target. The laser rangefinder was aligned with the viewing direction of the IR camera. Both devices were mounted on the rotary stage to ensure that the laser rangefinder is always pointing at the center of the IR camera's field of view. When there is an interested human target detected in the field of view, the rotary stage will adjust the orientation of the IR subsystem to center the interested target. Thus, the position of the interested target, relative to the sensor system's platform, can be obtained.

Fig. 7. Assembled multimodal IR and RF based mixed sensor system.

2.3 Sensor System Integration

Hardware System Assembly

Before deploying the MIRRF system onto the UAV platform. The IR subsystem and the RF subsystem were integrated together and mounted on to the all-terrain robot platform to perform ground tests. As shown in Fig. 7, the system includes the following components:

1) The Radar sensor developed by Intelligent Fusion Technology, Inc.
2) The FLIR Boson 320 IR camera from Teledyne.
3) The SF30/C laser rangefinder from Lightware.
4) The HWT905 IMU sensor from Wit-Motion.
5) X-RSW series motorized rotary stage from Zaber.
6) The IG42 all-terrain robot from SuperDroid Robots.

All the cables are organized and extended to the inside of the robot, in which two 12 V batteries were used to generate the 24 V DC power for the rotary stage and the wheels of the robot. Only one USB cable is needed to connect with the USB hub on the robot to establish the communication between the host computer and all the sensors and the rotary stage.

Software Package

In order to communicate with all the devices in the sensor system and have a better display of the detection results, we design a graphical user interface (GUI) software package for easier user control and key data display, as presented in Fig. 8. The communication between all the hardware devices and the host computer are realized in the GUI software.

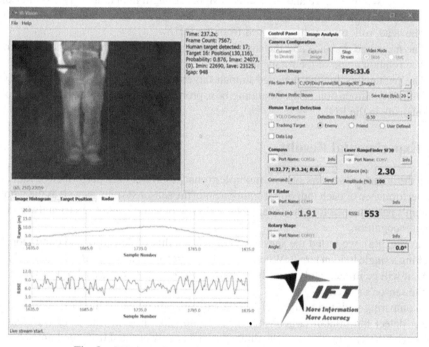

Fig. 8. GUI for online human target detection and tracking.

The GUI software has been developed with the following capabilities:

1) Display the image acquired from IR camera.
2) Configure the YOLO detector for human target detection.
3) Receive and display the measurement results from the IMU sensor.
4) Receive and display the measurement results from the Laser rangefinder.
5) Send the control command to the rotary stage.
6) Receive and display the measurement results from the radar.
7) Perform real-time target tracking to follow the interested target.

The measurement results from different sensors are sent back to the host computer at different data update rates. The measurement results from different sensors were synchronized in the GUI software to calculate the position of the tracking target. In the MIRRF system, the IR camera is the major sensor for human target detection and tracking. Therefore, all the other received sensors' measurements are synchronized with the update rate of the IR camera. In our test, the real time human target detection can run continuously at about 35 frames per second (fps) when the laptop computer (CPU: Intel Core i9-11900H; GPU: Nvidia RTX-3060 laptop GPU) was plugged into a power source, and about 28 fps when the laptop computer was running solely on battery. Each time a new frame of IR image is received in the image acquisition thread, the software will update the measured data from all the sensors, including:

1) The measured distance and the received signal strength indicator (RSSI) value from the RF subsystem.
2) The head, roll, pitch value received from the IMU.
3) The measured distance and the signal strength received from the laser rangefinder.
4) The rotation angle of the rotary stage.
5) The parameters for the detected human target by IR-image analysis with YOLO, including, dimension of the bounding box, probability, tracked target ID, maximum intensity in the ROI.

3   Experimental Results

With the integrated sensors system, multiple ground tests have been performed to validate the performance of each individual component in the sensor system, as well as the whole system's performance for human target detection, geolocation, and LOS friendly human target recognition.

In Fig. 9(a), we tested the sensor system's capability of detecting and continuously tracking a single human target. When the human target appears in the IR camera's field of view, it will be immediately identified (marked with the red bounding box) and tracked by the sensor's system.

Comparing with the traditional EO camera, one advantage of the IR camera is that it can detect the human target when there is no illumination. The long wavelength Infrared (LWIR) camera detect the direct thermal energy emitted from the human body. Figure 9(b) shows that the MIRRF system can function correctly even in the dark environment.

In Fig. 9(c), we tested the measurement accuracy of the radar subsystem. When the friendly human target is detected by our system, his distance to the platform is measured by both radar subsystem and the laser rangefinder. The measurement results verified that the radar subsystem can provide accurate distance information of the friendly target, with the error of less than 0.3 m when comparing with the laser rangefinder.

In the last test, as shown in Fig. 9(d), there are two human targets. The one holding the IR emitter (a heat source) is the friendly target. The other is the non-friendly target. The system was configured to tracking only the non-friendly target. When both targets came into the IR camera's field of view, the sensor system immediately identified them, and marked the friendly target with green bounding box and the non-friendly target with red box. Moreover, the sensor system immediately started to continuously track and follow the non-friendly target.

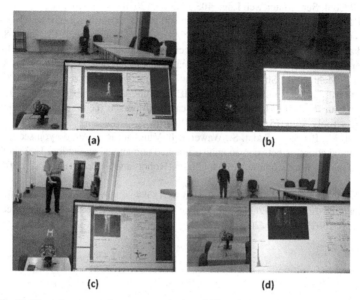

(a)   (b)

(c)   (d)

Fig. 9. Experiments to demonstrate the capability of the MIRRF sensor system.

4 Conclusion

In this paper, a multimodal IR and RF based sensor system was proposed and developed. This mixed sensor system was designed to be deployed onto the UAV to perform real-time human target detection, recognition, and geolocation. This paper is focused on the hardware and software integration of the whole MIRRF sensor system. Before the integration, the communication between each device and the host computer, as well as the functionality of each individual device have been tested accordingly. After that, through multiple ground tests, it has been verified that this cooperative MIRRF sensor system can consistently detect and recognize human targets in different environments. With the small size and light weight of the whole system, it is promising that the integrated MIRRF sensor system can be deployed to the UAV to perform real-time human target detection tasks.

In the future, we will deploy and test the MIRRF system on the UAV. we will also leverage the concept of the DDDAS to re-train the YOLO model with the new images that will be acquired from the IR camera on the UAV. Thus, it is anticipated that the MIRRF system would be able to perform the same human target detection, recognition, and geolocation tasks that have been verified from the ground tests.

References

1. Perera, F., Al-Naji, A, Law, Y., Chahl, J.: Human detection and motion analysis from a quadrotor UAV. IOP Conf. Ser.: Mater. Sci. Eng., **405**, 012003, (2018)
2. Rudol, P., Doherty, P.: Human body detection and geolocalization for uav search and rescue missions using color and thermal imagery. In: Aerospace Conference, IEEE, pp. 1–8, (2008)
3. Andriluka, M., et al.: Vision based victim detection from unmanned aerial vehicles. Intelligent Robots and Systems (IROS), IEEE/RSJ International Conference on, pp. 1740–1747 (2010)
4. Oreifej, O., Mehran, R., Shah, M.: Human identity recognition in aerial images. Computer Vision and Pattern Recognition (CVPR). In: IEEE Conference on, pp 709–716, (2010)
5. Yeh, M., Chiu, H., Wang, J.: Fast medium-scale multiperson identification in aerial videos. Multimedia Tools Appl. **75**, 16117–16133 (2016)
6. Monajjemi, M., Bruce, J., Sadat, S., Wawerla, J., Vaughan, R.: UAV, do you see me? Establishing mutual attention between an uninstrumented human and an outdoor UAV in flight. In: IEEE/RSJ International Conference on Intelligent Robots and Systems (IROS), pp. 3614–3620, (2015)
7. Wang, C., Bochkovskiy, A., Liao, H.: Scaled-YOLOv4: scaling cross stage partial network. In: Proceedings of the IEEE/CVF Conference on Computer Vision and Pattern Recognition (CVPR), pp. 13029–13038, (2021)

Learning Interacting Dynamic Systems with Neural Ordinary Differential Equations

Song Wen$^{(\boxtimes)}$, Hao Wang, and Dimitris Metaxas

Rutgers University, New Brunswick, USA
`song.wen@rutgers.edu, dnm@cs.rutgers.edu`

Abstract. Interacting Dynamic Systems refer to a group of agents which interact with others in a complex and dynamic way. Modeling Interacting Dynamic Systems is a crucial topic with numerous applications, such as in time series forecasting and physical simulations. To accurately model these systems, it is necessary to learn the temporal and relational dimensions jointly. However, previous methods have struggled to learn the temporal dimension explicitly because they often overlook the physical properties of the system. Furthermore, they often ignore the distance information in the relational dimensions. To address these limitations, we propose a Dynamic Data Driven Application Systems (DDDAS) approach called Interacting System Ordinary Differential Equations (ISODE). Our approach leverages the latent space of Neural ODEs to model the temporal dimensions explicitly and incorporates the distance information in the relational dimensions. Moreover, we demonstrate how our approach can dynamically update an agent's trajectory when obstacles are introduced, without requiring retraining. Our experimental studies reveal that our ISODE DDDAS approach outperforms existing methods in prediction accuracy. We also illustrate that our approach can dynamically adapt to changes in the environment by showing our agent can dynamically avoid obstacles. Overall, our approach provides a promising solution to modeling Interacting Dynamic Systems that can capture the temporal and relational dimensions accurately.

Keywords: Interacting Dynamic Systems · Ordinary Differential Equations · ISODE · DDDAS

1 Introduction

Interacting Dynamic Systems are prevalent in various applications, including autonomous driving, traffic flow forecasting, and swarm formation movements. However, modeling these systems is challenging due to nonlinear and complex agent interactions in dynamic environments. To accurately model these systems, it is essential to learn the temporal and relational dimensions jointly. Previous data-driven approaches have achieved promising results in learning these two dimensions. However, there are still limitations because they fail to learn the temporal and relational dimensions explicitly. For instance, they often use graph

E. Blasch et al. (Eds.): DDDAS 2022, LNCS 13984, pp. 217–226, 2024.
https://doi.org/10.1007/978-3-031-52670-1_21

neural networks [18], social pooling [1], or Transformers [30] to model the relational dimension. These methods often overlook the distance information among agents, resulting in reduced forecasting accuracy and collisions if sudden obstacles occur. Moreover, previous methods use RNNs or Transformers [8,24,30] to learn the temporal dimension, which ignores the continuous property of Interacting Dynamic Systems. They also fail to consider the physical property and cannot explain how agent interactions drive the change of agent states. These limitations hinder the methods' ability to dynamically adapt to changes in the environment.

To address the limitations mentioned above, we observe that an agent's next state depends on the current state and its interactions with other agents. For example, when driving a vehicle, if another vehicle approaches closely with a high velocity, we tend to change our vehicle's velocity and direction quickly. We model the relational dimension as agent interactions and the temporal dimension as temporal dynamics. We propose a novel DDDAS approach, called Interacting System Ordinary Differential Equations (ISODE), which explicitly learns the agent interactions and temporal dynamics. Our approach uses Ordinary Differential Equations (ODEs) to learn the continuous temporal dynamics and incorporates distance information in agent interactions.

Our proposed ISODE is an encoder-decoder architecture based on Variational Autoencoders (VAE). The encoder module transforms the inputted previous trajectories into latent vectors, while the decoder module recovers the inputted previous trajectories and predicts future trajectories based on the latent vectors. In the latent space, We use neural ODEs to model the continuous temporal dynamics for each agent, which explicitly shows how the state of each agent changes based on the current state and agent interactions. In the ODE, the derivative of the agent state depends on agent interactions and the current state. Additionally, we consider the distance information among agents to model agent interactions, where shorter distances result in more significant agent interactions. By using neural ODEs to learn temporal dynamics and incorporating distance information into agent interactions, our proposed ISODE can improve prediction accuracy and enable agents to deal with unforeseen obstacles without retraining.

Our paper's main contributions are as follows:

- Our ISODE framework utilizes Neural ODEs to learn an agent's continuous temporal dynamics in the latent space, explicitly explaining how an agent's state changes depending on its current state and other agents. Our approach considers the continuous property of the system, leading to a better interpretation and understanding of the dynamics.
- Our proposed ISODE incorporates distance information into agent interactions. This approach enhances the interpretation of agent interactions.
- We demonstrate that ISODE can effectively adapt an agent's motion without retraining when sudden obstacles appear. Our approach is based on DDDAS, enabling us to model dynamic environments where obstacles can appear and disappear dynamically.

– We conducted extensive experiments on various traffic datasets to evaluate the effectiveness of our proposed ISODE framework. The results demonstrate that our model achieves superior accuracy in trajectory forecasting compared to state-of-the-art approaches. Moreover, our approach significantly reduces collision rates when introducing sudden obstacles, making it highly effective in dynamic environments.

2   Related Work

2.1   Neural Ordinary Differential Equations

Chen et al. introduced Neural ODE [3], a continuous-time neural network for modeling sequences. Rubanova et al. proposed Latent ODE [23] to model irregularly-sampled time series. Yildiz et al. proposed ODE2VAE [29] for modeling complex non-interacting systems using the latent space to model a second-order ODE. Dupont et al. proposed Augmented Neural ODE [5] to increase the model's capacity by preserving the input space's topology. Brouwer et al. proposed GRU-ODE-Bayes [4] to process sporadic observations in continuous time series. Other works have adapted Neural ODEs for density estimation [6,13,14,20,22,28], trajectory modeling or planning [19,25,27], theoretical analysis [9,10], and continuous-time video generation [21].

Building on these approaches, we propose ISODE, a novel framework based on a Latent ODE for modeling inputting trajectories and underlying temporal dynamics, which can also infer future trajectories. Different from the Latent ODE, our approach models agent interactions with other agents and obstacles in the ODE, which reveal how the state of each agent change based on the current state and other agents.

2.2   Interacting Systems

Various deep learning models have been proposed for trajectory forecasting in interacting dynamic systems with Neural Ordinary Differential Equations. Alahi et al. introduced Social LSTM [1], which applies social pooling to the hidden state of LSTM to model social interactions among agents. Gupta et al. extended Social LSTM by proposing Social GAN [12], which uses global social pooling and GANs to generate trajectories consistent with the input. Sun et al. proposed Graph-VRNN [26], a graph network and RNN-based approach that models the relational dimension. To represent underlying agent interactions in the latent space, Kipf et al. used graphs [15], which were later extended by Graber to dynamic neural relational inference [8]. Salzmann et al. introduced Trajectron++ [24], a graph-structured model with LSTM that accounts for environmental information. Li et al. proposed EvolveGraph [18], which uses dynamic relational reasoning through a latent interaction graph to forecast trajectories. Yuan et al. introduced AgentFormer [30], a Transformer-based approach that jointly models social and temporal dimensions. Gu et al. proposed DenseTNT [11], based on

VectorNet [7], which encodes all agents as vectors in a graph. These approaches have demonstrated state-of-the-art performance in trajectory forecasting, especially in modeling the relational dimension.

Our proposed ISODE differs from previous methods by utilizing neural ODEs to learn temporal dynamics in latent space, which provides greater interpretability for continuous-time sequences and dynamic adaptive interactions among agents. Additionally, we model agent interactions by incorporating distance information among agents, leveraging the intuitive understanding that shorter distances indicate stronger interactions. Furthermore, ISODE is based on DDDAS, enabling us to model dynamic environments where new obstacles and other agents can dynamically appear and disappear.

3   Methodology

3.1   Problem Formulation and Notation

Our objective is to predict the future states of multiple agents simultaneously, denoted as $\mathbf{X}_{T_h+1:T_h+T_f} = \mathbf{x}^i_{T_h+1}, \mathbf{x}^i_{T_h+2}, ..., \mathbf{x}^i_{T_h+T_f}, i = 1, ..., N$, based on their past states, represented as $\mathbf{X}_{0:T_h} = \mathbf{x}^i_0, \mathbf{x}^i_1, ..., \mathbf{x}^i_{T_h}, i = 1, ..., N$. Here, N represents the number of agents, while T_h and T_f represent the number of states in the current and future states, respectively. The state of all agents at time t is denoted as X_t, and the state of agent i at time t, which includes its position and velocity, is represented by $\mathbf{x}^i_t$. In our application, we use two-dimensional vectors to represent the position and velocity.

3.2   ISODE

The proposed ISODE method is an Encoder-Decoder architecture based on Variational Autoencoders (VAE). It utilizes the Neural ODE, which is a continuous-time, latent-variable approach, to model the state sequence based on agent interactions and the current state. Figure 1 illustrates the two main components of the model: the Encoder and the Decoder.

The encoder component of ISODE encodes the previous trajectories of each agent $\mathbf{X}^i_{0:T_h}$ into a latent space. For each agent, the output of the encoder is the initial state in the latent space, which serves as the initial value for the ordinary differential equation. Unlike Latent ODEs, our encoder module connects a spatial attention module and a temporal attention module in parallel, and concatenates their output. This enables the model to learn both spatial and temporal features in the inputted previous trajectories.

The proposed ISODE's decoder component solves ODEs to generate latent trajectories, which are then decoded back into the real space, including positions and velocities. The encoder generates a latent vector, and an ODE is constructed based on agent interactions and the current state of each agent in the latent space. The model explicitly incorporates distance information to model agent interactions. Given the initial state from the Encoder in the latent space, the

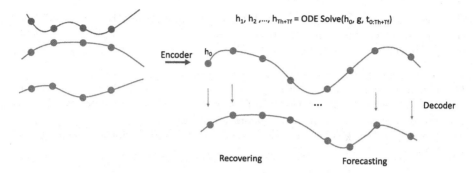

Fig. 1. The architecture of our proposed ISODE approach, which comprises two main components: an encoder and a decoder. Specifically, the encoder takes the previous trajectories as input and projects it into the latent space. In contrast, the decoder uses an ODE solver to generate the latent trajectories in the latent space, which are then decoded back into real space trajectories.

ODE is solved to generate latent trajectories. These latent trajectories are then recovered back to realistic trajectories in real space.

For agent i, our model can be represented as:

$$\mu_{h^i}, \sigma_{h^i} = g_{enc}(x^i_{0:T_h}, x^j_{0:T_h}), \; j \neq i \tag{1}$$

$$\mathbf{h}^i_0 \sim N(\mu_{h^i}, \sigma_{h^i}), \tag{2}$$

$$\mathbf{h}^i_0, \mathbf{h}^i_1, ..., \mathbf{h}^i_{T_h+T_f} = \text{ODESolve}(\mathbf{h}^i_0, g_\theta, t_{0:T_h+T_f}) \tag{3}$$

$$\text{each } \hat{x}^i_t \sim p(\hat{x}^i_t | \mathbf{h}^i_t), \tag{4}$$

where Eq. 1 represents the encoder, while Eqs. 2 to 4 denote the decoder of our proposed ISODE framework. To solve the differential equation $\frac{dh}{dt} = g_\theta$ with an initial value of $\mathbf{h}^i_0$, we use a numerical ODE solver called ODESolver, which can be Euler method or Runge-Kutta methods.

To explicitly model temporal dynamics during prediction, our proposed ISODE encodes the state sequence of each agent as initial values in the latent space and uses agent interactions and the current state to build the ODE. This enables us to model how the state of each agent changes with the current state and other agents. In the latent space, we can build the ODE as follows:

$$\frac{dh_i(t)}{dt} = g_\theta(h_i(t), h_j(t)) \tag{5}$$

$$= \sum_{j \neq i} \frac{1}{||h_i - h_j||} f_1(h_i(t), h_j(t)) + f_2(h_i(t)), \tag{6}$$

where $h_i(t)$ denotes the latent vector of agent i in time t. Besides, $||h_i - h_j||$ denotes the distance between agent i and agent j in latent space, while $f_1(h_i(t), h_j(t))$ is the interaction intensity between two agents. Equation 6 means that the derivative of the state depends on agent interactions (the first term) and the current state (the second term). The agent interactions in ISODE consist of two components:

Interaction Intensity: To estimate how agent j affects the dynamics of agent i, we concatenate the latent vectors of both agents (h_t^i and h_t^j) and pass them through a fully connected neural network to obtain $f_1(h_i(t), h_j(t))$. The output of the neural network represents the strength of the interaction between agents i and j.

Distance: The distance component explicitly models the influence of the distance between two agents on each other. In the latent space, we represent this relationship as $\frac{1}{||h_i - h_j||}$, where i denotes the agent being modeled and j denotes other agents. The L2 distance $||h_i - h_j||$ between two agents in the latent space corresponds to the real agent distance in the real space. As agent j approaches agent i, the dynamics of agent i are affected, causing $\frac{dh_i(t)}{dt}$ to increase accordingly. A shorter distance between two agents in the latent space results in a larger interaction effect.

In the above ODE, we incorporate the distance information in the agent interactions and explicitly model how the state of each agent changes with the current state and interactions with other agents. By using an ODE solver and the initial values given by the encoder, we can estimate $h_{t_1}, h_{t_2}, .., h_{t_n}$. Finally, we use fully connected neural networks to decode the latent vectors into realistic states $x_{t_1}, x_{t_2}, .., x_{t_n}$ for each agent. This allows us to dynamically estimate the future trajectories of multiple agents based on their previous trajectories and incorporate the interactions among them, including the influence of distance.

4   Experiments

To evaluate the performance of our proposed ISODE model, we conducted experiments on three traffic datasets, namely inD [2], rounD [17], and highD [16]. We measured the accuracy of trajectory forecasting and evaluated the model's ability to adapt to new scenarios by finetuning it on inputting previous trajectories. Additionally, we assessed the collision rate when introducing sudden obstacles to demonstrate the effectiveness of our approach in dynamic environments.

Forecasting Accuracy. We utilized ADE (Average Displacement Error) as the evaluation metric to assess the performance of our model. ADE quantifies the mean square error between the ground truth and the prediction, and it is calculated as

$$ADE = \frac{1}{T} \sum_{t=1}^{T} ||x_t^i - \hat{x}_t^i||, \tag{7}$$

where x_t^i represents the ground truth and $\hat{x}_t^i$ denotes the predicted trajectory.

We compared our ISODE with four state-of-the-art methods: Social LSTM [1], Social GAN [12], DenseTNT [11], and AgentFormer [30]. The comparison results in Table 1 demonstrate that ISODE outperforms several state-of-the-art methods, including Social LSTM, Social GAN, and DenseTNT. To further evaluate the effectiveness of ISODE in dealing with more complex trajectories, we classified trajectories into two categories, "curve" and "straight", and found that ISODE consistently achieved the best performance on curved trajectories. These results highlight the superior performance of ISODE in modeling and forecasting complex trajectories.

Table 1. Performance of ISODE and other methods on inD, rounD and highD traffic datasets. The best-performing method is highlighted in bold.

Method	length	inD		highD		rounD	
		straight	curve	straight	curve	straight	curve
Social LSTM	4 s	0.7973	3.1463	0.9525	3.5364	0.7268	2.6473
Social GAN		0.7861	3.1583	0.8367	3.4637	0.7483	2.6940
DenseTNT		0.7794	3.1578	0.7431	3.1778	0.6543	2.4764
AgentFomer		**0.7604**	3.1483	**0.6814**	3.1527	**0.5924**	2.4748
ISODE		0.7742	**3.1428**	0.6930	**3.1523**	0.6073	**2.4742**

Finetuning Our Model in Historical Trajectories. To further enhance the accuracy of our model, we finetuned our pre-trained model on the historical trajectories in the test dataset. During the test period, we used the historical trajectory X_{T_h} to estimate the future trajectory X_{T_f} without ground truth data, which allowed us to improve the model's ability to recover input trajectories accurately from the latent space in the test dataset. The finetuned model and ODE solver were then used to predict the future trajectory. As shown in Table 2, using the test data without ground truth significantly improved the prediction accuracy.

Table 2. Performance of finetuning on test data for the inD, rounD, and highD traffic datasets with a forecasting length of 4 s.

Method	inD		highD		rounD	
	straight	curve	straight	curve	straight	curve
ISODE	0.7742	3.1428	0.6930	3.1523	0.6073	2.4742
ISODE(finetuned)	0.7626	3.1409	0.6852	3.1499	0.5971	2.4720

Sudden Obstacle. We conducted experiments to evaluate the performance of our proposed ISODE approach in dynamic environments. Specifically, we examined the ability of the agent to adjust its trajectory when a sudden obstacle appears in its path, which is not seen during the training phase. We introduced static or moving agents as obstacles in the original trajectory to model dynamic environments. To assess the collision avoidance performance of ISODE and other methods, we established a collision criterion where a collision occurs when the distance between an agent and an obstacle is less than 0.5 m. The collision rates for all methods are presented in Table 3. The results of our experiments indicate that ISODE achieved the lowest collision rate when avoiding both static and moving obstacles, showcasing the effectiveness of our approach in modifying an agent's trajectory to adapt to dynamic environments.

Table 3. The collision rates of different methods by introducing sudden obstacles in the trajectory. The bold values indicate the best performance among the compared methods.

Method	Social LSTM	Social GAN	DenseTNT	AgentFormer	ISODE
Static obstacle	28.6%	29.6%	22.8%	28.4%	**11.4%**
Moving obstacle	32.4%	35.2%	32.6%	33.0%	**17.2%**

5   Conclusion

In this paper, we present ISODE, a DDDAS approach that explicitly models and learns both the relational and temporal dimensions, which correspond to agent interactions and the underlying temporal dynamics, respectively. ISODE incorporates distance information to represent agent interactions and learns the dynamics using a Neural ODE in latent space. The ODE is based on the current state and agent interactions, providing an explanation of how each agent's state changes with these factors. Extensive experiments using traffic datasets demonstrate the superior forecasting accuracy of ISODE in complex environments compared to previous methods. Moreover, ISODE allows for the dynamic insertion of obstacles and agents during an agent's trajectory without requiring retraining, while still dynamically adapting to changes in the environment. Our approach achieves a lower collision rate when sudden dynamic obstacles appear on or near an agent's trajectory.

Acknowledgements. Research partially funded by research grants to Metaxas from NSF: 1951890, 2003874, 1703883, 1763523 and ARO MURI SCAN.

References

1. Alahi, A., Goel, K., Ramanathan, V., Robicquet, A., Fei-Fei, L., Savarese, S.: Social LSTM: Human trajectory prediction in crowded spaces. In: Proceedings of the IEEE Conference On Computer Vision And Pattern Recognition, pp. 961–971 (2016)
2. Bock, J., Krajewski, R., Moers, T., Runde, S., Vater, L., Eckstein, L.: The IND dataset: A drone dataset of naturalistic road user trajectories at german intersections. In: 2020 IEEE Intelligent Vehicles Symposium (IV), pp. 1929–1934 (2020). https://doi.org/10.1109/IV47402.2020.9304839
3. Chen, R.T., Rubanova, Y., Bettencourt, J., Duvenaud, D.K.: Neural ordinary differential equations. In: Advances in Neural Information Processing Systems, vol. 31 (2018)
4. De Brouwer, E., Simm, J., Arany, A., Moreau, Y.: Gru-ode-bayes: continuous modeling of sporadically-observed time series. In: Advances in Neural Information Processing Systems 32 (2019)
5. Dupont, E., Doucet, A., Teh, Y.W.: Augmented neural odes. In: Advances in Neural Information Processing Systems, vol. 32 (2019)
6. Durkan, C., Bekasov, A., Murray, I., Papamakarios, G.: Neural spline flows. Advances in Neural Information Processing Systems, vol. 32 (2019)
7. Gao, J., Sun, C., Zhao, H., Shen, Y., Anguelov, D., Li, C., Schmid, C.: Vectornet: Encoding hd maps and agent dynamics from vectorized representation. In: Proceedings of the IEEE/CVF Conference on Computer Vision and Pattern Recognition, pp. 11525–11533 (2020)
8. Graber, C., Schwing, A.: Dynamic neural relational inference for forecasting trajectories. In: Proceedings of the IEEE/CVF Conference on Computer Vision and Pattern Recognition Workshops, pp. 1018–1019 (2020)
9. Gruenbacher, S., et al.: Gotube: Scalable stochastic verification of continuous-depth models. arXiv preprint arXiv:2107.08467 (2021)
10. Grunbacher, S., Hasani, R., Lechner, M., Cyranka, J., Smolka, S.A., Grosu, R.: On the verification of neural odes with stochastic guarantees. In: Proceedings of the AAAI Conference on Artificial Intelligence. vol. 35, pp. 11525–11535 (2021)
11. Gu, J., Sun, C., Zhao, H.: Densetnt: End-to-end trajectory prediction from dense goal sets. In: Proceedings of the IEEE/CVF International Conference on Computer Vision, pp. 15303–15312 (2021)
12. Gupta, A., Johnson, J., Fei-Fei, L., Savarese, S., Alahi, A.: Social gan: socially acceptable trajectories with generative adversarial networks. In: Proceedings of the IEEE Conference on Computer Vision and Pattern Recognition, pp. 2255–2264 (2018)
13. Hasani, R., Lechner, M., Amini, A., Rus, D., Grosu, R.: Liquid time-constant networks. arXiv preprint arXiv:2006.04439 (2020)
14. Jia, J., Benson, A.R.: Neural jump stochastic differential equations. In: Advances in Neural Information Processing Systems, vol. 2 (2019)
15. Kipf, T., Fetaya, E., Wang, K.C., Welling, M., Zemel, R.: Neural relational inference for interacting systems. In: International Conference on Machine Learning, pp. 2688–2697. PMLR (2018)
16. Krajewski, R., Bock, J., Kloeker, L., Eckstein, L.: The highd dataset: A drone dataset of naturalistic vehicle trajectories on German highways for validation of highly automated driving systems. In: 2018 21st International Conference on Intelligent Transportation Systems (ITSC), pp. 2118–2125 (2018). https://doi.org/10.1109/ITSC.2018.8569552

17. Krajewski, R., Moers, T., Bock, J., Vater, L., Eckstein, L.: The round dataset: a drone dataset of road user trajectories at roundabouts in Germany. In: 2020 IEEE 23rd International Conference on Intelligent Transportation Systems (ITSC), pp. 1–6 (2020). https://doi.org/10.1109/ITSC45102.2020.9294728

18. Li, J., Yang, F., Tomizuka, M., Choi, C.: Evolvegraph: multi-agent trajectory prediction with dynamic relational reasoning. Adv. Neural. Inf. Process. Syst. **33**, 19783–19794 (2020)

19. Liang, Y., Ouyang, K., Yan, H., Wang, Y., Tong, Z., Zimmermann, R.: Modeling trajectories with neural ordinary differential equations. In: IJCAI, pp. 1498–1504 (2021)

20. Liebenwein, L., Hasani, R., Amini, A., Rus, D.: Sparse flows: pruning continuous-depth models. In: Advances in Neural Information Processing Systems, vol. 34 (2021)

21. Park, S., Kim, K., Lee, J., Choo, J., Lee, J., Kim, S., Choi, E.: Vid-ode: continuous-time video generation with neural ordinary differential equation. In: Proceedings of the AAAI Conference on Artificial Intelligence. vol. 35, pp. 2412–2422 (2021)

22. Quaglino, A., Gallieri, M., Masci, J., Koutník, J.: Snode: spectral discretization of neural odes for system identification. arXiv preprint arXiv:1906.07038 (2019)

23. Rubanova, Y., Chen, R.T., Duvenaud, D.K.: Latent ordinary differential equations for irregularly-sampled time series. In: Advances in Neural Information Processing Systems, vol. 32 (2019)

24. Salzmann, T., Ivanovic, B., Chakravarty, P., Pavone, M.: Trajectron++: dynamically-feasible trajectory forecasting with heterogeneous data. In: Vedaldi, A., Bischof, H., Brox, T., Frahm, J.-M. (eds.) Computer Vision – ECCV 2020: 16th European Conference, Glasgow, UK, August 23–28, 2020, Proceedings, Part XVIII, pp. 683–700. Springer International Publishing, Cham (2020). https://doi.org/10.1007/978-3-030-58523-5_40

25. Shi, R., Morris, Q.: Segmenting hybrid trajectories using latent odes. In: International Conference on Machine Learning, pp. 9569–9579. PMLR (2021)

26. Sun, C., Karlsson, P., Wu, J., Tenenbaum, J.B., Murphy, K.: Stochastic prediction of multi-agent interactions from partial observations. arXiv preprint arXiv:1902.09641 (2019)

27. Vorbach, C., Hasani, R., Amini, A., Lechner, M., Rus, D.: Causal navigation by continuous-time neural networks. In: Advances in Neural Information Processing Systems, vol. 34 (2021)

28. Yan, H., Du, J., Tan, V.Y., Feng, J.: On robustness of neural ordinary differential equations. arXiv preprint arXiv:1910.05513 (2019)

29. Yildiz, C., Heinonen, M., Lahdesmaki, H.: Ode2vae: Deep generative second order odes with bayesian neural networks. In: Advances in Neural Information Processing Systems, vol. 32 (2019)

30. Yuan, Y., Weng, X., Ou, Y., Kitani, K.M.: Agentformer: agent-aware transformers for socio-temporal multi-agent forecasting. In: Proceedings of the IEEE/CVF International Conference on Computer Vision, pp. 9813–9823 (2021)

Relational Active Feature Elicitation
for DDDAS

Nandini Ramanan[1], Phillip Odom[2], Erik Blasch[3], Kristian Kersting[4],
and Sriraam Natarajan[1]([envelope])

[1] University of Texas, Dallas, USA
sriraam.natarajan@utdallas.edu
[2] Georgia Institute of Technology, Atlanta, USA
[3] Air Force Research Lab, Dayton, USA
[4] TU Darmstadt, Darmstadt, Germany

Abstract. Dynamic Data Driven Applications Systems (DDDAS) utilize data augmentation for system performance. To enhance DDDAS systems with domain experts, there is a need for interactive and explainable active feature elicitation in relational domains in which a small subset of data is fully observed while the rest of the data is minimally observed. The goal is to identify the most informative set of entities for whom acquiring the relations would yield a more robust model. Assuming the presence of a human expert who can interactively score the relations, there is a need for an explainable model designed using the Feature Acquisition via Interaction in Relational domains (FAIR) algorithm. FAIR employs a relational tree-based distance metric to identify the most diverse set of relational examples (entities) to obtain more relational feature information for user refinement. The model that is learned iteratively is usable, interpretable, and explainable.

1 Introduction

Feature-value acquisition is a necessary step in successful deployment of DDDAS paradigms. It has long been pursued in standard domains (i.e., the ones that can be easily described by a flat vector representation) from the perspective of active learning [5,15]. Such approaches, while successful, cannot be directly extended to relational domains such as networks or hyper-graphs [1]. Acquisition in these domains require reasoning at different levels of abstraction – individual feature values, sub-groups of entities/relations and at the level of the set of all objects. Statistical Relational Learning (SRL) [4,6,12] naturally combines the power of relational representations such as first-order logic with the ability of statistical/probabilistic models to handle uncertainty. Specifically, actively acquiring features in relational data.

While previous research in relational data for active learning have mainly considered acquiring the labels in the context of link prediction or relation extraction [1,9,10], a different yet related task is that of identifying the best entities

E. Blasch et al. (Eds.): DDDAS 2022, LNCS 13984, pp. 227–232, 2024.
https://doi.org/10.1007/978-3-031-52670-1_22

(examples) to acquire more feature information on. This is quite natural in settings such as networks where a subset of the network is fully observed while in the other part of the network, only the entities are known with partial relations. This poses several challenges in relational settings – First, a need for *active selection strategy* of the elicitable (relational) examples on whom acquiring additional features will improve the classifier performance. Acquiring diverse set of examples will help improve generalization. The second challenge is that of computing distances between relational examples. While in Natarajan et al. [11], it was easily computed using a divergence metric such as KL-divergence [8], in relational models, the distance calculation is not straightforward. Thus the notion of first-order tree-based distance due to Khot et al. [7] for computing the most diverse set of examples is a better strategy. Third challenge is that of *feature-subspace selection* for each example, to avoid uninformative/redundant features. Feature sub-space selection in relational domains is intractable due to the number of groundings. Inspired by the ideas of subjective understanding in explanation-based learning literature [13,14], it is necessary to perform inference on the current model based on the chosen examples from the tree-based distances. This inference step **explains the most important aspects of the current model to the human learner**. The base learner is a *relational gradient-boosted tree* that is combined to a single tree at each explanation step. The algorithm picks the most uncertain part of this tree for the current example and elicits the additional features from the human. The final challenge in relational domains is identifying the appropriate level of abstraction – should the query be over an instance (an individual co-author), a sub-group (area of research) or the population (over all the journals)? We note that the asbtraction challenge is automatically handled by our **explainable, interpretable and elicitable** base learner which is a (combined) relational decision-tree. Identifying the appropriate paths will automatically present the set of features to the human expert. The presence of human expert makes our work distinctly different from the previous active learning work in relational data. It reinforces the need of **explainable** model and we precisely addresses this issue.

To summarize we make the following contributions: (1) Adressing the problem of training-time feature acquisition in the presence of a human-expert. (2) Developing the first **F**eature **A**cquisition via **I**nteraction in **R**elational domains (*FAIR*) algorithm that actively solicits features on the most informative examples. (3) Adapting a relational tree-based distance metric for identifying the most diverse set of examples.

2   Interactive Feature Elicitation for DDDAS

Traditional ML methods typically assume that all features are provided for each example. However, not all features require the same amount of effort to obtain. For example, medical applications depend on multiple modalities of data such as imagery, text, family history, and demographic or epigenetic information. While family history may be collected from patients, diagnostic procedures, which produce different data modalities, vary in terms of their cost and invasiveness. Therefore, selecting the correct procedure, and consequently the most informative feature subspace can be critical in improving decision-making while minimizing the cost of acquiring features.

Identifying the best subset of features to elicit from the experts is intractable, especially in relational domains. FAIR is an approximate method that first selects a set of examples about whom to query and then identifies relevant features for those examples. Formally, we assume that a given dataset, $\mathbf{D} = \mathbf{D}^b \cup \mathbf{D}^o$, is composed of examples with only baseline features $(\mathbf{D}^b)$ and examples which have additional observed features that have been previously queried $(\mathbf{D}^o)$. For each previously elicited example, $\langle (\mathbf{x}_i^b, \mathbf{x}_i^o, y_i) \rangle \in \mathbf{D}^o$, $\mathbf{x}_i^b$ represents the base features while $\mathbf{x}_i^{o_i} \subseteq \mathbf{x}_i^e$ represents any features from the elicitable set that have been previously acquired for example i, whose label is y_i. We denote the set of observed features for all examples as $\mathbf{x}^o$. Note that each example may have a different set of observed features. Interactive feature elicitation iteratively expands $\mathbf{x}^o$ in order to improve the model. Thus, the learning problem is as follows:

Given: A relational data base $\mathbf{D}$, Relational database schema $\mathcal{R}$, Query budget B and The expert, E.

To Do: Identify the most useful set of examples $\mathbf{y}_b \in \mathbf{D}^b$ for which to obtain more attribute or relations $\mathbf{x}_i^{o_i} \subseteq \mathbf{x}_i^e$ in order to improve classifier performance.

Our proposed approach (*FAIR*), interatively identifies a representative subset of query-able examples (1), learns an explanation model to identify subset of features from elicitable set (2) and elicits the features from experts (3). Intuitively, *FAIR* aims to acquire additional features for a diverse set of examples which could not be classified using the available features. By prioritizing examples/features, *FAIR* efficiently utilizes the resources for feature elicitation. While previous work focused on propositional distance measures [11], we faithfully capture the underlying relational data via a tree-based distance measure to capture the semantic similarity which uses the path similarity in relational decision trees [7].

The two key components of *FAIR* are: 1) **Example Subset Selection**, which relies on a relational distance measure over the full data, $\mathbf{D}$, to pick the most diverse and informative examples and 2) **Feature Sub-Space Selection**, which leverages a relational base model on the observed data $\mathbf{D}_o$, to generate robust explanation which can be used to do informed feature subspace elicitation for the queryable examples. The advantage of using a relational representation is that it succinctly capture probabilistic (noisy) dependencies among the attributes of different objects, leading to a compact representation of learned models. SRL is ideal for this tasks due to two key reasons: their ability to handle *generalized, noisy* knowledge and data and their capability to produce *explainable and interpretable* hypotheses.

It can be strongly argued that *FAIR* has a strong potential for impact in DDDAS paradigms. For instance, consider the system designed by Blasch et al. [2]. The physical systems, sensors and the simulators can easily be modeled using a hybrid relational model. The advantage of such a model is that

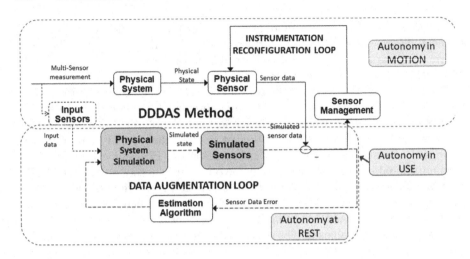

Fig. 1. Typical DDDAS paradigm (as shown by Blasch et al. [2])

it can allow for information fusion from multiple data sources – raw sensor data, database schemas describing domain knowledge, natural language text and images, to name a few. All the different data sources can be easily encapsulated using a rich relational description.

Once such a rich relational structure can be obtained, it is fairly straightforward to incorporate the *FAIR* framework for this task. The sensor management part can be replaced by a higher-order information management module that can handle multiple modalities of the data. The features can then be aquired effectively byt the *FAIR* framework by interacting with the human experts. The relational system then can serve to enable efficient learning of SRL models as presented in Fig. 2. The use case of this system could be in the context of a conversational AI agent that communicates actively with the human expert to identify potentially interesting scenarios based on both sensor data and natural language interactions.

2.1   Preliminary Results

FAIR was compared with a random algorithm for eliciting features on a **NELL:Sport** data set [3]. This data set consists of relations about players and teams generated by the Never Ending Language Learner (NELL). The task is to predict the relation **TeamPlaysSport** i.e., whether a team plays a particular sport. The data has fixed train and test set and the results were averaged over 3 runs.

As can be seen in Table 1, with only 10 query instances *FAIR* is able to achieve considerably higher performance in all the metrics – area under ROC and PR curves, F1 scores and the all important recall in domains with class imbalance. While increasing the number of query instances to 100 reduces the difference between the methods, it is clear that *FAIR* is superior. This is particularly

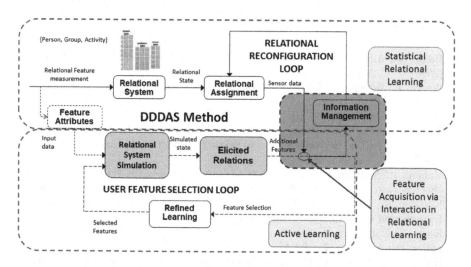

Fig. 2. *FAIR* integrated into the DDDAS framework. The proposed system alters the one due to Blasch et al. (Fig. 1). The key change is that physical system is absorbed by a relational simulator that allows for rich multimodal inputs and the *FAIR*framework is then used to obtain the appropriate labels.

Table 1. Comparison of *FAIR* with a random selection strategy on NELL data set. The results are presented with only 10 query instances and with 100 instances for both the methods.

Metric	*FAIR* 10 instances	Random 10 instances	*FAIR* 100 instances	Random 100 instances
AUC ROC	0.93	0.87	0.96	0.94
AUC PR	0.8	0.75	0.9	0.87
F1	0.85	0.75	0.86	0.84
Recall	0.93	0.74	0.97	0.9

significant in lower query instances since the goal is to minimize human's (domain expert's) effort in providing appropriate features.

3   Conclusion

We considered the problem of acquiring features in relational domains through interaction with human expert. To this effect we proposed the *FAIR* algorithm that is both interactive and explainable. Our distance metric uses tree-based distances that are interpretable as well. Rigorous evaluations on larger domains is the immediate next step. Extending the algorithm to dynamic domains where the feature sets can change consistently can result in interesting insights and potentially create interesting scenarios for building *conversational AI agents*. Finally, performing large scale human evaluation for explainability and interpretability remains interesting direction of future research.

References

1. Bilgic, M., Mihalkova, L., Getoor, L.: Active learning for networked data. In: ICML (2010)
2. Blasch, E., Ravela, S., Aved, A.: Handbook of Dynamic Data Driven Application Systems. Springer (2018)
3. Carlson, A., Betteridge, J., Kisiel, B., Settles, B., Hruschka, E., Jr.: and T. Mitchell. Toward an architecture for never-ending language learning, In AAAI (2010)
4. Getoor, L., Taskar, B.: Introduction to Statistical Relational Learning. MIT Press, (2007)
5. Kanani, P., Melville, P.: Prediction-time active feature-value acquisition for cost-effective customer targeting. Workshop on Cost Sensitive Learning at NIPS (2008)
6. Kersting, K., De Raedt, L.: Bayesian logic programming: Theory and tool. In: An Introduction to Statistical Relational Learning (2007)
7. Khot, T., Natarajan, S., Shavlik, J.W.: Relational one-class classification: A non-parametric approach. In: AAAI (2014)
8. Kullback, S., Leibler, R.A.: On information and sufficiency. The annals of mathematical statistics (1951)
9. Kuwadekar, A., Neville, J.: Relational active learning for joint collective classification models. In: ICML (2011)
10. Macskassy, S.A.: Using graph-based metrics with empirical risk minimization to speed up active learning on networked data. In: KDD (2009)
11. Natarajan, S., Das, S., Ramanan, N., Kunapuli, G., Radivojac, P.: On whom should i perform this lab test next? an active feature elicitation approach. In: IJCAI (2018)
12. Raedt, L., Kersting, K., Natarajan, S., Poole, D.: Statistical relational artificial intelligence: Logic, probability, and computation. Morgan Claypool (2016)
13. Shavlik, J.W., Towell, C.G.: Combining explanation-based learning and artificial neural networks. In: Proceedings of the Sixth International Workshop on Machine Learning. Elsevier (1989)
14. Tadepalli, P.: A formalization of explanation-based macro-operator learning. In: IJCAI (1991)
15. Thahir, M., Sharma, T., Ganapathiraju, M.K.: An efficient heuristic method for active feature acquisition and its application to protein-protein interaction prediction. In: MC Proc (2012)

Explainable Human-in-the-Loop Dynamic Data-Driven Digital Twins

Nan Zhang[1,2], Rami Bahsoon[2], Nikos Tziritas[4],
and Georgios Theodoropoulos[1,3(✉)]

[1] Department of Computer Science and Engineering, Southern University of Science and Technology (SUSTech), Shenzhen, China
theogeorgios@gmail.com
[2] School of Computer Science, University of Birmingham, Birmingham, UK
[3] Research Institute for Trustworthy Autonomous Systems, Shenzhen, China
[4] Department of Informatics and Telecommunications, University of Thessaly, Volos, Greece

Abstract. Digital Twins (DT) are essentially dynamic data-driven models that serve as real-time symbiotic "virtual replicas" of real-world systems. DT can leverage fundamentals of Dynamic Data-Driven Applications Systems (DDDAS) bidirectional symbiotic sensing feedback loops for its continuous updates. Sensing loops can consequently steer measurement, analysis and reconfiguration aimed at more accurate modelling and analysis in DT. The reconfiguration decisions can be autonomous or interactive, keeping human-in-the-loop. The trustworthiness of these decisions can be hindered by inadequate explainability of the rationale, and utility gained in implementing the decision for the given situation among alternatives. Additionally, different decision-making algorithms and models have varying complexity, quality and can result in different utility gained for the model. The inadequacy of explainability can limit the extent to which humans can evaluate the decisions, often leading to updates which are unfit for the given situation, erroneous, compromising the overall accuracy of the model. The novel contribution of this paper is an approach to harnessing explainability in human-in-the-loop DDDAS and DT systems, leveraging bidirectional symbiotic sensing feedback. The approach utilises interpretable machine learning and modelling to explainability, and considers trade-off analysis of utility gained. We use examples from smart warehousing to demonstrate the approach.

Keywords: Digital Twins · DDDAS · Explainability · Human-in-the-loop

1 Introduction

Digital Twins (DT) are data-driven models that serve as real-time symbiotic "virtual replicas" of real-world systems. Digital Twin modelling can leverage

E. Blasch et al. (Eds.): DDDAS 2022, LNCS 13984, pp. 233–243, 2024.
https://doi.org/10.1007/978-3-031-52670-1_23

principles of Dynamic Data-Driven Applications Systems (DDDAS) in its realisation and refinement. DT feedback loops, for example, can be designed following the DDDAS paradigm to continuously steer the measurement, modelling and analysis [2]. As is the case in some DDDAS applications, some DT decisions cannot be fully autonomous and need to be enacted by human operators [8–10]. There can be decisions that are counter-intuitive, which may confuse the human operators. Humans need to understand the patterns or rationale behind the system's decision-making logic in order to trust and follow the instructions or take further decisions. Providing explanations to humans (e.g., operators, developers or users) can facilitate their understanding of the rationale behind the decisions and their rightfulness for a given context. Explanations can also provide assurance for humans to trust the autonomous adaptation by DT and its underlying DDDAS. Humans are also able to examine and inspect the operation of DDDAS. In case of any anomaly or fault identified from the explanations, humans are then able to intervene in the system to assist or improve its decision-making.

The paper makes the following novel contribution:

(1) It motivates the need for providing explainability in human-in-the-loop Digital Twins, leveraging DDDAS principles and its design for feedback loops; (2) As a prerequisite for explanation is the identification of areas within the system, where the provision of explanation is crucial for trustworthy service; the paper describes a data-centric strategy to identify areas within DDDAS and DT that require explanations; (3) It provides an enriched reference architecture model for DT, building on DDDAS and supported with explainability; the model extends the authors' previous work on digitally Twinning for intelligent systems and illustrates the utility gained for supporting explainability; (4) It discusses the trade-offs and cost of providing explainability.

The rest of the paper is structured as follows. Section 2 provides some background on explainability. A motivating example of smart warehouse with autonomous agents is presented in Sect. 3. Section 4 presents a reference architecture of explainable Digital Twin systems, and categorises the areas in a DDDAS-based Digital Twin system where explainability can be applied. Section 5 discusses the trade-off analysis of models in DTs. Section 6 concludes the paper and outlines future work.

2 Background and Related Work

In recent years, there has been increasing interest in the explainability and interpretability of Artificial Intelligence (AI), denoted as XAI (eXplainable Artificial Intelligence) [13]. The empirical success of deep learning models leads to the rise of opaque decision systems such as Deep Neural Networks, which are complex black-box models. Explanations supporting the output of a model are crucial in providing transparency of predictions for various AI stakeholders [1]. Explanation relates to how intelligent systems explain their decisions. An Explanation refers to causes, and is an answer to a *why-question* [13].

In Digital Twins, decisions can be a product of autonomous reasoning processes and require explanation. A decision model of DT may suggest locally sub-optimal decisions, but also has the risk of making mistakes. A typical example is DTs that employ Reinforcement Learning (RL), which is a sequential decision-making model. RL aims at deciding a sequence of actions that maximises long-term cumulative rewards. However, an action that is optimal in the long term may at present be sub-optimal or even temporally lead to undesired system states. The lack of explanation for sub-optimal actions can lead to confusion and difficulty for humans to distinguish whether the autonomous system has made a mistake or the "faulty" decision is intentional for the long-term benefit. With explanations, human operators and stakeholders can understand the rationale behind the decisions, and increase trust in the DDDAS and DT systems they utilise for decision support.

Preliminary works have been presented in discussing the explainable decision-making in dynamic data-driven Digital Twins. In [6] explanations for dynamically selecting a Digital Twin model from a library of physics-based reduced-order models are provided. Based on sensor measurements, a trained optimal classification tree is used to match a physics-based model that represents the structural damage state at the moment of measurement. The approach is demonstrated in a case study of unmanned aerial vehicle (UAV).

The work presented in this paper differs in that it involves the human factor in justifying explainability. Explanations can also be provided in different areas within the system architecture.

3   Motivating Example

In this section, a motivating example based on our previous work in [16,17] is presented. The example assumes a Digital Twin for an intelligent system. The intelligent system is composed of multiple collaborating computing nodes, each of which is controlled by an onboard intelligent agent. Each agent learns by accumulating knowledge of its environment and interaction with other agents. Each agent acts autonomously based on its knowledge, but its quality of decision-making may be restricted by the processing capability of the node or the power consumption if it is battery-powered. Therefore, the agents can act autonomously most of the time, but may need a Digital Twin of the entire system of agents for global optimisation of trade-off analysis between different goals and requirements. The resultant control decisions can be in the form of changing the decision algorithms/models of the agents [17], or modifying the parameters [12]. Smart warehouse is an application domain where the above-mentioned system can be applied. Multiple Autonomous Mobile Robots (AMRs) can travel within the warehouse for collaborative picking-up and delivery tasks. Each AMR is controlled by an agent that decides which task to take and which other AMRs to collaborate with. Each agent uses rules to decide its actions, but the optimality of rules can be contextual. Rules need to be adapted in order to reach optimal performance in different runtime contexts. However, it is risky for agents to self-adapt the rules, since each agent only has partial knowledge about a subset of the entire environment in space and time. The quality of self-adaptation is threatened by the lack of knowledge of other parts of the environment. A Digital Twin of the warehouse is able to provide more informed decisions on modifying the rules of agents, with a global view of the entire system and what-if predictions of various possible scenarios.

The Digital Twin utilises the DDDAS feedback loop to dynamically ensure both the state and knowledge of the intelligent system are accurately modelled [17]. Based on the model, predictions of various what-if scenarios can then further inform decision-making to optimise the behaviour of the system. Due to the nature of autonomy and high complexity within the system, decisions made by the DT may still be imperfect or even faulty. Human supervision and intervention are necessary, which requires explanations from the DT for humans to identify potentially undesired decisions.

4   A Reference Architecture for Explainable Digital Twins Leveraging DDDAS Principles

To address the issue of explainability in the motivating example, we present a novel reference architecture for an explainable Digital Twin system, leveraging DDDAS feedback loops in Fig. 1. The architecture extends on our previous work [17] to provide refinements of the architecture informed by DDDAS feedback loop design and enrichment with primitives for data-driven explainability. The architecture utilises interpretable machine learning and goal-oriented requirement modelling to provide explanations of decisions that are driven or influenced by data. Our architecture adopts the classical design of DDDAS computational feedback loops: (1) *sensor reconfiguration*, which guides the sensor to enhance the information content of the collected data, and (2) *data assimilation*, which uses the sensor data error to improve the accuracy of the simulation model [2]. Additionally, we refine the above two loops with human-in-the-loop inputs, as explainability is essentially a human-centric concern. Utilising the simulation model and analysis, the system behaviour can be dynamically controlled in a feedback manner.

In Fig. 1, the physical space at the bottom contains the system that is composed of intelligent agents, as in the motivating example. The Digital Twin contains a simulation model that replicates the physical space. The agents in the model are identical software programs to the agents in the physical space. There are three controllers that utilise feedback loops for autonomous decision-making: *Sensor Re-configurator*, *Model Updater*, and *Behaviour Optimiser*. Each controller incorporates interpretable machine learning and goal models to provide explanations to human operators. The three control components can also interact with each other. For instance, the *Behaviour Optimiser* may lack high fidelity data to justify and validate its decision, then it can notify *Sensor Re-configurator* asking for further enhanced sensory data.

4.1   Explainable Decisions for Human-in-the-Loop Digital Twins

Human factors can be present in the two computational feedback loops in DDDAS-based DT and can be captured in an additional physical human-in-

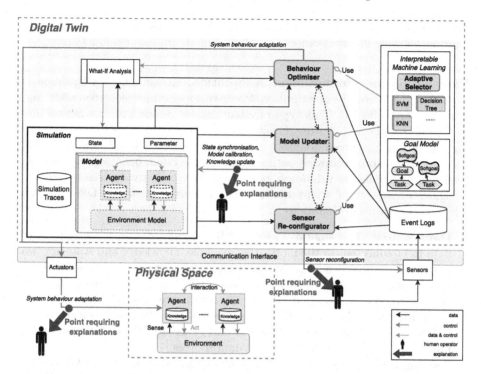

Fig. 1. A novel reference architecture for explainable human-in-the-loop DDDAS-inspired Digital Twins

the-loop feedback process, as illustrated in Fig. 1. Explanations are needed to enhance the trust of humans in the system and/or allow humans to find anomalies or faults in the autonomous process in order to improve the decisions made by the system. With explanations provided, humans can trace the input data and how the data are assimilated to generate the adaptation decision. Inadequate decisions can be identified and their cause can be located: because of faulty input data or during the reasoning process.

Explanations are closely linked to decisions that are steered by data, accumulated knowledge or computations by data assimilated into the system. Our architecture aims at providing explanations on how and why decisions or adaptations are made, linked to data: the data that support a decision need to be traced, since decision is a result of data input and how the data is then processed and/or manipulated. In the architecture, the adaptations of the three controllers represent three categories of decisions that can be made by the DDDAS-based Digital Twin: (1) *measurement adaptation*, (2) *model adaptation*, and (3) *system behaviour adaptation*. For each type of adaptation decision, explanations are required to increase the transparency of decisions for humans (shown as "Point requiring explanations" in Fig. 1). Goal models help in presenting the satisfaction of system requirements by their utility. Each controller uses interpretable machine learning for its decision-making to directly provide explanations.

Explanation for Measurement Adaptation. Measurement adaptation is a part of the sensor reconfiguration feedback loop, which controls what and how data are collected. This type of adaptation is decided by the *Sensor Reconfigurator* in the architecture. Explanations are essential to answer: what is the purpose of a measurement adaptation, and how is it beneficial? Examples of reasons can be to more accurately estimate the state, system behaviour adaptation requiring more data for validation, or cost-effective concerns such as energy.

Different aspects of adaptation on the measurement are as follows. Each adaptation requires explanations.

- **Data source**: The Digital Twin can dynamically decide what sources of data to be collected. In the motivating example, knowledge and state are two sources of data. When identifying discrepancies between the model and the real system, comparing knowledge between the two can lead to more cost-efficient results than comparing states in certain scenarios [17].
- **Sampling rate**: The frequency of sensing can be adjusted. A sensor can be put into sleep mode for the reason of energy saving, if its readings can be predicted by historical data or other data sources [14].
- **Sensing fidelity**: Sensors can send back imagery data in high or low resolution, based on the trade-off between data enhancement due to higher fidelity observation vs. cost [2].
- **Sensor placement**: The locations of sensors can be adjusted to monitor areas of interest possibly with multiple sensors from different perspectives.
- **Human monitoring**: Humans can follow instructions from the Digital Twin to monitor and inspect the system and environment on the spot and report data back.

Explanation for Model Adaptation. The aim of model adaptation is to increase the fidelity of the simulation model. This adaptation decision is made by the *Model Updater* in Fig. 1. Explanations should justify why the adaptations are beneficial in improving fidelity. The adaptation can be state replication and estimation, parameter calibration, and knowledge update.

Data assimilation is a common approach for state estimation, which aims at finding a "best" estimate of the (non-observable) state of the system. However, the motivating example is different from the conventional application scenarios of data assimilation such as Kalman Filter. The system is discrete-event and involves complex interactions between different entities, which cannot be easily characterised by mathematical properties and assumptions with numerical models. Although a data assimilation framework for discrete event simulation models has been proposed [5], the example system itself is driven by inherent computation and knowledge of the agents, which poses an additional challenge for the model to replicate not only the state but also the knowledge of the system.

There have already been some research efforts in providing explanations for model adaptation. Interpretable machine learning can be used to select within a library of models based on real-time observation [6]. Model adaptation can be driven by the discrepancies between sensor data and the data from the Digital Twin model. In [3], a Gaussian Mixture Model-based discrepancy detector is first designed to identify anomalies. Then the detected anomalies are further classified into different types by Hidden Markov Model. The classification of anomalies can serve as the explanation for discrepancies to support later decisions. However, their work does not involve feedback loops that correct the anomalies after classification.

Explanation for System Behaviour Adaptation. System behaviour adaptation refers to how the Digital Twin model is used for simulating different what-if scenarios to inform decisions on optimising the system behaviour. This type of decision is made by the *Behaviour Optimiser* in Fig. 1.

What-if analysis can be self-explainable, since *what if*-questions are essentially asking explanatory questions of *how*, and are just a contrast case analysing what would happen under a different situation or with an alternative choice [13]. Nevertheless, the process of exploring all different what-if scenarios can be expensive thus requiring to be optimised by e.g. searching algorithms.

Explanations can also be achieved by providing the satisfaction of the system requirements during operation [15]. These requirements are imposed by experts during the design time of the Digital Twin system, which can involve the task goals (e.g. delivery efficiency in the warehouse example) or system performance (e.g. computational load of the Digital Twin). Goal models can be utilised to provide explanations [15], which is also incorporated in our architecture.

5   Discussion

5.1   Trade-Off Analysis

Our architecture has taken a data-driven stance to explainability of decisions in the system. However, dynamism in data sensing, assimilation and processing can make unnecessary explainability costly. Although explanation is beneficial for transparency, trustworthiness and understandability, the trade-off between performance and explainability needs to be considered in dynamic and data-driven systems. Additionally, decision-making algorithms and models differ in quality of decisions; henceforth, the levels of explanation needed [1] can vary with the quality of decisions induced by these models. The empirical evidence that evaluates how end users perceive the explainability of AI algorithms has been established [4]. The trade-off is situational and depends on the complexity of the problem. Decision models may rank differently in performance vs. explainability when solving different problems, along with the varying complexity of input and training data (e.g. tabular data or images) [4]. For highly dynamic data-driven open systems, the data sensing can vary in complexity, quantity and refresh rate. The type and sophistication of the decisions can also vary. There is a need for DT that self-adaptively decides on when to use which decision models and the level of explanation required. In the reference architecture in Fig. 1, an *Adaptive Selector* is proposed to dynamically switch between the models in different runtime situations. One possible solution to adaptive selection is to abstract the system into different states. Each state can be pre-calculated for its requirement on the level of explanation. The *Adaptive Selector* monitors the state to make selections.

The choice of decision models may further influence the trust of human operators and their willingness to enact the decisions. DT's dynamic trade-off analysis demands awareness of the cost due to the explanation's impact on humans' behaviour. An adaptive cost-aware approach has been proposed in [11], which uses probabilistic model checking to reason about when to provide explanations in a case study of web services. This study assumes that explanation can improve humans' probability of successfully performing a task, but will incur delayed human actions, because humans need time to comprehend the explanation.

5.2   Architecture Evaluation

The inclusion of explainability is viewed as an architecture design decision. Well-established architectural evaluation methods may be used to evaluate the extent to which explainability can improve the utility of the architecture under design, against other quality attributes of interest that need to be satisfied within the architecture.

Scenario-based evaluation methods, such as ATAM [7] may be used during the inception and development stages of the architecture to inform the design of explainability. Scenarios are human-centric and context-dependent. Types of scenarios can be typical, exploratory, or stress scenarios. The extent to which the design for explainability is fit depends on the choice of scenarios. Scenarios may focus on issues such as the need, effectiveness and added value of having explanation embedded as part of the architecture. Examples of the added value of supporting explainability in the data-driven cycle include: ensuring compliance (safety, security, etc.) and avoiding accidental violation of compliance as data is streamed, analysed, processed and/or decisions are actuated; tracing back the cause of issues attributed to data and decisions, mainly for debugging and forensic investigation; transparency of decisions and control by explaining data that can drive control; explaining the risk of critical data-driven decisions, among the many others.

Considering the *Updater* as an example, the parameter calibration can be triggered by policies and the measured discrepancy between the simulation and the real system. Calibration should be triggered only when necessary, since estimating the parameters requires excessive computational resources. However, issues such as sensor anomaly may trigger the calibration by mistake, causing unnecessary computational costs. Explanations can inform humans to trace the need for calibration and identify such a measurement anomaly. Frequent explanations contribute to cautious anomaly identification by always involving humans, but can instead cause annoyance and decreased quality of experience (QoE) of the humans when even trivial decisions need confirmation from humans. The trade-off analysis between computational cost and QoE is essential in evaluation.

6   Conclusion

This paper has aspired to investigate the problem of explainability in human-in-the-loop dynamic data-driven Digital Twin systems. We have contributed to a novel reference architecture of DT, which draws inspiration from the design of DDDAS feedback loops and enrich it with primitives for explainability. The architecture is designed to identify and elaborate on where and when explanations can support decisions that are data-driven or influenced by data assimilated and/or computed by the system.

The areas that need explanations are identified by focusing on three types of adaptation decisions, traced to data. We have refined and enriched the architecture to provide explainability linked to the decisions of three controllers. The controllers use interpretable machine learning and goal models to provide explanations. Future work will focus on further refinements and implementation of variants of explainable Digital Twin architectures for autonomous systems. We will look at cases from smart warehouse logistics and manufacturing, where humans and machines collaborate to inform the design and evaluate the effectiveness of adaptive explanation. Special attention will be paid to the impact of explanation on humans' behaviour. The *Adaptive Selector* in the architecture will be instantiated by algorithms that use state-based approaches to dynamically switch the decision-making models. The work is ongoing and a prototype agent-based modelling testbed of the motivating example has already been implemented in AnyLogic[1].

Acknowledgements. This research was supported by: Shenzhen Science and Technology Program, China (No. GJHZ20210705141807022); SUSTech-University of Birmingham Collaborative PhD Programme; Guangdong Province Innovative and Entrepreneurial Team Programme, China (No. 2017ZT07X386); SUSTech Research Institute for Trustworthy Autonomous Systems, China; and EPSRC/EverythingConnected Network project on Novel Cognitive Digital Twins for Compliance, UK.

---

[1] https://www.anylogic.com/.

References

1. Barredo Arrieto, A., et al.: Explainable Artificial Intelligence (XAI): Concepts, taxonomies, opportunities and challenges toward responsible AI. Inform. Fusion **58**, 82–115 (2020). https://doi.org/10.1016/j.inffus.2019.12.012
2. Blasch, E.: DDDAS advantages from high-dimensional simulation. In: 2018 Winter Simulation Conference (WSC), pp. 1418–1429 (2018). https://doi.org/10.1109/WSC.2018.8632336
3. Gao, C., Park, H., Easwaran, A.: An anomaly detection framework for digital twin driven cyber-physical systems. In: Proceedings of the ACM/IEEE 12th International Conference on Cyber-Physical Systems, pp. 44–54. ICCPS '21, ACM, New York, NY, USA (2021). https://doi.org/10.1145/3450267.3450533
4. Herm, L.V., Heinrich, K., Wanner, J., Janiesch, C.: Stop ordering machine learning algorithms by their explainability! A user-centered investigation of performance and explainability. Int. J. Inform. Manage. 102538 (Jun 2022). https://doi.org/10.1016/j.ijinfomgt.2022.102538
5. Hu, X., Wu, P.: A data assimilation framework for discrete event simulations. ACM Trans. Model. Comput. Simul. **29**(3) (2019). https://doi.org/10.1145/3301502
6. Kapteyn, M.G., Knezevic, D.J., Willcox, K.: Toward predictive Digital Twins via component-based reduced-order models and interpretable machine learning. In: AIAA Scitech 2020 Forum, pp. 1–19. American Institute of Aeronautics and Astronautics, Reston, Virginia (Jan 2020). https://doi.org/10.2514/6.2020-0418
7. Kazman, R., Klein, M., Clements, P.: ATAM: Method for architecture evaluation. Tech. Rep. CMU/SEI-2000-TR-004, Software Engineering Institute, Carnegie Mellon University, Pittsburgh, PA (2000). https://resources.sei.cmu.edu/library/asset-view.cfm?AssetID=5177
8. Kennedy, C., Theodoropoulos, G.: Intelligent management of data driven simulations to support model building in the social sciences. In: Computational Science - ICCS 2006, pp. 562–569. Springer, Berlin Heidelberg, Berlin, Heidelberg (2006). https://doi.org/10.1007/11758532_74
9. Kennedy, C., Theodoropoulos, G., Sorge, V., Ferrari, E., Lee, P., Skelcher, C.: AIMSS: an architecture for data driven simulations in the social sciences. In: Shi, Y., van Albada, G.D., Dongarra, J., Sloot, P.M.A. (eds.) ICCS 2007. LNCS, vol. 4487, pp. 1098–1105. Springer, Heidelberg (2007). https://doi.org/10.1007/978-3-540-72584-8_144
10. Kennedy, C., Theodoropoulos, G., Sorge, V., Ferrari, E., Lee, P., Skelcher, C.: Data driven simulation to support model building in the social sciences. J. Algorithm. Comput. Technol. **5**(4), 561–581 (2011). https://doi.org/10.1260/1748-3018.5.4.561
11. Li, N., Cámara, J., Garlan, D., Schmerl, B.: Reasoning about when to provide explanation for human-involved self-adaptive systems. In: 2020 IEEE International Conference on Autonomic Computing and Self-Organizing Systems (ACSOS), pp. 195–204 (2020). https://doi.org/10.1109/ACSOS49614.2020.00042
12. McCune, R.R., Madey, G.R.: Control of artificial swarms with DDDAS. Procedia Comput. Sci. **29**, 1171–1181 (2014). https://doi.org/10.1016/j.procs.2014.05.105
13. Miller, T.: Explanation in artificial Intelligence: insights from the social sciences. Artif. Intell. **267**, 1–38 (2019). https://doi.org/10.1016/j.artint.2018.07.007
14. Minerva, R., Lee, G.M., Crespi, N.: Digital twin in the IoT context: a survey on technical features, scenarios, and architectural models. Proc. IEEE **108**(10), 1785–1824 (2020). https://doi.org/10.1109/JPROC.2020.2998530

15. Welsh, K., Bencomo, N., Sawyer, P., Whittle, J.: Self-explanation in adaptive systems based on runtime goal-based models. In: Kowalczyk, R., Nguyen, N.T. (eds.) Transactions on Computational Collective Intelligence XVI. LNCS, vol. 8780, pp. 122–145. Springer, Heidelberg (2014). https://doi.org/10.1007/978-3-662-44871-7_5

16. Zhang, N., Bahsoon, R., Theodoropoulos, G.: Towards engineering cognitive digital twins with self-awareness. In: 2020 IEEE International Conference on Systems, Man, and Cybernetics (SMC), pp. 3891–3891. IEEE (Oct 2020). https://doi.org/10.1109/SMC42975.2020.9283357

17. Zhang, N., Bahsoon, R., Tziritas, N., Theodoropoulos, G.: Knowledge equivalence in Digital Twins of intelligent systems. arXiv:2204.07481 [cs, eess] (Apr 2022). https://arxiv.org/abs/2204.07481

Plenary Presentations - Tracking

Transmission Censoring and Information Fusion for Communication-Efficient Distributed Nonlinear Filtering

Ruixin Niu[✉]

Virginia Commonwealth University, Richmond, VA 23284, USA
rniu@vcu.edu

Abstract. A transmission censoring and information fusion approach is proposed for distributed nonlinear system state estimation in Dynamic Data Driven Applications Systems (DDDAS). In this approach, to conserve communication resources, based on the Jeffreys divergence between the prior and posterior probability density functions (PDFs) of the system state, only local posterior PDFs that are sufficiently different from their corresponding prior PDFs will be transmitted to a fusion center. To further reduce the communication cost, the local posterior PDFs are approximated by Gaussian mixtures, whose parameters are learned by an expectation-maximization algorithm. At the fusion center, the received PDFs will be fused via a generalized covariance intersection algorithm to obtain a global PDF. Numerical results for a multi-senor radar target tracking example are provided to demonstrate the effectiveness of the proposed censoring approach.

Keywords: Information fusion · Particle filter · Censoring · Jeffreys divergence · Gaussian mixtures

1 Introduction

In Dynamic Data Driven Applications Systems (DDDAS), state estimation/tracking, information fusion, and subsequent sensor management play a critical role [4,8]. The inherent resource constraints in networked DDDAS, such as limited energy/power of the distributed agents and constrained communication bandwidth, present challenges for distributed estimation and tracking algorithms in such systems. It is therefore desirable to limit the amount of the communication between the distributed sensors/agents and a central fusion center (FC). This can be achieved through sensor censoring, where based on its assessment on the importance of the local information, each sensor makes its decision on whether to send the local information to a FC or not.

Sensor censoring has been investigated for distributed detection problems [2] and distributed estimation problems [13]. For system state estimation/filtering

---

This work was supported in part by the AFOSR Dynamic Data and Information Processing Portfolio under Grant FA9550-22-1-0038.

E. Blasch et al. (Eds.): DDDAS 2022, LNCS 13984, pp. 247–254, 2024.
https://doi.org/10.1007/978-3-031-52670-1_24

problems, in [15] a strategy for censoring distributed sensors' *measurements* was proposed based on the normalized innovation squared (NIS) obtained by a Kalman filter or an extended Kalman filter (EKF).

In many distributed sensing systems, the distributed sensors/agents only share/transmit their local state estimates, since the storage, management, and transmission of the local raw sensor measurements can be expensive and cumbersome. Motivated by this, in [6,7], state estimate censoring schemes were proposed for a distributed target tracking system, where an information theoretic metric, Jeffreys divergence, is evaluated for making censoring decisions by the distributed sensors.

In [6,7], the local posterior probability density functions (PDFs) are approximated by Gaussian PDFs either through an EKF or a Gaussian particle filter. Hence, after passing the local censoring test, each distributed sensor only transmits the mean and covariance of the approximated Gaussian PDF. At the FC, the received Gaussian PDFs are fused according to the optimal track fusion approach [5,10].

Even though the Gaussian approximation reduces the number of parameters transmitted and simplifies track fusion significantly, it will unavoidably lead to information loss and approximation errors. To address this issue, in this paper, at each distributed sensor, the posterior PDF of the system state is approximated by a Gaussian mixture model (GMM), whose parameters are learned using an expectation-maximization (EM) algorithm. The censored local GMM PDFs received at the FC will be fused by an approximate covariance intersection approach.

2   Jeffreys Divergence Based Sensor Censoring

2.1   Jeffreys Divergence

The goal of censoring is to save communication costs between the sensors and the FC, while at the same time maintaining a reasonably accurate fused state estimate at the FC. In a distributed tracking system, each sensor runs its own nonlinear filter to estimate the system state based on local sensor measurements. To judge whether or not significant amount of new information has been brought by the newly arrived sensor measurements, one can compare the posterior and prior PDFs of the system state. A widely used metric measuring dissimilarity between two PDFs is the Kullback-Leibler divergence (KLD). The KLD between PDFs $p(\mathbf{x})$ and $q(\mathbf{x})$ is defined as

$$D\left(p(\mathbf{x})\|q(\mathbf{x})\right) = \int p(\mathbf{x}) \log \frac{p(\mathbf{x})}{q(\mathbf{x})} \tag{1}$$

However, the KLD is an asymmetric measure, and its symmetrized version, Jeffreys divergence, is adopted to measure the distance between $p(\mathbf{x})$ and $q(\mathbf{x})$:

$$J\left(p(\mathbf{x})\|q(\mathbf{x})\right) = D\left(p(\mathbf{x})\|q(\mathbf{x})\right) + D\left(q(\mathbf{x})\|p(\mathbf{x})\right)$$

$$= \int [p(\mathbf{x}) - q(\mathbf{x})] \log \frac{p(\mathbf{x})}{q(\mathbf{x})} d\mathbf{x} \tag{2}$$

2.2   Evaluation of Jeffreys Divergence

Denoting the system state at time k as $\mathbf{x}_k$, the sensor measurements up to time k as $\mathbf{z}_{1:k}$, the posterior PDF of $\mathbf{x}_k$ can be written as

$$p(\mathbf{x}_k|\mathbf{z}_{1:k}) = \frac{p(\mathbf{z}_k|\mathbf{x}_k)p(\mathbf{x}_k|\mathbf{z}_{1:k-1})}{p(\mathbf{z}_k|\mathbf{z}_{1:k-1})} \tag{3}$$

From (3), it can be proved that the Jeffreys divergence between $p(\mathbf{x}_k|\mathbf{z}_{1:k})$ and $p(\mathbf{x}_k|\mathbf{z}_{1:k-1})$ is:

$$J\left(p(\mathbf{x}_k|\mathbf{z}_{1:k})\|p(\mathbf{x}_k|\mathbf{z}_{1:k-1})\right) = \int \left[p(\mathbf{x}_k|\mathbf{z}_{1:k}) - p(\mathbf{x}_k|\mathbf{z}_{1:k-1})\right] \log p(\mathbf{z}_k|\mathbf{x}_k)d\mathbf{x}_k \tag{4}$$

In this paper, it is assumed that each local sensor runs a particle filter to estimate the system state. In a sequential importance sampling (SIS) particle filter, at time k and before the possible particle re-sampling, the system state prior and posterior PDFs can be approximated by

$$p(\mathbf{x}_k|\mathbf{z}_{1:k-1}) \approx \sum_{i=1}^{N_p} w_{k-1}^i \delta(\mathbf{x}_k - \mathbf{x}_k^i)$$

$$p(\mathbf{x}_k|\mathbf{z}_{1:k}) \approx \sum_{i=1}^{N_p} w_k^i \delta(\mathbf{x}_k - \mathbf{x}_k^i) \tag{5}$$

where N_p is the total number of particles, $\{\mathbf{x}_k^i\}_{i=1}^{N_p}$ is a set of particles, and w_{k-1}^i w_k^i are the particle weights before and after the weight update, respectively. Plugging these approximations into (4), the Jeffreys divergence becomes

$$J\left(p(\mathbf{x}_k|\mathbf{z}_{1:k})\|p(\mathbf{x}_k|\mathbf{z}_{1:k-1})\right) \approx \sum_{i=1}^{N_p} (w_k^i - w_{k-1}^i) \log p(\mathbf{z}_k|\mathbf{x}_k^i) \tag{6}$$

In the special case when the particle filter chooses the prior as the importance density, (6) reduces to

$$J\left(p(\mathbf{x}_k|\mathbf{z}_{1:k})\|p(\mathbf{x}_k|\mathbf{z}_{1:k-1})\right) \approx \sum_{i=1}^{N_p} (w_k^i - w_{k-1}^i) \log\left(\frac{w_k^i}{w_{k-1}^i}\right) \tag{7}$$

which coincides with the result obtained in our previous work [7]. Therefore, in this paper, a more rigorous and general derivation for evaluating the Jeffreys divergence in particle filters is provided.

For censoring, if the calculated Jeffreys divergence exceeds a pre-specified threshold, the local posterior PDF will be transmitted to the FC.

3   Estimation of Posterior PDFs via GMM

The transmission of all the particles and their associated weights obtained by a local particle filter requires a prohibitive amount of communications especially when the filter uses a large number of particles. A more practical solution is to approximate the posterior PDFs by Gaussian mixtures, and transmit the mixture parameters to the FC. A GMM is quite flexible and can be applied to approximate an arbitrary PDF. In this paper, the parameters of a GMM are learned via the widely used EM algorithm [9].

4   Fusion of Gaussian Mixtures

At the FC, the received local posterior PDFs (GMM PDFs) need to be fused to obtain a more accurate global PDF. Fusion of local posterior PDFs from local sensors can be addressed by the optimal fusion rule for distributed nonlinear filtering [5], or the optimal channel filter [14]. However, these fusion approaches do not have closed-form solutions for nonlinear problems, and Monte-Carlo importance sampling based fusion approaches have been proposed to fuse Gaussian mixtures [1].

Alternatively, a popular approach called covariance intersection (CI) [12], provides a powerful and general means for fusing information in arbitrary networks. The CI approach can robustly fuse correlated tracks or state estimates without the knowledge of the cross-covariance between them. In [11], the CI was generalized to fuse Gaussian mixtures, based on a first order approximation of the Chernoff information. This approach is adopted here due to its computational efficiency and its robust fusion performance for highly nonlinear problems. A brief summary of the generalized CI algorithm is provided below.

Suppose the FC aims to fuse two Gaussian mixtures:

$$p_a(\mathbf{x}) = \sum_{i=1}^{N_a} p_i \mathcal{N}(\mathbf{x}; \mathbf{a}_i, \mathbf{A}_i)$$

$$p_b(\mathbf{x}) = \sum_{i=1}^{N_b} q_i \mathcal{N}(\mathbf{x}; \mathbf{b}_i, \mathbf{B}_i) \tag{8}$$

where N_a and N_b are the numbers of Gaussian components in the GMMs a and b respectively, p_i and q_i are the weights for the ith Gaussian components in the two models, $\mathbf{a}_i$ and $\mathbf{b}_i$ denote the means for the ith Gaussian components, and $\mathbf{A}_i$ and $\mathbf{B}_i$ represent the covariance matrices for the ith Gaussian components. The fused GMM has totally $N_c = N_a N_b$ components:

$$p_c(\mathbf{x}) = \sum_{i=1}^{N_a} \sum_{j=1}^{N_b} r_{ij} \mathcal{N}(\mathbf{x}; \mathbf{c}_{ij}, \mathbf{C}_{ij}) \tag{9}$$

The parameters of the fused GMM can be obtained by

$$\mathbf{C}_{ij}^{-1} = \omega \mathbf{A}_i^{-1} + (1 - \omega) \mathbf{B}_j^{-1}$$
$$\mathbf{c}_{ij} = \mathbf{C}_{ij} \left[\omega \mathbf{A}_i^{-1} \mathbf{a}_i + (1 - \omega) \mathbf{B}_j^{-1} \mathbf{b}_j \right]$$
$$r_{ij} = \frac{p_i^\omega q_j^{(1-\omega)}}{\sum_i \sum_j p_i^\omega q_j^{(1-\omega)}} \tag{10}$$

where ω is an optimization variable. As a result, both the mean and covariance matrix of the fused GMM in (9) are clearly functions of ω. The optimal ω used can be found by minimizing the covariance matrix (its determinant or trace) of the fused GMM.

To take advantage of the global prior information, the fused GMM based on local PDFs will be further fused with the global prior PDF, via the same generalized CI algorithm. As observed in our numerical experiment, fusion with the global prior is especially advantageous when only one local PDF is received by the FC.

5   Numerical Results for a Radar Tracking Example

5.1   System Model

The target's motion follows a discrete-time white noise acceleration model [3]:

$$\mathbf{x}_k = \mathbf{F}\mathbf{x}_{k-1} + \mathbf{v}_k \tag{11}$$

where $\mathbf{x}_k = [\xi_k \ \dot{\xi}_k \ \eta_k \ \dot{\eta}_k]$ is the target state at time k, consisting of positions and velocities along ξ and η axes in a 2D Cartesian coordinate system. $\mathbf{F}$ is the state transition matrix:

$$\mathbf{F} = \begin{bmatrix} 1 & T & 0 & 0 \\ 0 & 1 & 0 & 0 \\ 0 & 0 & 1 & T \\ 0 & 0 & 0 & 1 \end{bmatrix} \tag{12}$$

in which T is the time interval between two adjacent samples. $\mathbf{v}_k$ denotes the zero-mean white Gaussian process noise with covariance matrix $\mathbf{Q}$

$$\mathbf{Q} = \tilde{q} \begin{bmatrix} \frac{T^3}{3} & \frac{T^2}{2} & 0 & 0 \\ \frac{T^2}{2} & T & 0 & 0 \\ 0 & 0 & \frac{T^3}{3} & \frac{T^2}{2} \\ 0 & 0 & \frac{T^2}{2} & T \end{bmatrix} \tag{13}$$

where $\tilde{q}$ is the power spectral density of the process noise in the original continuous-time kinematic model before its discretization.

The radar sensor provides nonlinear range and bearing measurements:

$$\mathbf{z}_k = h(\mathbf{x}_k) + \mathbf{w}_k \tag{14}$$

where

$$h(\mathbf{x}_k) = \begin{bmatrix} \sqrt{(\xi_k - \xi_s)^2 + (\eta_k - \eta_s)^2} \\ \arctan\left(\frac{\eta_k - \eta_s}{\xi_k - \xi_s}\right) \end{bmatrix} \tag{15}$$

and (ξ_s, η_s) is the sensor coordinates.

5.2   Performance Evaluation in a Two-Radar Network

In this experiment, there are two radars tracking a target, as illustrated in Fig. 1, in which $T=2$ s, $\tilde{q}=5$, and the standard deviations for range and bearing measurement noises are 10 m and 3°, respectively. With these parameters, for each radar without collaboration, the tracking problem is highly nonlinear due to its poor bearing accuracy. Hence, a particle filter is used at each radar.

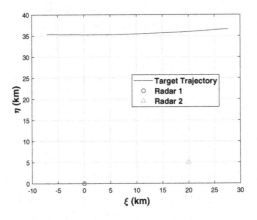

Fig. 1. Target trajectory and radars.

The tracking performance of the proposed censoring strategy is compared with a random data selection scheme, where each sensor randomly transmits its local posterior PDF with a transmission probability of 1/2. For a fair comparison, in the proposed censoring strategy, the threshold for the Jeffreys divergence is set in such a way that the censoring and random selection strategies incur the same amount of communications.

The positional root mean square error (RMSE) is obtained via 100 Monte-Carlo trials and shown in Fig. 2. It is clear that information fusion at the FC improves tracking performance dramatically, even at half of the communication rate, since it takes advantage of the spatial diversity of the sensors. During the first 15 s, the proposed censoring strategy leads to a significantly better tracking performance than the random selection strategy. As time goes on, the tracking performance in both strategies converges to a level close to that of the full-rate communication strategy. The velocity RMSE results are similar and skipped due to limited space.

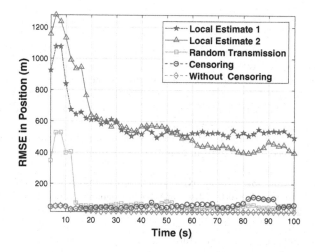

Fig. 2. Positional target tracking performance for various transmission strategies.

In Table 1, the average number of local PDF transmissions is provided for the censoring scheme over the time. Clearly, the censoring scheme is adaptive to the tracking performance over time, which allocates more communication resources at the beginning when more sensor measurement information is needed, and gradually reduces the communications when the tracking result improves and converges to a steady state.

Table 1. Average number of PDF transmissions in censoring strategy

Radar Scan #	1	2	3	4	5	6	10	20	30	40	50
PDF Transmissions	2.00	2.00	1.97	1.90	1.73	1.33	0.97	0.73	0.86	0.84	0.85

6 Conclusion

A transmission censoring and information fusion approach was proposed for communication-efficient distributed nonlinear system state estimation. Transmission censoring is performed based on the Jeffreys divergence between the prior and posterior PDFs of the system state. The local posterior PDFs are approximated by Gaussian mixtures via an EM algorithm to further reduce communication costs. At the fusion center, the received PDFs are fused via a fast generalized covariance intersection algorithm. Numerical results for a multi-senor radar target tracking example were provided to show that the proposed censoring scheme allocates communication resources adaptively to achieve significantly better tracking performance. In the future, this work will be extended to other network structures, such as those with feedback and decentralized networks.

References

1. Ahmed, N.R.: Decentralized Gaussian mixture fusion through unified quotient approximations. ArXiv:1907.04008 (2019)
2. Appadwedula, S., Veeravalli, V.V., Jones, D.L.: Decentralized detection with censoring sensors. IEEE Trans. Signal Process $56(4)$, 1362–1373 (2008)
3. Bar-Shalom, Y., Li, X.R., Kirubarajan, T.: Estimation with Applications to Tracking and Navigation. Wiley, New York (2001)
4. Blasch, E., et al.: DDDAS-based joint nonlinear manifold learning for target localization. In: Proceedings of International Workshop on Structural Health Monitoring Conference (September 2017)
5. Chong, C.Y., Mori, S., Chang, K.C.: Distributed multitarget multisensor tracking. In: Bar-Shalom, Y. (ed.) Multitarget-Multisensor Tracking: Advanced Applications, pp. 247–295. Artech House (1990)
6. Conte, A., Niu, R.: Censoring in distributed radar tracking systems with various feedback models. In: 2015 18th International Conference on Information Fusion (Fusion), pp. 476–483 (2015)
7. Conte, A.S., Niu, R.: Censoring distributed nonlinear state estimates in radar networks. In: Pham, K.D., Chen, G. (eds.) Sensors and Systems for Space Applications IX, vol. 9838, pp. 127–142. International Society for Optics and Photonics, SPIE (2016)
8. Blasch, E.P., Darema, F., Ravela, S., Aved, A.J. (eds.): Handbook of Dynamic Data Driven Applications Systems: Volume 1. Springer International Publishing, Cham (2022). https://doi.org/10.1007/978-3-030-74568-4
9. Dempster, A.P., Laird, N.M., Rubin, D.B.: Maximum likelihood from incomplete data via the EM Algorithm. J. Royal Stat. Society: Series B (Methodological) $39(1)$, 1–22 (1977). https://doi.org/10.1111/j.2517-6161.1977.tb01600.x
10. Govaers, F., Koch, W.: An Exact solution to track-to-track-fusion at arbitrary communication rates. IEEE Trans. Aerosp. Electron. Syst. $48(3)$, 2718–2729 (2012). https://doi.org/10.1109/TAES.2012.6237623
11. Julier, S.: An empirical study into the use of Chernoff information for robust, distributed fusion of Gaussian mixture models. In: 2006 9th International Conference on Information Fusion (Fusion) (July 2006)
12. Julier, S., Uhlmann, J.: A non-divergent estimation algorithm in the presence of unknown correlations. In: Proceedings of the American Control Conference, pp. 2369–2373 (1997)
13. Msechu, E.J., Giannakis, G.B.: Sensor-centric data reduction for estimation with WSNs via censoring and quantization. IEEE Trans. Signal Process. $60(1)$, 400–414 (2012)
14. Ong, L., Bailey, T., Durrant-Whyte, H., Upcroft, B.: Decentralised particle filtering for multiple target tracking in wireless sensor networks. In: Proceedings of 2008 11th International Conference on Information Fusion (June 2008)
15. Zheng, Y., Niu, R., Varshney, P.K.: Sequential Bayesian estimation with censored data for multi-sensor systems. IEEE Trans. Signal Process. $62(10)$, 2626–2641 (2014)

Distributed Estimation of the Pelagic Scattering Layer Using a Buoyancy Controlled Robotic System

Cong Wei[1]([⊠])[iD] and Derek A. Paley[1,2][iD]

[1] Institute for Systems Research, University of Maryland, College Park, MD 20742, USA
[2] Department of Aerospace Engineering, University of Maryland, College Park, MD 20742, USA
{weicong,dpaley}@umd.edu

Abstract. This paper formulates a strategy for Driftcam, an ocean-going robot system, to observe and track the motion of an ocean biological phenomenon called the pelagic scattering layer, which consists of organisms that migrate vertically in the water column once per day. Driftcam's horizontal motion is determined by the flow field and the vertical motion is regulated by onboard buoyancy control. In order to observe the evolution of the scattering layer, an ensemble Kalman filter is applied to estimate organism density; the density dynamics are propagated using the Perron-Frobenius operator. Multiple Driftcam are subject to depth regulation by open-loop and closed-loop controllers; a control strategy is proposed to track the peak of the density. Numerical simulations illustrate the efficacy of this strategy and motivate ongoing and future efforts to design a coordination formation algorithm for multi-agent Driftcam system to track the motion of the scattering layer, with implications for ocean monitoring.

Keywords: Density mapping · Ensemble Kalman filter · Scattering Layer

1 Introduction

This paper addresses the problem of underwater mapping of a marine biological system called the pelagic scattering layer. An ocean-going system called Driftcam is deployed to perform the mapping, ideally in a group. This robotic system is propelled by ocean currents and is used for counting and measuring organisms in the pelagic scattering layer with a high-definition low-light camera [22]. Depth is regulated through a piston pump engine, which pumps oil into an external bladder that can change buoyancy by its expandable volume [1].

The pelagic scattering layer, also referred to as the sound scattering layer, is a biological layer in the ocean consisting of a variety of marine animals[1]. An important feature of the pelagic scattering layer is the daily movement of organisms from the deep ocean during the day to a shallow depth during the night [20]. This diel vertical migration represents the largest biology movement on the planet in terms of biomass, number

---

[1] https://en.wikipedia.org/wiki/Deep_scattering_layer.

E. Blasch et al. (Eds.): DDDAS 2022, LNCS 13984, pp. 255–263, 2024.
https://doi.org/10.1007/978-3-031-52670-1_25

of individuals, and species[2] and plays a key role in structuring ecological and physic-ochemical processes, as well as the biological carbon pump of vast oceanic ecosystems [3].

Despite the importance of the scattering layer, we lack knowledge of its biological features and biogeochemical processes [10], because the scattering layer is difficult to sample and observe in real time. Most sampling and sensing systems either cause disturbances or collect indirect data, resulting in inaccuracy [2,16]. The untethered Driftcam system modeled in this paper provides a direct observation of the organisms living in the scattering layer [1]. This sampling problem is an example of a dynamic, data-driven application system in which the collection of measurements is used to update a model, which in turn is used to guide the collection of subsequent measurements.

Our work seeks a solution for long-endurance and large-scale sampling over mesopelagic[3] ocean space. The design and development of a long-endurance passive marine sensor system makes it possible to sample large-scale ocean environments for specified phenomena of interest [11,12,21]. Purely passive drifters like the Argo system are not able to automatically achieve the configurations that improves sampling efficiency [19]. Actuated vehicles like autonomous surface vehicles and autonomous underwater vehicles are used to achieve faster surveys in the dynamic marine environment [9], but they may rely on prior knowledge of the flow field and require a large number of vehicles, depending on the scale of the region of interest. Multi-agent autonomous drifters are also deployed in a depth-holding configuration to measure the internal waves near the shore [12], however, this configuration is more suitable to monitoring ocean dynamics or biological systems occurring at relative limited depth interval. Another ocean sampling method includes using actuated autonomous underwater vehicles to track a patch of interest tagged by Lagrangian drifters. [4]. An adaptive sampling strategy [5] uses feedback control to redesign paths in response to updated sensor measurements, such as for sampling ocean features [15,18].

This paper aims at identifying the depth and vertical distribution of the scattering layer by sampling the density of organisms using onboard cameras, which use image processing to measure organism density locally. The vertical depth dynamics of the scattering layer determines the density propagation over time. An estimator is used to recover the density field from discrete density measurements by one or more Driftcam. The goal is to collect data that minimizes the estimation error of scattering layer density, e.g., by tracking the density peak with a Driftcam using closed-loop control.

The contributions of this paper are (1) a dynamic model of organism density in the scattering layer using the Perron-Frobenius operator; (2) an estimation framework that assimilates discrete measurements taken by one or more Driftcam to reconstruct the density and identify its peak using an ensemble Kalman filter; (3) and an adaptive sampling strategy based on the recovered density map in which one or more Driftcam tracks the peak density using closed-loop control.

The paper is organized as follows. Section 2 gives a brief description of the Perron-Frobenius operator and the ensemble Kalman filter. Section 3 models the density prop-

---

[2] https://oceanexplorer.noaa.gov/technology/development-partnerships/21scattering-layer/features/scattering-layer/scattering-layer.html.

[3] https://ocean.si.edu/ecosystems/deep-sea/deep-sea.

agation based on the dynamics of diel vertical migration. The ensemble Kalman filter is applied to recover the density field from measurements collected by one or more Driftcam. Numerical simulations in Sect. 4 illustrate the performance of the sampling method. Section 5 summarizes the paper and ongoing and future work.

2 Preliminary

2.1 Perron-Frobenius Operator

The Perron-Frobenius (PF) operator is used in ergodic theory to study measure-theoretic characterization. This is a brief introduction; detailed information can be found in [6,8]. Let $\mathbb{X}$ be a compact manifold and $f : \mathbb{X} \to \mathbb{X}$ be a smooth time invariant vector field. Consider the following time invariant system

$$\dot{x} = f(x). \tag{1}$$

Let $\phi_f : \mathbb{R} \times \mathbb{X} \to \mathbb{X}$ be the solution of system (1), i.e., $x = \phi_f(t, x_0)$ satisfies (1) with initial condition $x(0) = x_0$.

Definition 1. *A semigroup of operator* $\mathcal{P}^\tau : \tau > 0$ *is said to be the PF operator if* $\mathcal{P}^\tau : L^1(\mathbb{X}) \to L^1(\mathbb{X})$ *is defined by [6]*

$$\mathcal{P}^\tau \rho(\cdot) = \rho \circ \phi_f(-\tau, \cdot)|\det(D_x \phi_f(-\tau, \cdot))|, \tag{2}$$

where D_x represents the Jacobian matrix with respect to state variable x.

If $\rho(\cdot)$ is a probability density (PDF) with respect to an absolutely continuous probability measure v, then $\mathcal{P}^\tau \rho$ is another PDF with respect to the absolutely continuous probability measure $v \circ \phi(-\tau, \cdot)$. Specifically,

$$\int_B \mathcal{P}^\tau \mathrm{d}v = \int_{\phi_f(-\tau, B)} \rho \mathrm{d}v, \tag{3}$$

for any v-measurable set B [14]. The PF operator transport a density function with time according to the flow of the system dynamics.

2.2 Ensemble Kalman Filter

The ensemble Kalman filter (EnKF) is an Monte Carlo approximation of the Kalman filter that stores, propagates and updates an ensemble of vectors to approximate the state distribution [13,17]. Consider a nonlinear dynamic system and a linear measurement equation,

$$x(t_{k+1}) = f(x_k) + w(t_{k+1}) \tag{4a}$$
$$y(t_{k+1}) = Hx(t_{k+1}) + v(t_{k+1}), \tag{4b}$$

where $x(t_k), w(t_k) \in \mathbb{R}^n$ and $y(t_k), v(t_k) \in \mathbb{R}^m$. Assume that $w(t_k)$ and $v(t_k)$ are zero-mean white noise with covariance matrices $Q(t_k)$ and $R(t_k)$, respectively. Moreover, $x(t_0), w(t_k)$ and $v(t_k)$ are uncorrelated. For ensemble Kalman filter, the distribution is replaced by a collection of realizations called an ensemble. Let

$$X = [x_1, \cdots, x_N] = [x_i] \tag{5}$$

be an $n \times N$ matrix, where x_i is a sample from prior distribution. Matrix X is the prior ensemble. In the same way, the distribution of measurement is represented by

$$Y = [y_1, \cdots, y_N] = [y_i]. \tag{6}$$

Here we show only the result of ensemble Kalman filter; detailed information can be found in [7, 13, 17]. The forecast of EnKF is

$$\tilde{X}(t_{k+1}) = f(\hat{X}(t_k)) + W(k) \tag{7a}$$

$$E(\tilde{X}(t_{k+1})) = \frac{1}{N} \sum_{i=1}^{N} x_i, \quad C = \frac{AA^T}{N-1}, \tag{7b}$$

where

$$A = \tilde{X}(t_{k+1}) - E(\tilde{X}(t_{k+1})). \tag{8}$$

The update is given by

$$\hat{X}(t_{k+1}) = \tilde{X}(t_{k+1}) + CH^T(HCH^T + R)^{-1}(Y - HX). \tag{9}$$

The EnKF implemented below has as state vector the density of the scattering layer over a range of discrete depths; it uses the PF operator to propagate those density estimates forward in time.

3   Estimation of the Scattering Layer

The depth ζ of the scattering layer at time t is modeled dynamically as follows:

$$\dot{\zeta} = -\omega(\zeta(0) - \zeta_0)\sin(\omega t) \tag{10a}$$

$$\dot{t} = 1 \tag{10b}$$

where ω indicates the frequency of vertical migration (one cycle per 24 h). To mimic the width contraction at the surface, assume $\zeta_0 = \alpha\zeta(0)$, $\alpha \in (0.5, 1)$. The dynamics (10) translate the organism density vertically. Define $\phi(t, \zeta(0)) : \mathbb{R} \times \mathbb{R}^+ \to \mathbb{R}$, as the flow map, where

$$\phi(t, \zeta(0)) = \zeta(0)[\alpha + (1 - \alpha)\cos\omega t]. \tag{11}$$

The density at a depth $\zeta_d \in \mathbb{R}^-$ can be predicted by the PF operator, i.e.,

$$\rho(t_k, \zeta_d) = \rho(0, \phi(-t_k, \zeta_d))D(t_k, 0) \tag{12a}$$

$$\rho(t_{k+1}, \zeta_d) = \rho(0, \phi(-t_{k+1}, \zeta_d))D(t_{k+1}, 0), \tag{12b}$$

where $D(t_k, 0) = 1/[\alpha + (1 - \alpha) \cos \omega t_k]$

$$D(t_{k+1}, t_k) \triangleq \frac{\alpha + (1 - \alpha) \cos \omega t_k}{\alpha + (1 - \alpha) \cos \omega t_{k+1}}. \tag{13}$$

The density propagation at ζ_d from t_k to t_{k+1} derived from (12) is

$$\rho(t_{k+1}, \zeta_d) = \Delta(t_{k+1}, t_k)\rho(t_k, \zeta_d). \tag{14}$$

where

$$\Delta(t_{k+1}, t_k) = D(t_{k+1}, t_k)\frac{\rho(0, \phi(-t_{k+1}, \zeta_d))}{\rho(0, \phi(-t_k, \zeta_d))}. \tag{15}$$

Consider M Driftcam at depths $[z_{D_1}, \cdots, z_{D_M}]^T$; each collects a noisy measurement at a constant time step $\tau_m > t_{k+1} - t_k$; the measurement is $y(t_k) \in \mathbb{R}^M$. Discretizing the full ocean depth into n_o levels, let $\mathbb{R}^{n_o} \ni x(t_k) \triangleq [\rho(t_k, \zeta_d(t_k))]$. Based on density propagation model (14), the density state-space model is

$$x(t_{k+1}) = \Delta(t_{k+1}, t_k)x(t_k) + w(t_{k+1}), \quad w(t_{k+1}) \sim \mathcal{N}(0, Q(t_{k+1})) \tag{16a}$$

$$t_{k+1} = t_k + \tau_p \tag{16b}$$

$$y(t_{k+1}) = H(t_{k+1})x(t_{k+1}) + v(t_{k+1}), \quad v(t_{k+1}) \sim \mathcal{N}(0, R(t_{k+1})), \tag{16c}$$

where

$$H_{lj}(t_k) = \begin{cases} 1 & \text{if } z_{D_l}(t_k) = \zeta_d(t_k) \\ 0 & \text{otherwise.} \end{cases}$$

We design an unbiased estimator following the steps from (5)–(9). The result is

$$\hat{\mathbf{X}}(k) = [\hat{x}_1(k), \hat{x}_2(k), \cdots, \hat{x}_N(k)] \in \mathbb{R}^{n_o \times N}.$$

The EnKF forecast step obtains

$$\tilde{x}_i(t_{k+1}) = \Delta(t_{k+1}, t_k)\hat{x}_i(t_k) + w_i(t_{k+1}), \quad w_i(t_{k+1}) \sim \mathcal{N}(0, Q(t_{k+1})).$$

Forecast covariance is calculated from (7b). The observation $y(k + 1)$ here forms a matrix, $\tilde{\mathbf{Y}} \in \mathbb{R}^{M \times N}$, given by

$$\tilde{\mathbf{Y}}(t_{k+1}) = [\tilde{y}_1(t_{k+1}), \cdots, \tilde{y}_N(t_{k+1})],$$

where $\tilde{y}_i(t_{k+1}) = y_i(t_{k+1}) + v(t_{k+1})$. The updated ensemble $\hat{\mathbf{X}}$ is calculated as follows [13],

$$\hat{\mathbf{X}}(t_{k+1}) = \tilde{\mathbf{X}}(t_{k+1}) + K[\tilde{\mathbf{Y}}(t_{k+1}) - H(t_k)\tilde{\mathbf{X}}(t_{k+1})],$$

where

$$K = CH^T(t_{k+1})[H(t_{k+1})CH^T(t_{k+1}) + R(t_{k+1})]^{-1}.$$

The following section illustrates the scattering layer modeling and estimation framework.

4   Numerical Results

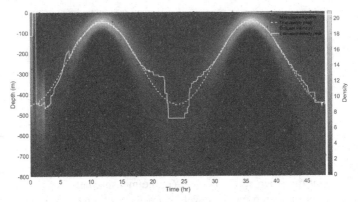

(a) Density over time estimated by a single Driftcam.

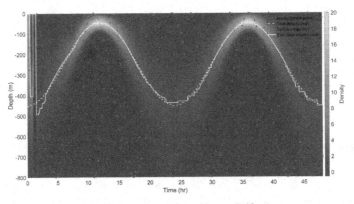

(b) Density over time estimated by two Driftcam.

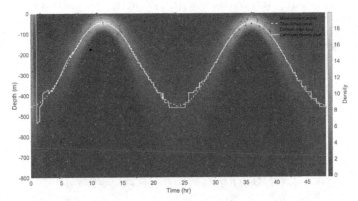

(c) Density over time estimated by two Driftcam; one is equipped with a closed-loop controller that tracks the peak of the estimate, when the ensemble covariance is sufficiently low.

Fig. 1. Estimation of scattering layer density over time, with one or two Driftcam

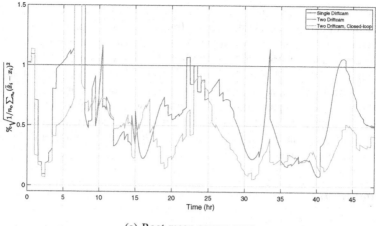

(a) Root-mean-square error

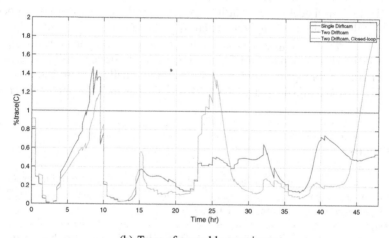

(b) Trace of ensemble covariance

Fig. 2. Performance evaluation for open- and closed-loop Driftcam control. All the data is normalized by that from the single Driftcam case.

This section presents three cases of scattering layer sampling: a single Driftcam, two Driftcam, and two Driftcam with a closed-loop tracking controller. All the cases share the same period of diel migration (24 h), the same number of ensemble members $N = 100$, and depth resolution $n_o = 100$.

In the first case, one Driftcam is commanded to perform sinusoidal motion from -800 m to 0 m with frequency 0.15/h. The Driftcam collects a density measurement every 30min, shown as red dots on Fig. 1(a). The average vertical speed is 7 cm/s. Fig. 1(a) shows that a single Dirftcam is able to map the density field and identify the density peak as indicated by the white line in Fig. 1(a). However, performance improves with a second Driftcam as shown next.

The second sampling result is for the case with two Driftcam. The second Driftcam is commanded to perform sinusoidal motion from −800 m to 0 m. From Fig. 1(b), the quality of mapping is improved compared to a single Driftcam, especially when the organisms are sparsely distributed at the bottom. In the last case, after being deployed for a certain time (10 h), a feedback controller is activated when the trace of ensemble covariance is below a threshold (20). This controller drives one Driftcam to track the estimated density peak. Figure 2(a) and Fig. 2(b) suggest that more Driftcam deployed in the region of interest can reduce estimation error and its covariance. The estimation error is the smallest in the closed-loop control case; however, the trace of the ensemble covariance becomes higher.

5   Conclusion

This work presents an ocean-sampling strategy for a buoyancy-driven underwater vehicle designed to observe the density field of the pelagic scattering layer. The strategy is designed based on modeling the density dynamics by the Perron-Frobenius operator. Samples of the density field are assimilated by an ensemble Kalman filter. This paper reveals two strategies to improve the estimation performance either by deploying more Driftcam or by closed-loop control. In ongoing work, the strategy will be extended to multiple Driftcam for higher spatial dimensions. Additionally, a coordinated controller may be incorporated into the sampling strategy to regulate the Driftcam motion relative to one another.

References

1. Berkenpas, E.J., et al.: A buoyancy-controlled Lagrangian camera platform for in situ imaging of marine organisms in midwater scattering layers. IEEE J. Oceanic Eng. 43(3), 595–607 (2018). https://doi.org/10.1109/JOE.2017.2736138
2. Blanluet, A., et al.: Characterization of sound scattering layers in the Bay of Biscay using broadband acoustics, nets and video. PLoS ONE 14(10), 1–19 (2019). https://doi.org/10.1371/journal.pone.0223618
3. Brodeur, R., Pakhomov, E.: Nekton. In: Cochran, J.K., Bokuniewicz, H.J., Yager, P.L. (eds.) Encyclopedia of Ocean Sciences (Third Edition), pp. 582–587. Academic Press, Oxford, third edn. (2019)
4. Das, J., et al.: Coordinated sampling of dynamic oceanographic features with underwater vehicles and drifters. Int. J. Robot. Res. 31(5), 626–646 (2012). https://doi.org/10.1177/0278364912440736
5. Fiorelli, E., Leonard, N.E., Bhatta, P., Paley, D.A., Bachmayer, R., Fratantoni, D.M.: Multi-auv control and adaptive sampling in Monterey Bay. IEEE J. Oceanic Eng. 31, 935–948 (10 2006). https://doi.org/10.1109/JOE.2006.880429
6. Froyland, G., Padberg, K.: Almost-invariant sets and invariant manifolds-connecting probabilistic and geometric descriptions of coherent structures in flows. Physica D 238(16), 1507–1523 (2009). https://doi.org/10.1016/j.physd.2009.03.002
7. Gillijns, S., Mendoza, O., Chandrasekar, J., De Moor, B., Bernstein, D., Ridley, A.: What is the ensemble Kalman filter and how well does it work? In: Proceedings of the IEEE American Control Conference, pp. 4448–4453 (2006). https://doi.org/10.1109/ACC.2006.1657419

8. Goswami, D., Thackray, E., Paley, D.A.: Constrained Ulam dynamic mode decomposition: Approximation of the Perron-Frobenius operator for deterministic and stochastic systems. IEEE Control Syst. Lett. **2**(4), 809–814 (2018). https://doi.org/10.1109/LCSYS.2018.2849552

9. Hansen, J., Manjanna, S., Li, A.Q., Rekleitis, I., Dudek, G.: Autonomous marine sampling enhanced by strategically deployed drifters in marine flow fields. In: Proceedings of the OCEANS MTS/IEEE Charleston, pp. 1–7 (2018). https://doi.org/10.1109/OCEANS.2018.8604873

10. Haëntjens, N., et al.: Detecting mesopelagic organisms using biogeochemical-argo floats. Geophys. Res. Lett. **47**(6), e2019GL086088 (2020). https://doi.org/10.1029/2019GL086088

11. Hsieh, M.A., et al.: Small and adrift with self-control : Using the environment to improve autonomy. In: Proceedings of the International Symposium on Robotics Research, pp. 1–16 (2015). https://doi.org/10.1007/978-3-319-60916-4_22

12. Jaffe, J.S., et al.: A swarm of autonomous miniature underwater robot drifters for exploring submesoscale ocean dynamics. Nat. Commun. **8**, 1–8 (2017). https://doi.org/10.1038/ncomms14189

13. Katzfuss, M., Stroud, J.R., Wikle, C.K.: Understanding the ensemble kalman filter. Am. Stat. **70**(4), 350–357 (2016). https://doi.org/10.1080/00031305.2016.1141709

14. Klus, S., et al.: Data-driven model reduction and transfer operator approximation. J. Nonlinear Sci. **28**(4), 985–1010 (2018)

15. Kularatne, D., Hsieh, A.: Tracking attracting lagrangian coherent structures in flows. In: Proceedings of Robotics: Science and Systems. Rome, Italy (July 2015). https://doi.org/10.15607/RSS.2015.XI.021

16. Lavery, A.C., Chu, D., Moum, J.N.: Measurements of acoustic scattering from zooplankton and oceanic microstructure using a broadband echosounder. ICES J. Marine Sci. **67**(2), 379–394 (10 2009). https://doi.org/10.1093/icesjms/fsp242

17. Mandel, J.: A brief tutorial on the Ensemble Kalman Filter (2009). https://doi.org/10.48550/ARXIV.0901.3725, https://arxiv.org/abs/0901.3725

18. Michini, M., Hsieh, M.A., Forgoston, E., Schwartz, I.B.: Robotic tracking of coherent structures in flows. IEEE Trans. Rob. **30**(3), 593–603 (2014). https://doi.org/10.1109/TRO.2013.2295655

19. Roemmich, D., et al.: On the future of argo: a global, full-depth, multi-disciplinary array. Front. Marine Sci. **6** (2019). https://doi.org/10.3389/fmars.2019.00439

20. Seki, M.P., Polovina, J.J.: Ocean gyre ecosystems. In: Cochran, J.K., Bokuniewicz, H.J., Yager, P.L. (eds.) Encyclopedia of Ocean Sciences (Third Edition), pp. 753–758. Academic Press, Oxford, third edition edn. (2019)

21. Subbaraya, S., et al.: Circling the seas: design of Lagrangian drifters for ocean monitoring. IEEE Robot. Autom. Mag. **23**, 42–53 (2016)

22. Suitor, R., Berkenpas, E., Shepard, C.M., Abernathy, K., Paley, D.A.: Dynamics and control of a buoyancy-driven underwater vehicle for estimating and tracking the scattering layer. In: Proceedings of the IEEE American Control Conference (2022)

Towards a Data-Driven Bilinear Koopman Operator for Controlled Nonlinear Systems and Sensitivity Analysis

Damien Guého[(✉)] and Puneet Singla

The Pennsylvania State University, University Park, PA 16802, USA
damien.gueho@gmail.com

Abstract. A Koopman operator is a linear operator that can describe the evolution of the dynamical states of any arbitrary uncontrolled dynamical system in a lifting space of infinite dimension. In practice, analysts consider a lifting space of finite dimension with a guarantee to gain accuracy on the state prediction as the order of the operator increases. For controlled systems, a bilinear description of the Koopman operator is necessary to account for the external input. Additionally, bilinear state-space model identification is of interest for two main reasons: some physical systems are inherently bilinear and bilinear models of high dimension can approximate a broad class of nonlinear systems. Nevertheless, no well-established technique for bilinear system identification is available yet, even less in the context of Koopman. This paper offers perspectives in identifying a bilinear Koopman operator from data only. Firstly, a bilinear Koopman operator is introduced using subspace identification methods for the accurate prediction of controlled nonlinear systems. Secondly, the method is employed for sensitivity analysis of nonlinear systems where it is desired to estimate the variation of a measured output given the deviation of a constitutive parameter of the system. The efficacy of the methods developed in this paper are demonstrated on two nonlinear systems of varying complexity.

Keywords: Koopman operator · Bilinear system identification · Sensitivity analysis

1 Introduction

Data-driven analysis and control of dynamic systems is central for dynamic data-driven applications systems (DDDAS) and the ability to understand and model nonlinear dynamical systems in presence of control actions remains a challenge in the system identification and reduced-order modeling (ROM) community. A first step to describe nonlinear dynamics utilizes the Koopman operator framework where the dynamical states are lifted in a higher-dimensional measurement space. The theory to identify a Koopman operator of any order given some data from an unforced dynamical system is well developed [2,6,11,12,16]. For majority of

© The Author(s), under exclusive license to Springer Nature Switzerland AG 2024
E. Blasch et al. (Eds.): DDDAS 2022, LNCS 13984, pp. 264–271, 2024.
https://doi.org/10.1007/978-3-031-52670-1_26

the problems, the exact Koopman operator that would describe the evolution of the dynamical states in a lifting space linearly is of infinite dimension. In practice, analysts consider a lifting space of finite dimension (adequately with given requirements and computing capabilities) giving rise to a truncated Koopman operator. As presented in earlier work [3], increasing the order of the operator offers a guarantee to gain accuracy on the state prediction. When subspace identification methods like the eigensystem realization algorithm (ERA) and dynamic mode decomposition (DMD) [4,5,7,13–15] for time-invariant systems or the time-varying eigensystem realization algorithm (TVERA) [9,10] for time-varying systems are used to find a time-invariant Koopman operator (TIKO) or time-varying Koopman operator (TVKO), there is no difficulties as the dynamics that govern the evolution of lifting functions of the state are expressed with respect to these lifting functions themselves. These subspace identification methods identify the most controllable and observable linear subspace onto which the dominant dynamics evolve, given by a singular value decomposition (SVD) of the collected data. Additionally, selecting a basis of a function space as lifting functions provides the guarantee of the closure of the lifting space, under the dynamics considered. Now, let us consider the controlled version of the dynamics $\dot{x} = x^2$, such that $\dot{x} = x^2 + u$. Note that the controlled action appears linearly. If the state x is considered as a measurement, so does x^2 and u since $\dot{x} = x^2 + u$. If x^2 is considered as a measurement, so does x^3 and xu since $\dot{x^2} = 2\dot{x}x = 2x^3 + 2xu$. This new term, xu, presents two difficulties. The first major difficulty is that the dynamics in a lifted space is *not* linear anymore, nor it is control affine and approximating the controlled nonlinear system by a linear system with affine control would yield poor results. Some previous attempts in this direction showed very mixed results [2,6] and the theory around the Koopman operator for controlled system is not mature enough. Other research works have considered introducing a control input to model chaos dynamics in certain settings [1] but only for autonomous nonlinear systems. Secondly, one could argue that a new type of lifting functions could be introduced, function of both the state and the control vectors. This would lead to identify a Koopman operator that would not only predict future values of the state but also future values of the control input, which is not desirable. Instead, this paper introduces the concept of bilinear Koopman operator. Bilinear state-space model identification is of interest for two main reasons. Some physical systems are inherently bilinear and bilinear models of high dimension can approximate a broad class of nonlinear systems. Nevertheless, no well-established technique for bilinear system identification is available yet, even less in the context of Koopman. The aim of this paper is to offer some perspectives and advances for bilinear system identification, working towards a bilinear Koopman operator.

2   Continuous Bilinear System Identification with Specialized Input

The algorithm presented in [8] relies on the central observation that the bilinear system of equations becomes a linear time invariant system upon the application of constant forcing functions. The authors exhibited the solution of the bilinear system of equations and showed that while the general input output behavior is indeed nonlinear, one can generate an analytical solution for a set of specified inputs. The generic algorithm is presented in details in [8] and the reader can refer to it for more details. In this section, it is desired to present an adapted version of this algorithm when the response of the system is from a set of initial condition $x_0 \neq 0$, by identifying matrices A_c, N_{c_i}, C and the initial condition x_0 for a bilinear dynamical system of the form

$$\dot{x} = A_c x + \sum_{i=1}^{r} N_{c_i} x u_i, \quad x_0 \neq 0, \quad y = C x. \tag{1a}$$

Additionally, the algorithm outlined in this section uses a set of step inputs when other methods in the literature are using pulses, with the only requirement that the step inputs have to go to zero at some point in time and be zero for a few time steps. This weaker condition on the input provides more flexibility to the analyst when adjust the control action (also pulses are very difficult to apply to real mechanical systems for example, with instances where pulse inputs can impair the system). The step by step algorithm is presented below.

1. We perform a set of $N_1 + N_1 \times N_2$ experiments. This set is comprised of N_1 random initial condition response experiments from arbitrary x_0, and for each of them, an additional set of N_2 forced response experiments is performed with step inputs. The requirement is that the step inputs have to go to zero at some point in time and be zero for a few time steps, but can have any profile before or after. We will assume that the input is nonzero at time step k_0 and is zero for p time steps after that. Throughout the description of the procedure we will give conditions on N_1 and N_2.

2. Perform ERA/DMD [4,5,7,13–15] on the first set of N_1 experiments. This allows to obtain a realization of the pair $(\hat{A}, \hat{C})$ (and hence, $\hat{A}_c$) as well as the observability matrix $\hat{O}^{(p)}$. For the identification to capture the full dynamics, it is required that $N_1 \geq n$.

3. For each group of N_2 experiments, build matrices

$$Y_{k_0} = \begin{bmatrix} y_{k_0}^{\#1} & y_{k_0}^{\#2} & \cdots & y_{k_0}^{\#N_2} \end{bmatrix}, \quad Y_{k_0}^{(N_2)} = \begin{bmatrix} y_{k_0}^{\#1} & y_{k_0}^{\#2} & \cdots & y_{k_0}^{\#N_2} \\ y_{k_0+1}^{\#1} & y_{k_0+1}^{\#2} & \cdots & y_{k_0+1}^{\#N_2} \\ \vdots & \vdots & \ddots & \vdots \\ y_{k_0+p-1}^{\#1} & y_{k_0+p-1}^{\#2} & \cdots & y_{k_0+p-1}^{\#N_2} \end{bmatrix}. \tag{2a}$$

Calculate the identified state at time k_0: $\hat{x}_{k_0} = \hat{C} Y_{k_0}$ and the matrix

$$F = \frac{1}{\Delta t} \log \left(\hat{O}^{(p)\dagger} Y_{k_0}^{(N_2)} \hat{x}_{k_0}^{\dagger} \right) - \hat{A}_c. \tag{3}$$

4. Repeat the procedure N_1 times and populate the matrix

$$N_c = \begin{bmatrix} F^{\#1} & F^{\#2} & \cdots & F^{\#N_1} \end{bmatrix}. \tag{4}$$

In parallel, build the matrix

$$V_{k_0}^{(N_2)} = \begin{bmatrix} Iu_{1,k_0}^{\#1} & Iu_{1,k_0}^{\#2} & \cdots & Iu_{1,k_0}^{\#N_2} \\ Iu_{2,k_0}^{\#1} & Iu_{2,k_0}^{\#2} & \cdots & Iu_{2,k_0}^{\#N_2} \\ \vdots & \vdots & \ddots & \vdots \\ Iu_{r,k_0}^{\#1} & Iu_{r,k_0}^{\#2} & \cdots & Iu_{r,k_0}^{\#N_2} \end{bmatrix}. \tag{5}$$

The identified bilinear matrices $\hat{N}_{c_i}$ are: $\begin{bmatrix} \hat{N}_{c_1} & \hat{N}_{c_2} & \cdots & \hat{N}_{c_r} \end{bmatrix} = N_c V_{k_0}^{(N_2)\dagger}$. The matrix $V_{k_0}^{(N_2)}$ is invertible if full rank hence $N_2 \geq r$ and a rich input.

5. Initial condition $\hat{x}_0$ can then be identified similarly as in the ERA procedure solving a least-squares problem.

That procedure will be used in subsequent sections for sensitivity analysis. Figure 1 summarizes the overall procedure for bilinear system identification.

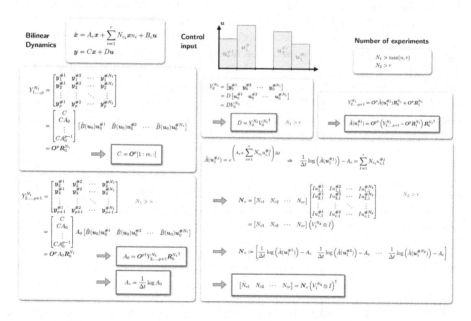

Fig. 1. Overview of the bilinear system identification framework

3   Numerical Simulations

3.1   Hovering Helicopter

The example of a hovering helicopter under wind disturbance as well as model parameter uncertainties [24,28] is considered. The dynamics of the system are given by

$$\dot{x} = Ax + B\delta + B_w u_w \tag{6}$$

where

$$x = \begin{bmatrix} u_h & q_h & \theta_h & y \end{bmatrix}^T, \quad A = \begin{bmatrix} p_1 & p_2 & -g & 0 \\ p_3 & p_4 & 0 & 0 \\ 0 & 1 & 0 & 0 \\ 1 & 0 & 0 & 0 \end{bmatrix}, \quad B = \begin{bmatrix} p_5 \\ p_6 \\ 0 \\ 0 \end{bmatrix}, \quad B_w = \begin{bmatrix} -p_1 \\ -p_3 \\ 0 \\ 0 \end{bmatrix}, \tag{7}$$

u_h [ft/s] represents the horizontal velocity of the helicopter, θ_h [$\times 10^{-2}$ rad] represents the pitch angle, q_h [$\times 10^{-2}$ rad/s] represents the pitch angular velocity and y [ft] represents the horizontal perturbation from a ground point reference. g corresponds to the acceleration due to gravity and is equal to 0.322. δ represents the control input to the system. u_w represents the wind disturbance on the helicopter and is modeled as a zero mean Gaussian white noise with variance $\sigma_w^2 = 18$. The model comprises of six model parameters p_1 to p_6. The first four parameters p_1 to p_4 represent the aerodynamic stability derivatives while the parameters p_5 and p_6 represent the aerodynamic control derivatives. For identification purposes, initial conditions to the system are assumed to be zero: $x_0 = 0$. The control law implemented is that of a full state feedback [24] where $\delta = -Kx$, and $K = \begin{bmatrix} 1.9890 & 0.2560 & 0.7589 & 1 \end{bmatrix}^T$. On substituting the control law in the original system, one obtains the closed-loop stochastic system

$$\dot{x} = A_c x + B_w u_w \tag{8}$$

where $A_c = A - BK$. Similar to Ref. [24], it is assumed that parameters $p = \begin{bmatrix} p_1 & p_2 & p_3 & p_4 \end{bmatrix}^T$ are uncertain. Equation 8 can be re-written in the form of a bilinear system

$$\dot{x} = \tilde{A}_c x + \sum_{i=1}^{4} N_{c_i} x u_i + B_c u, \tag{9}$$

with the augmented input vector being $u = \begin{bmatrix} p_1 & p_2 & p_3 & p_4 & p_1 u_w & p_3 u_w \end{bmatrix}^T$. Continuous time-invariant system matrices $\tilde{A}_c$, B_c and N_{c_i} can be derived using Eq. (6) to Eq. (8). For testing, the parameters p_1 to p_4 are monotonically varied between the lower and upper bounds

$$p_{lb} = \begin{bmatrix} -0.0488 & 0.0013 & 0.126 & -3.3535 \end{bmatrix}^T, \quad p_{ub} = \begin{bmatrix} -0.0026 & 0.0247 & 2.394 & -0.1765 \end{bmatrix}^T. \tag{10}$$

Prediction errors come out to be at 10^{-9} with eigenvalues of A_c and bilinear matrices N_c matching up to machine precision,

$$\left\|\lambda\left(A_c\right) - \lambda\left(\hat{A}_c\right)\right\| \simeq 10^{-12} \quad \cdot \quad \left\|\lambda\left(N_c\right) - \lambda\left(\hat{N}_c\right)\right\| \simeq 10^{-12} \quad (11)$$

The identified bilinear model is able to reproduce the dynamics of the true model and Fig. 2 shows the sensitivity of the state vector with respect to the first parameter p_1.

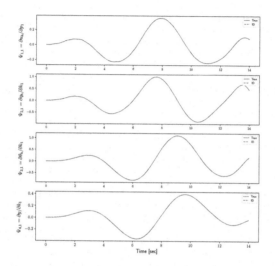

Fig. 2. Sensitivity accuracy for the hovering helicopter

3.2 Controlled Duffing Oscillator

This example corresponds to the controlled nonlinear Duffing oscillator governed by the following equations

$$\dot{x} = y + g_1 u_1, \qquad \dot{y} = -\delta y - \alpha x - \beta x^3 + g_2 u_2 \qquad (12a)$$

with parameters $\alpha = 1$, $\delta = -0.1$, $g_1 = 0$ and $g_2 = 1$. For this example, we want to study the capabilities of a bilinear system to approximate nonlinear dynamics in presence of an external input, using the Koopman framework. When the value of β is significant, the nonlinear term in Eq. (12)() has a huge impact on the approximation capabilities. A classic bilinear approach is valid for small nonlinearities in contained domains, but has the domain of interest grows and the nonlinear term becomes more and more significant, an other approach is desired. One could augment the measurement vector with additional lifting functions, giving rise to a bilinear Koopman operator. Figures 3(a) to 3(f) present the approximation capabilities with increasing order of the bilinear Koopman operator. Reaching order 6, or a dimension of the operator of 27, the approximation is excellent, confirming that a bilinear system identification approach in conjunction with the Koopman framework is a valid method to approximate controlled nonlinear systems.

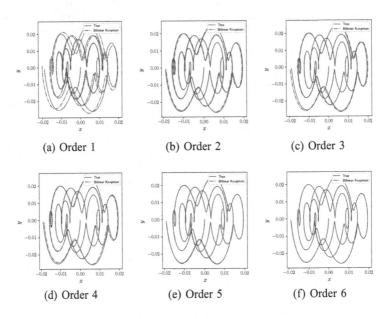

(a) Order 1   (b) Order 2   (c) Order 3

(d) Order 4   (e) Order 5   (f) Order 6

Fig. 3. Koopman bilinear representation of the nonlinear Duffing oscillator, $\beta = 5$

4   Conclusion

This paper has introduced the concept of bilinear Koopman operator. Controlled systems in the context of Koopman yield bilinear dynamics in a lifted space. Since some physical systems are inherently bilinear and bilinear models of high dimension can approximate a broad class of nonlinear systems, this paper has offered some perspectives and advances for bilinear system identification, working towards a bilinear Koopman operator. Several numerical simulations confirm the growing interest in bilinear system identification and validate the methods and algorithms presented. The same framework is also successfully employed for sensitivity analysis of nonlinear systems where it is desired to estimate the variation of a measured output given the deviation of a constitutive parameter of the system.

Acknowledgement. This material is based upon work supported by the AFOSR grant FA9550-15-1-0313 and FA9550-20-1-0176.

References

1. Brunton, S.L., Brunton, B.W., Proctor, J.L., Kaiser, E., Kutz, J.N.: Chaos as an intermittently forced linear system. Nature Commun. **8**(19) (2017). https://doi. org/10.1038/s41467-017-00030-8

2. Brunton, S.L., Brunton, B.W., Proctor, J.L., Kutz, J.N.: Koopman invariant subspaces and finite linear representations of nonlinear dynamical systems for control. PLoS ONE **11**(1), e0150171 (2016)

3. Guého, D.: Data-Driven Modeling for Analysis and Control of Dynamical Systems. Ph.D. thesis, The Pennsylvania State University (2022)

4. Juang, J.N., Cooper, J.E., Wright, J.R.: An eigensystem realization algorithm using data correlation (era/dc) for modal parameter identification. Control Theory Adv. Technol. **4**(1), 5–14 (1988)

5. Juang, J.N., Pappa, R.S.: An eigensystem realization algorithm (era) for modal parameter identification and model reduction. J. Guid. Control. Dyn. **8**(5), 620–627 (1985). https://doi.org/10.2514/3.20031

6. Korda, M., Mezic, I.: Linear predictors for nonlinear dynamical systems: Koopman operator meets model predictive control. Automatica **93**, 149–160 (2018)

7. Kutz, J.N., Brunton, S.L., Brunton, B.W., Proctor, J.L.: Dynamic Mode Decomposition: Data-Driven Modeling of Complex Systems. SIAM (2016)

8. Majji, M., Juang, J.N., Junkins, J.L.: Continuous time bilinear system identification using repeated experiments (2009)

9. Majji, M., Juang, J.N., Junkins, J.L.: Observer/kalman-filter time-varying system identification. J. Guid. Control. Dyn. **33**(3), 887–900 (2010). https://doi.org/10. 2514/1.45768

10. Majji, M., Juang, J.N., Junkins, J.L.: Time-varying eigensystem realization algorithm. J. Guid. Control. Dyn. **33**(1), 13–28 (2010). https://doi.org/10.2514/1. 45722

11. Mezić, I.: Spectral properties of dynamical systems, model reduction and decompositions. Nonlinear Dyn. **41**, 309–325 (2005)

12. Mezić, I., Banaszuk, A.: Comparison of systems with complex behavior. Physica D **197**, 101–133 (2004)

13. Rowley, C.W., Mezic, I., Bagheri, S., Schlatter, P., Henningson, D.: Spectral analysis of nonlinear flows. J. Fluid Mech. **641**, 115–127 (2009)

14. Schmid, P.J.: Dynamic mode decomposition of numerical and experimental data. J. Fluid Mech. **656**, 5–28 (2010)

15. Tu, J.H., Rowley, C.W., Luchtenburg, D.M., Brunton, S.L., Kutz, J.N.: On dynamic mode decomposition: theory and applications. J. Comput. Dyn. **1**(2), 391–421 (2014)

16. Williams, M.O., Kevrekidis, I.G., Rowley, C.W.: A data-driven approximation of the koopman operator: extending dynamic mode decomposition. J. Nonlinear Sci. **25**, 1307–1346 (2015)

Main-Track Plenary Presentations - Security

Tracking Dynamic Gaussian Density with a Theoretically Optimal Sliding Window Approach

Yinsong Wang[1]([⊠]) [iD], Yu Ding[2] [iD], and Shahin Shahrampour[1] [iD]

[1] Northeastern University, Boston, MA 02115, USA
wang.yinso@northeastern.edu
[2] Texas A&M University, College Station, TX 77843, USA

Abstract. Dynamic density estimation is ubiquitous in many applications, including computer vision and signal processing. One popular method to tackle this problem is the "sliding window" kernel density estimator. There exist various implementations of this method that use heuristically defined weight sequences for the observed data. The weight sequence, however, is a key aspect of the estimator affecting the tracking performance significantly. In this work, we study the exact mean integrated squared error (MISE) of "sliding window" Gaussian Kernel Density Estimators for evolving Gaussian densities. We provide a principled guide for choosing the optimal weight sequence by theoretically characterizing the exact MISE, which can be formulated as constrained quadratic programming. We present empirical evidence with synthetic datasets to show that our weighting scheme indeed improves the tracking performance compared to heuristic approaches.

Keywords: Dynamic Density Tracking · Kernel Density Estimator · Time Series

1 Introduction

Dynamic density estimation is an important topic in various applications, such as manufacturing, sensor networks, and traffic control. One popular choice for tackling this problem is the "sliding window" kernel density estimator [5,7,11,12]. This class of estimators has shown impressive performance in practice, and continuous studies on improving the method have contributed to its success. However, to the best of our knowledge, most existing works focus on the kernel function itself. For example, M-kernel method [12] merges data points to the closest previously defined grid points. Cluster kernel and resampling techniques [5] further improve the merging performance, assuming exponentially decaying importance for older data points. Local region kernel density estimator [2] varies the kernel bandwidth across different regions in the density support. Adaptive bandwidth method [1] updates the kernel bandwidth sequence as new data points are observed. All these studies put a heavy emphasis on the kernel function but use heuristic approaches in weighting the observed data points.

A recent development in natural language processing introduced the attention mechanism [8] for sequential data modeling, which has seen immense success in various

© The Author(s), under exclusive license to Springer Nature Switzerland AG 2024
E. Blasch et al. (Eds.): DDDAS 2022, LNCS 13984, pp. 275–282, 2024.
https://doi.org/10.1007/978-3-031-52670-1_27

fields. The method highlights the importance of correlations between sequential data points, which can be effectively modeled with a weight sequence that captures the importance of each data point. For example, the state-of-the-art natural language processing model BERT [4] represents each word as a weighted average of other words in a sentence to make predictions. Graph attention networks [9] represent each node feature as a weighted average of other nodes features to capture structural information. These developments motivate us to revisit "sliding window" kernel density estimators and improve upon heuristic weighting schemes.

In this work, we investigate the theoretical aspect of dynamic Gaussian density tracking using a "sliding window" Gaussian kernel density estimator. We calculate the exact estimation accuracy in terms of mean integrated squared error (MISE) and show the impact of the weight sequence on the MISE. We prove that MISE can be formulated as a constrained quadratic programming that can lead us to a unique optimal weight sequence. We provide numerical experiments using synthetic dynamic Gaussian datasets to support our theoretical claim that this weighting scheme indeed improves the tracking performance compared to heuristic approaches.

2   Problem Formulation and Algorithm

We focus on dynamic data arriving in batches, where at each time t we observe a batch of n_t data points $\{x_j^{(t)}\}_{j=1}^{n_t}$. This data structure applies to many real-world time-series datasets [3]. We assume that at time t, the data points are sampled from a Gaussian distribution, i.e., the true density has the following form

$$p_t(x) = \phi_{\gamma_t}(x - \mu_t) \triangleq \frac{1}{\sqrt{2\pi}\gamma_t} e^{\frac{-(x-\mu_t)^2}{2\gamma_t^2}}.$$

The density evolution can then be uniquely identified with sequences of means and standard deviations. For a dataset $\mathcal{D}$ over a time span of m, we have m batches, i.e., $\mathcal{D} = \{\mathcal{X}_i\}_{i=1}^m$, where $\mathcal{X}_i = \{x_j^{(i)} \sim N(\mu_i, \gamma_i^2)\}_{j=1}^{n_i}$ and $N(\mu_i, \gamma_i^2)$ denotes the Gaussian distribution with mean μ_i and standard deviation γ_i.

"Sliding Window" Kernel Density Estimator: Let us restrict our attention to Gaussian kernel density estimators where

$$K_\sigma(x - x') = \phi_\sigma(x - x') \triangleq \frac{1}{\sqrt{2\pi}\sigma} e^{\frac{-(x-x')^2}{2\sigma^2}},$$

where σ denotes the kernel bandwidth. Then, the dynamic density estimator has the following form

$$\hat{h}_t(x) = \sum_{i=1}^T \alpha_i \hat{p}_i(x) = \sum_{i=1}^T \frac{\alpha_i}{n_i} \sum_{j=1}^{n_i} K_\sigma(x - x_j^{(i)}), \tag{1}$$

where a window size of T previous batches is taken into account. The weights $\alpha^{(t)} \triangleq [\alpha_1^{(t)}, \ldots, \alpha_T^{(t)}]^\top$ vary over time t, but the superscript (t) is omitted for the presentation simplicity. Each "sliding window" kernel density estimator is essentially a weighted average of kernel density estimators, and using $\sum_{i=1}^{T} \alpha_i = 1$, we can ensure that the estimator (1) is a proper density function.

Mean Integrated Squared Error: MISE is a popular metric to characterize the accuracy of density estimators [6, 10]. For any density estimator $\hat{h}(x)$ and the true density $p(x)$, the MISE is formally defined as

$$MISE(\hat{h}(x)) \triangleq \int \mathbb{E}[(\hat{h}(x) - p(x))^2]dx,$$

where the expectation is taken over the randomness of the data samples (sampled from density $p(x)$), and the integral over x accounts for the accumulated error over the support of the density function. MISE is often decomposed into *bias* and *variance* terms as follows

$$
\begin{aligned}
MISE(\hat{h}(x)) &= \int \mathbb{E}[(\hat{h}(x) - \mathbb{E}[\hat{h}(x)])^2] + (\mathbb{E}[\hat{h}(x)] - p(x))^2 dx \\
&= \int V(\hat{h}(x)) + B^2(\hat{h}(x))dx = IV(\hat{h}(x)) + IB^2(\hat{h}(x)),
\end{aligned}
\tag{2}
$$

where $V(\cdot)$ and $B^2(\cdot)$ are called the variance and squared bias, and $IV(\cdot)$ and $IB^2(\cdot)$ are called the integrated variance and integrated squared bias, respectively.

3 Theoretical Result: Optimal Weight Sequence

We now present our main theorem, which states that the exact MISE of the "sliding window" kernel density estimators of evolving Gaussian densities can be calculated, and it is a quadratic function of the weight sequence $\boldsymbol{\alpha}$.

Theorem 1. *Estimating the evolving Gaussian density $p_t(x)$ with the "sliding window" Gaussian Kernel Density Estimator (1) results in the following exact mean integrated squared error*

$$MISE(\hat{h}_t(x)) = \boldsymbol{\alpha}^\top \boldsymbol{\Lambda} \boldsymbol{\alpha} - 2\boldsymbol{\theta}^\top \boldsymbol{\alpha} + \frac{1}{2\gamma_t \sqrt{\pi}},$$

where $\boldsymbol{\Lambda} = \boldsymbol{\Phi} + \mathbf{D}$, and $\boldsymbol{\Phi} \in \mathbb{R}^{T \times T}$ is such that $[\boldsymbol{\Phi}]_{ij} = \phi_{(\sigma^2 + \gamma_i^2 + \sigma^2 + \gamma_j^2)^{1/2}}(\mu_i - \mu_j)$. $\mathbf{D} \in \mathbb{R}^{T \times T}$ is a diagonal matrix with $[\mathbf{D}]_{ii} = \frac{1}{n_i 2\sqrt{\pi}}(\frac{1}{\sigma} - \frac{1}{\sqrt{\sigma^2 + \gamma_i^2}})$, and the i-th entry of vector $\boldsymbol{\theta}$ is $\phi_{(\sigma^2 + \gamma_i^2 + \gamma_t^2)^{1/2}}(\mu_i - \mu_t)$. To connect with (2), we have the following

$$IB^2(\hat{h}(x)) = \boldsymbol{\alpha}^\top \boldsymbol{\Phi} \boldsymbol{\alpha} - 2\boldsymbol{\theta}^\top \boldsymbol{\alpha} + \frac{1}{2\gamma_t \sqrt{\pi}},$$

$$IV(\hat{h}(x)) = \boldsymbol{\alpha}^\top \mathbf{D} \boldsymbol{\alpha}.$$

The proof of Theorem 1 can be found in the Appendix. We are now able to propose the following corollary, which suggests that one can find the optimal weight sequence for the estimator (1) by optimizing MISE over weights.

Corollary 1. *The optimal weight sequence under MISE for the dynamic density estimation is determined by the following constrained quadratic programming*

$$\min_{\boldsymbol{\alpha}} \quad \boldsymbol{\alpha}^\top \Lambda \boldsymbol{\alpha} - 2\boldsymbol{\theta}^\top \boldsymbol{\alpha} + \frac{1}{2\gamma_t \sqrt{\pi}}$$

$$s.t. \quad \mathbb{1}^\top \boldsymbol{\alpha} = 1, \quad \alpha_i \geq 0,$$

where $\mathbb{1}$ *is the vector of all ones.*

4   Empirical Results

In this section, we investigate the empirical behavior of the dynamic kernel density estimator for different weighting methods. We first design synthetic dynamic datasets with evolving Gaussian densities following the foundation of our theoretical claim. Then, we measure the MISE of the density estimation over different experimental settings to validate our theoretical results.

4.1   Synthetic Dataset Design

We consider a synthetic dynamic dataset for a time span of 100 batches. Each batch of data points is sampled from an evolving Gaussian density. The design principles of the synthetic dataset are as follows:

- The evolution in the mean of the Gaussian distribution follows a random walk model with the starting point of 0, i.e.,

$$\mu_{t+1} = \mu_t + \epsilon_1^{(t)},$$

  where $\mu_0 = 0$, and $\epsilon_1^{(t)}$ denotes a sample from the uniform distribution on $[-1, 1]$.
- The evolution in the standard deviation of the Gaussian distribution follows a lower bounded random walk model with the starting point of 1, i.e.,

$$\gamma_{t+1} = \max\{\gamma_t + \epsilon_{0.2}^{(t)}, \gamma_0\},$$

  where $\gamma_0 = 1$, and $\epsilon_{0.2}^{(t)}$ denotes a sample from the uniform distribution on $[-0.2, 0.2]$.
- Each batch of data randomly contains 3 to 20 data points, which limits the amount of available data at each time to justify the use of dynamic density estimators and also highlights the impact of data availability in the weightings.

4.2   Experimental Settings

In our experiment, we compare the optimal weighting sequence proposed in Corollary 1 (denoted by "dynamic") with three other popular baseline methods.

- **Current:** The traditional kernel density estimator (KDE), which only uses data points from the current batch for estimation. Basically, the last element of the weight sequence $\boldsymbol{\alpha} \in \mathbb{R}^T$ is equal to 1, and all other elements are 0.

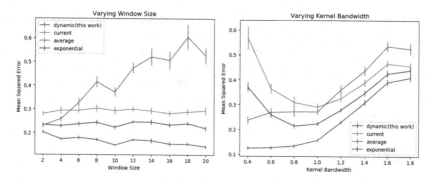

Fig. 1. Left: The estimated MISE versus different window sizes. **Right:** The estimated MISE versus different kernel bandwidths.

- **Average:** This method combines all data points in the window by equally weighting all batches. That is to say, all elements of $\alpha \in \mathbb{R}^T$ are equal to $1/T$.
- **Exponential:** This is one of the most popular weighting methods in common practice. It assumes the correlations between data points decay exponentially over time. The weight sequence $\alpha \in \mathbb{R}^T$ with a decay factor β is as follows

$$\alpha_1 = (1 - \beta)^{T-1}, \quad \{\alpha_i = (1 - \beta)^{T-i}\beta\}_{i=2}^T.$$

In our simulation, we set $\beta = 0.1$, which shows the best performance for this weighting method.

We perform the comparison over two variables, namely the window size T and the kernel bandwidth σ. When comparing different window sizes, we fix the kernel bandwidth for all estimators as $\sigma = 1$, which is a good kernel bandwidth choice for the baseline weighting methods according to our bandwidth comparison. When comparing different bandwidths, we fix the window size $T = 5$, which is also in favor of the baseline weighting methods according to our simulation.

We carry out 20 Monte-Carlo simulations to generate error bars, where we randomly create a brand new synthetic dataset for each Monte-Carlo simulation.

4.3 Performance

The estimated MISE for all methods is shown in Fig. 1. We observe that the optimal weighting sequence (dynamic) achieves the best performance (lowest MISE) in all cases.

For the window size comparison, the current weighting KDE serves as a baseline, where the error does not (statistically) change with different window sizes. We see that the average weighting has a significant error as window size increases. The exponential weighting is consistently better than the current weighting. However, the dynamic weighting achieves the best performance for all window sizes. This observation, together with our theoretical characterization, shed light on why the "sliding window" kernel density estimator works better than the traditional KDE for dynamic density estimation. It introduces a mild bias from previous data points in favor of reducing the variance (induced by low data volume) to improve the estimation accuracy.

For the kernel bandwidth comparison, the current weighting KDE again serves as a baseline, where the optimal performance occurs at $\sigma = 1$. We can see that (for $\sigma \leq 1$) all "sliding window" estimators are better than KDE under a relatively small window size of 5 (low bias), and the optimal weighting method (dynamic) is consistently better than all other benchmark methods.

5   Conclusion

In this work, we theoretically characterized the MISE of the "sliding window" kernel density estimator for evolving Gaussian densities as a quadratic function of the weight sequence. Our result underscores the important role of the weight sequence in the estimation accuracy. We also provided numerical experiments to support our theoretical claim. For future directions, we would like to expand this theory to Gaussian mixtures, which can potentially apply to any dynamic density tracking problem.

Acknowledgements. The authors gratefully acknowledge the support of NSF Award #2038625 as part of the NSF/DHS/DOT/NIH/USDA-NIFA Cyber-Physical Systems Program.

6   Appendix: Proof of Theorem 1

We first state a (standard) property of Gaussian densities in the following lemma.

Lemma 1. *For Gaussian density functions the following relationship holds*

$$\phi_a(x - \mu_a)\phi_b(x - \mu_b) = \phi_{(a^2+b^2)^{1/2}}(\mu_a - \mu_b)\phi_{\frac{ab}{(a^2+b^2)^{1/2}}}(x - \mu_{ab}),$$

where $\mu_{ab} = \frac{a^2\mu_b + b^2\mu_a}{a^2+b^2}$.

Using (1), we can write the bias of the estimator in (2) as

$$B(\hat{h}_t(x)) \triangleq \mathbb{E}[\hat{h}_t(x) - p_t(x)] = \mathbb{E}\left[\sum_{i=1}^{T} \frac{\alpha_i}{n_i} \sum_{j=1}^{n_i} K_\sigma(x - x_j^{(i)}) - p_t(x)\right]$$

$$= \sum_{i=1}^{T} \alpha_i \int K_\sigma(x - y)p_i(y)dy - p_t(x) = \sum_{i=1}^{T} \alpha_i (K_\sigma * p_i)(x) - p_t(x),$$

$$\tag{3}$$

where $*$ denotes the convolution, and $p_i(\cdot)$ is the true density of batch i. The estimator variance in (2) can also be calculated as

$$V(\hat{h}_t(x)) = \sum_{i=1}^{T} \alpha_i^2 V(\hat{p}_i(x)),$$

due to the independence, where

$$V(\hat{p}_i(x)) = \frac{1}{n_i}\left((K_\sigma^2 * p_i)(x) - (K_\sigma * p_i)^2(x)\right). \tag{4}$$

Given the expressions of bias (3) and variance (4), to calculate the exact MISE, we need to characterize the quantities $(K_\sigma^2 * p_i)(x)$ and $(K_\sigma * p_i)(x)$. First, we consider $(K_\sigma * p_i)(x)$ and use Lemma 1 to get

$$(K_\sigma * p_i)(x) = \int \phi_\sigma(x-y)\phi_{\gamma_i}(y-\mu_i)dy = \phi_{(\sigma^2+\gamma_i^2)^{1/2}}(x-\mu_i).$$

We further characterize $(K_\sigma^2 * p_i)(x)$ as following

$$\begin{aligned}(K_\sigma^2 * p_i)(x) &= \int \phi_\sigma^2(x-y)\phi_{\gamma_i}(y-\mu_i)dy \\ &= \frac{1}{2\sigma\sqrt{\pi}}\int \phi_{\sigma/\sqrt{2}}(x-y)\phi_{\gamma_i}(y-\mu_i)dy \\ &= \frac{1}{2\sigma\sqrt{\pi}}\phi_{(\sigma^2/2+\gamma_i^2)^{1/2}}(x-\mu_i).\end{aligned}$$

Having the above expressions, we can calculate the integrated variance term as

$$IV(\hat{p}_i(x)) = \frac{1}{n_i 2\sqrt{\pi}}\left(\frac{1}{\sigma} - \frac{1}{\sqrt{\sigma^2+\gamma_i^2}}\right),$$

using Lemma 1. The exact MISE depends also on the integrated bias square (2), which takes the following expression

$$IB(\hat{h}_t(x))^2 = \int\left(\sum_{i=1}^{T}\alpha_i\phi_{(\sigma^2+\gamma_i^2)^{1/2}}(x-\mu_i) - \phi_{\gamma_t}(x-\mu_t)\right)^2 dx.$$

There exist three types of terms in the above expression, which we examine one by one. First, we look at the interaction terms of the following form

$$\int \alpha_i\phi_{(\sigma^2+\gamma_i^2)^{1/2}}(x-\mu_i)\alpha_j\phi_{(\sigma^2+\gamma_j^2)^{1/2}}(x-\mu_j)dx = \alpha_i\alpha_j\phi_{(\sigma^2+\gamma_i^2+\sigma^2+\gamma_j^2)^{1/2}}(\mu_i-\mu_j),$$

which follows from Lemma 1. The second type is the interactions terms where

$$\int \alpha_i\phi_{(\sigma^2+\gamma_i^2)^{1/2}}(x-\mu_i)\phi_{\gamma_t}(x-\mu_t)dx = \alpha_i\phi_{(\sigma^2+\gamma_i^2+\gamma_t^2)^{1/2}}(\mu_i-\mu_t).$$

The last term is the square of $p_t(x)$, for which we have $\int p_t^2(x)dx = \frac{1}{2\gamma_t\sqrt{\pi}}$.

Given the above expressions, we can write the square integrated bias as a quadratic function of weight sequence α as follows

$$IB(\hat{h}_t(x))^2 = \alpha^\top \Phi\alpha - 2\theta^\top\alpha + \frac{1}{2\gamma_t\sqrt{\pi}},$$

where the matrix $\boldsymbol{\Phi} \in \mathbb{R}^{T \times T}$ is such that $[\boldsymbol{\Phi}]_{ij} = \phi_{(\sigma^2 + \gamma_i^2 + \sigma^2 + \gamma_j^2)^{1/2}}(\mu_i - \mu_j)$, and the i-th entry of vector $\boldsymbol{\theta}$ is $\phi_{(\sigma^2 + \gamma_i^2 + \gamma_t^2)^{1/2}}(\mu_i - \mu_t)$. By the same token, we can write the variance term as

$$IV(\hat{h}_t(x)) = \boldsymbol{\alpha}^\top \mathbf{D} \boldsymbol{\alpha},$$

where $\mathbf{D}$ is a diagonal matrix with $[\mathbf{D}]_{ii} = \frac{1}{n_i 2\sqrt{\pi}}\left(\frac{1}{\sigma} - \frac{1}{\sqrt{\sigma^2 + \gamma_i^2}}\right)$. This completes the proof of Theorem 1.

References

1. Amiri, A., Dabo-Niang, S.: Density estimation over spatio-temporal data streams. Econometrics Stat. **5**, 148–170 (2018)
2. Boedihardjo, A.P., Lu, C.T., Chen, F.: A framework for estimating complex probability density structures in data streams. In: Proceedings of the 17th ACM Conference on Information and Knowledge Management, pp. 619–628 (2008)
3. Dau, H.A., et al.: Hexagon-ML: the UCR time series classification archive, October 2018. https://www.cs.ucr.edu/~eamonn/time_series_data_2018/
4. Devlin, J., Chang, M.W., Lee, K., Toutanova, K.: BERT: pre-training of deep bidirectional transformers for language understanding. arXiv preprint arXiv:1810.04805 (2018)
5. Heinz, C., Seeger, B.: Cluster kernels: resource-aware kernel density estimators over streaming data. IEEE Trans. Knowl. Data Eng. **20**(7), 880–893 (2008)
6. Marron, J.S., Wand, M.P.: Exact mean integrated squared error. Ann. Stat. **20**(2), 712–736 (1992)
7. Qahtan, A., Wang, S., Zhang, X.: KDE-track: an efficient dynamic density estimator for data streams. IEEE Trans. Knowl. Data Eng. **29**(3), 642–655 (2016)
8. Vaswani, A., et al.: Attention is all you need. In: Advances in Neural Information Processing Systems, vol. 30 (2017)
9. Veličković, P., Cucurull, G., Casanova, A., Romero, A., Lio, P., Bengio, Y.: Graph attention networks. arXiv preprint arXiv:1710.10903 (2017)
10. Wand, M.P., Jones, M.C.: Kernel Smoothing. CRC Press, Boston (1994)
11. Wang, Y., Ding, Y., Shahrampour, S.: TAKDE: temporal adaptive kernel density estimator for real-time dynamic density estimation. arXiv preprint arXiv:2203.08317 (2022)
12. Zhou, A., Cai, Z., Wei, L., Qian, W.: M-kernel merging: towards density estimation over data streams. In: Proceedings of the Eighth International Conference on Database Systems for Advanced Applications, (DASFAA 2003), pp. 285–292 (2003)

Dynamic Data-Driven Digital Twins for Blockchain Systems

Georgios Diamantopoulos[1,2], Nikos Tziritas[3], Rami Bahsoon[1],
and Georgios Theodoropoulos[2(✉)]

[1] School of Computer Science, University of Birmingham, Birmingham, UK
[2] Department of Computer Science and Engineering and Research Institute
for Trustworthy Autonomous Systems, Southern University of Science
and Technology (SUSTech), Shenzhen, China
gxd192@student.bham.ac.uk
[3] Department of Informatics and Telecommunications, University of Thessaly,
Volos, Greece

Abstract. In recent years, we have seen an increase in the adoption of blockchain-based systems in non-financial applications, looking to benefit from what the technology has to offer. Although many fields have managed to include blockchain in their core functionalities, the adoption of blockchain, in general, is constrained by the so-called trilemma trade-off between decentralization, scalability, and security. In our previous work, we have shown that using a digital twin for dynamically managing blockchain systems during runtime can be effective in managing the trilemma trade-off. Our Digital Twin leverages DDDAS feedback loop, which is responsible for getting the data from the system to the digital twin, conducting optimisation, and updating the physical system. This paper examines how leveraging DDDAS feedback loop can support the optimisation component of the trilemma benefiting from Reinforcement Learning agent and a simulation component to augment the quality of the learned model while reducing the computational overhead required for decision making.

1 Introduction

Blockchain's rise in popularity is undeniable; many non-financial applications have adopted the technology for its increased transparency, security and decentralisation [19]. Supply chain, e-government, energy management, IoT [3,11,15,21] are among the many systems benefiting from blockchain.

Two main types of blockchain exist, namely, Public and Private [9] with Consortium [14] being a hybrid of the two. From the above, private blockchain systems lend themselves easier to a dynamic control system. In a private blockchain, participating nodes, communicate through a peer-to-peer (P2P) network and hold a personal (local) ledger storing the transactions that take place in the system. This network is private and only identified users can participate. The set of individual ledgers can be viewed as a distributed ledger, denoting the

E. Blasch et al. (Eds.): DDDAS 2022, LNCS 13984, pp. 283–292, 2024.
https://doi.org/10.1007/978-3-031-52670-1_28

global state of the system. A consensus protocol is used to aid in validating and ordering the transactions and through it, a special set of nodes called block producers, vote on the order and validity of recent transactions, and arrange them in blocks. These blocks are then broadcasted to the system, for the rest of the nodes to update their local ledger accordingly. With the above working correctly, the nodes of the system vote on the new state of the ledger, and under the condition of a majority, each local ledger, and thus the global state of the system, is updated to match the agreed new system state. The above eliminates the need for a central authority to update the system state and assures complete transparency.

Despite the potential of Blockchain in many different domains, factors such as low scalability and high latency have limited the technology's adoption, especially in time-critical applications, while in general, blockchain suffers from the so-called trilemma trade-off that is between decentralisation, scalability, and security [27].

The most notable factor affecting the performance of the blockchain, excluding external factors we cannot control such as the system architecture, network, and workload, is the consensus protocol, with system parameters such as block time, and block interval getting a close second. The trilemma trade-off in combination with blockchains time-varying workloads makes the creation of robust, general consensus protocols extremely challenging if not impossible, creating a need for other solutions [8]. Although no general consensus protocol exists, existing consensus protocols perform best under specific system conditions [5,10,13,22]. Additionally, blockchain systems support and are influenced by dynamic changes to the system parameters (including the consensus protocol) during runtime. Thus there is a need for dynamic management of blockchain systems.

Digital Twins and DDDAS have been utilised in autonomic management of computational infrastructures [2,7,16,20] and the last few years have witnessed several efforts to bring together Blockchain and Digital Twins. However, efforts have focused on utilising the former to support the latter; a comprehensive survey is provided in [23]. Similarly, in the context of DDDAS, Blockchain technology has been utilised to support different aspects of DDDAS operations and components [4,25,26].

In our previous work [6], we presented a Digital Twin architecture for the dynamic management of blockchain systems focusing on the optimisation of the trilemma trade-off and we demonstrated its use to optimise a blockchain system for latency. The novel contribution of this work is enriching Digital Twins design for blockchain-based systems with DDDAS-inspired feedback loop. We explore how DDDAS feedback loop principles can support the design of info-symbiotic link connecting the blockchain system with the simulation and analytic environment to dynamically manage the trilemma. As part of the loop, we contribute to a control mechanism that uses Reinforcement Learning agent (RL) and combined with our existing simulation framework. The combination overcomes the limitations of just relying on RL while relaxing the computational overhead required when relying solely on simulation.

The rest of the paper is structured as follows: Sect. 2 discusses the utilisation of Digital Twins for the management of Blockchain systems and provides an overview of a Digital Twin framework for this purpose. Section 3 delves into the DDDAS feedback loop at the core of the Digital Twin and examines its different components. As part of the loop, it proposes a novel optimisation approach based on the combination of an RL and what-if analysis. Section 4 presents a quantitative analysis of the proposed optimisation approach. Finally, Sect. 5 concludes this paper.

2   Digital Twins for Blockchain Systems

For this paper, we consider a generic permissioned blockchain system illustrated as 'Physical System' in Fig. 1 with K nodes denoted as:

$$P = \{p_1, p_2, ..., p_K\} \tag{1}$$

M of which are block producers denoted as:

$$B = \{b_1, b_2, ..., b_M\}, \ B \subset P \tag{2}$$

which take part in the Consensus Protocol (CP) and are responsible for producing the blocks [6]. Additionally, each node $p \in P$ holds a local copy of the Blockchain(BC) while the block producers $b \in B$ also hold a transaction pool (TP) which stores broadcasted transactions.

2.1   Consensus

In the above-described system, nodes produce transactions, which are broadcasted to the system, and stored by the block producers in their individual transaction pools. Each block producer takes turns producing and proposing blocks in a round-robin fashion. Specifically, when it's the turn of a block producer to produce a new block, it first gathers the oldest transactions in the pool, verifies them and packs them into a block, signs the block with its private key and initiates the consensus protocol. The consensus protocol acts as a voting mechanism for nodes to vote on the new state of the system, the new block in this case, and as mentioned earlier, is the main factor affecting the performance of the blockchain. It is pertinent to note that producing blocks in a round robin fashion is simple to implement albeit inefficient due to "missed cycles" caused by invalid blocks or offline nodes [1]. Other alternative implementations are possible, such as having every block producer produce a new block or leaving the selection up to the digital twin.

Although consensus protocols have been studied for many years, due to their use in traditional distributed systems for replication, blockchains larger scale, in combination with the unknown network characteristics of the nodes, make the vast majority of existing work incompatible. Recent works have focused on adapting some of these traditional protocols for the permissioned blockchain, achieving good performance but so far no one has managed to create a protocol achieving good performance under every possible system configuration [8]. With several specialized consensus protocols available, a dynamic control system is a natural solution for taking advantage of many specialised solutions while avoiding their shortcomings.

The idea of trying to take advantage of many consensus protocols is not new, similar concepts already exist in the literature, in the form of hybrid consensus algorithms [12,17] which combine 2 protocols in one to get both benefits of both. Although fairly successful in achieving their goal of getting the benefits of two consensus protocols, hybrid algorithms also combine the shortcomings of the algorithms and are usually lacking in performance or energy consumption. In contrast, a dynamic control system allows for the exploitation of the benefits of the algorithms involved, with the cost of additional complexity in the form of the selection mechanism.

In our previous work [6], we focused on minimizing latency by dynamically changing between 2 consensus protocols Practical Byzantine Fault Tolerance (PBFT) [5] and BigFoot [22]. PBFT acts as a robust protocol capable of efficiently achieving consensus when byzantine behaviour is detected in the system while BigFoot is a fast alternative when there are no signs of byzantine behaviour [22].

To achieve the above, we employed a Digital Twin (DT) coupled with a Simulation Module to conduct what-if analysis, based on which, an optimiser would compute the optimal consensus protocol for the next time step. Using a DT can overcome the shortcomings of relying on an RL agent alone since the simulation element and what-if analysis allow for the exploration of alternative future scenarios [24]. The complete architecture can be seen in Fig. 1.

3   The DDDAS Feedback Loop

The system described in the previous section closely follows the DDDAS paradigm, with the Digital Twin containing the simulation of the physical system, the node feeding data to the Digital Twin acting as the sensors, and the optimiser updating the system closing the feedback loop.

Interacting with the Blockchain System. Blockchains design, in combination with the communication protocol for the consensus, allows block producers to have access to or infer with high accuracy, a large amount of data about the state of the blockchain system. These block producers can act as the sensors of the physical system tasked with periodically sending state data to the digital twin. Specifically, every new transaction and block are timestamped and broadcasted to the system and thus are easily accessible. Using the list of historical transactions, we can develop a model of the workload used in the simulation. Although using queries to request the state of block producers requires a mechanism to overcome the Byzantine node assumption, blocks contain a large amount of data which could make the above obsolete. Each new block contains an extra

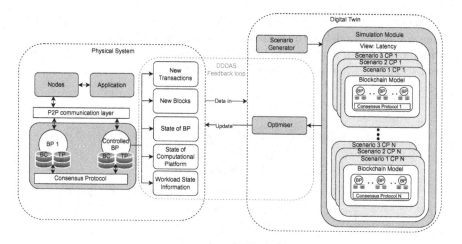

Fig. 1. Digital Twin Architecture and DDDAS feedback loop

data field in which the full timestamped history of the consensus process is stored and can be used to infer the state of the Block producers. Specifically, through blocks, we can learn the state of block producers (offline/online), based on whether the node participated in the consensus protocol or not, as well as develop a model of the block producers and their failure frequency. Additionally, using the relative response times we can infer a node's network state, and update it over time. With all of the above, a fairly accurate simulation of the blockchain system can be achieved.

Updating the Model and Controlling the Physical System. Relying on simulation to calculate the optimal system parameters is a computationally expensive approach [6]. As the optimisation tasks get more complicated, with multiple views taken into account (Fig. 1), smart contract simulation, harder to predict workloads, and especially once the decision-making process gets decentralised and replicated over many block producers, conducting what-if analysis becomes taxing on the hardware. Depending on the case i.e. energy aware systems or systems relying on low-powered/battery-powered nodes might not be able to justify such an expensive process or worst case, the cost of optimisation can start to outweigh the gains.

3.1 Augmenting Reinforcement Learning with Simulation

In this paper, we propose the use of a Reinforcement Learning (RL) agent in combination with simulation and what-if analysis to overcome the individual shortcomings of each respective technique. Reinforcement Learning trained on historical data cannot, on its own, provide a nonlinear extrapolation of future scenarios, essential in modelling complex systems such as blockchain [18], while simulation can be computationally expensive. By using the simulation module to augment the training with what-if generated decisions the agent can learn a more complete model of the system improving the performance of the agent. Additionally, what-if analysis can be used when the agent encounters previously unseen scenarios, avoiding the risk of bad decisions (Fig. 2).

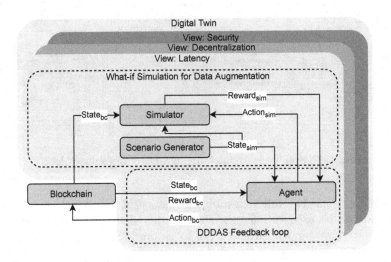

Fig. 2. General architecture for RL based control

For the optimisation of the trilemma trade-off, views are utilised with each view specialised in a different aspect of the trilemma [6]. In this case, the DDDAS component may be viewed as consisting of multiple feedback loops one for each aspect of optimisation. By splitting the DDDAS into multiple feedback loops, we can allow for finer control of both the data needed and the frequency of the updates. Additionally, moving away from a monolithic architecture allows for a more flexible, and scalable architecture. Specifically, each view consists of two components: the DDDAS feedback loop and the training data augmentation loop. The DDDAS feedback loop contains the RL agent which is used to update the system. The what-if simulation component includes the simulation module (or simulator) and the Scenario Generator. The data gathered from the physical system are used to update the simulation model while the scenario generator generates what-if scenarios, which are evaluated and used in the training of the agent. In Fig. 3 a high-level architecture of the proposed system can be seen.

4   Experimental Setup and Evaluation

Experimental Setup. To illustrate the utilisation of RL-based optimisation and analyse the impact of using simulation to enhance the RL agent, a prototype implementation of the system presented in Fig. 3 has been developed focusing on latency optimisation. More specifically we consider the average transaction latency defined as $\frac{\sum_i^{T_B} Time_B - Time_{T_i}}{T_B}$, with T_B denoting the number of transactions in the block B, T_i the i_{th} transaction in B and $Time_B$, $Time_{T_i}$ the time B and T_i were added to the system, respectively.

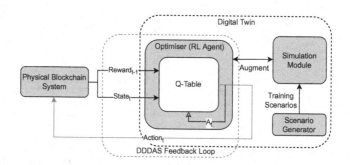

Fig. 3. An example instantiation of RL-based control: Latency View

For the experiments, a general permissioned blockchain system like the one described as "Physical System" in Fig. 1, was used with 5 nodes taking part in the consensus protocol. Two consensus algorithms were implemented specifically, PBFT and BigFoot which complement each other as shown in [6]. The scenario generator created instances of the above blockchain system by randomly generating values for the following parameters: (a) Transactions Per Second (TPS) which denotes the number of transactions the nodes generate per second; (b) T which denotes the size of the transactions; (c) Node State which signifies when and for how long nodes go offline; and (d) Network State which denotes how the network state fluctuates over time. Additionally, following our previous approach, we assume that the system does not change state randomly, but does so in time intervals of length TI. Finally, the digital twin updates the system in regular time steps of size TS.

A Q-Learning agent has been used. The state of the system S is defined as $S = (F, N_L, N_H)$ with F being a binary metric denoting whether the system contains a node which has failed, and N_L, N_H denoting the state of the network by represented by the lower and upper bounds of the network speeds in Mbps rounded to the closest integer. The action space is a choice between the two consensus protocols and the reward function is simply the average transaction latency of the optimised TS as measured in the physical system.

Results. For evaluating the performance of the proposed optimiser, the average transaction latency was used. Specifically, two workloads (WL1, and WL2) were generated using the scenario generator. WL1 was used for the training of the agent (Fig. 4a), while WL2 was used to represent the system at a later stage, where the state has evolved over time. Two approaches were used for the optimisation of WL2: (a) the agent on its own with no help from the simulator and (b) the agent augmented with simulation in the form of what-if analysis.

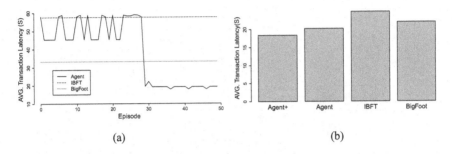

(a)   (b)

Fig. 4. Results of the experimental evaluation with (a) showing the training performance of the agent on WL1 (b) the performance of the agent and the agent + simulation (denoted as agent+) for WL2

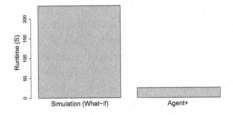

Fig. 5. Comparison of the runtimes of simulation-based optimisation and agent + simulation

As shown in Fig. 4 the agent achieves good training performance on WL1 managing to outperform both algorithms on their own. In WL2 the agent's performance is shown to decrease in comparison to that of the agent augmented with the simulation (agent+) (Fig. 4b). Additionally, Fig. 5 shows the runtime of the agent+ as compared to that of the what-if-based optimiser demonstrating the agent's efficiency. The increased performance in combination with the reduced computational overhead of the agent+, greatly increases the potential of the proposed framework to be used in low-powered/energy-aware systems.

5   Conclusions

Leveraging on our previous work on utilising Digital Twins for dynamically managing the trilemma trade-off in blockchain systems, in this paper we have focused on the DDDAS feedback loop that links the Digital twin with the blockchain system. We have elaborated on the components and challenges to implement the loop. A key component of the feedback loop is an optimiser and we have proposed a novel optimisation approach for the system. The optimiser combines Re-enforcement Learning and Simulation to take advantage of the efficiency of the agent with the accuracy of the simulation. Our experimental results confirm that the proposed approach not only can successfully increase the performance of the agent but do so more efficiently, requiring less computational overhead.

Acknowledgements. This research was supported by: Shenzhen Science and Technology Program, China (No. GJHZ20210705141807022); SUSTech-University of Birmingham Collaborative PhD Programme; Guangdong Province Innovative and Entrepreneurial Team Programme, China (No. 2017ZT07X386); SUSTech Research Institute for Trustworthy Autonomous Systems, China.

References

1. Byzantine fault tolerance round robin proposal. https://github.com/ethereum/EIPs/issues/650
2. Abar, et al.: Automated dynamic resource provisioning and monitoring in virtualized large-scale datacenter. In: 2014 IEEE 28th International Conference on Advanced Information Networking and Applications, pp. 961–970 (2014). https://doi.org/10.1109/AINA.2014.117
3. Andoni, M., et al.: Blockchain technology in the energy sector: a systematic review of challenges and opportunities. Renew. Sustain. Energy Rev. **100**, 143–174 (2019)
4. Blasch, et al.: A study of lightweight DDDAS architecture for real-time public safety applications through hybrid simulation. In: 2019 Winter Simulation Conference (WSC), pp. 762–773 (2019). https://doi.org/10.1109/WSC40007.2019.9004727
5. Castro, et al.: Practical byzantine fault tolerance. In: OSDI, vol. 99, pp. 173–186 (1999)
6. Diamantopoulos, G., Tziritas, N., Bahsoon, R., Theodoropoulos, G.: Digital twins for dynamic management of blockchain systems. arXiv preprint arXiv:2204.12477 (2022)
7. Faniyi, et al.: A dynamic data-driven simulation approach for preventing service level agreement violations in cloud federation. Proc. Comput. Sci. **9**, 1167–1176 (2012). https://doi.org/10.1016/j.procs.2012.04.126. https://www.sciencedirect.com/science/article/pii/S1877050912002475, proceedings of the International Conference on Computational Science, ICCS 2012
8. Giang-Truong, et al.: A survey about consensus algorithms used in blockchain. J. Inf. Process. Syst. **14**(1) (2018)
9. Guegan, D.: Public blockchain versus private blockhain (2017)
10. Guerraoui, et al.: The next 700 BFT protocols. In: Proceedings of the 5th European Conference on Computer Systems, EuroSys 2010, New York, NY, USA, pp. 363–376. Association for Computing Machinery (2010). https://doi.org/10.1145/1755913.1755950
11. Gürpinar, T., Guadiana, G., Asterios Ioannidis, P., Straub, N., Henke, M.: The current state of blockchain applications in supply chain management. In: 2021 The 3rd International Conference on Blockchain Technology, pp. 168–175 (2021)

12. Huang, et al.: Incentive assignment in hybrid consensus blockchain systems in pervasive edge environments. IEEE Trans. Comput. **71**, 2102–2115 (2021)

13. Kotla, et al.: Zyzzyva: speculative byzantine fault tolerance. In: Proceedings of Twenty-First ACM SIGOPS Symposium on Operating Systems Principles, pp. 45–58 (2007)

14. Li, et al.: Consortium blockchain for secure energy trading in industrial internet of things. IEEE Trans. Ind. Inform. **14**(8), 3690–3700 (2017)

15. Liang, X., Zhao, J., Shetty, S., Li, D.: Towards data assurance and resilience in IoT using blockchain. In: MILCOM 2017–2017 IEEE Military Communications Conference (MILCOM), pp. 261–266. IEEE (2017)

16. Liu, et al.: Towards an agent-based symbiotic architecture for autonomic management of virtualized data centers. In: Proceedings of the Winter Simulation Conference, WSC 2012, Winter Simulation Conference (2012)

17. Liu, et al.: Fork-free hybrid consensus with flexible proof-of-activity. Futur. Gener. Comput. Syst. **96**, 515–524 (2019)

18. Liu, et al.: Performance optimization for blockchain-enabled industrial internet of things (IIoT) systems: a deep reinforcement learning approach. IEEE Trans. Ind. Inform. **15**(6), 3559–3570 (2019)

19. Mansfield-Devine, S.: Beyond bitcoin: using blockchain technology to provide assurance in the commercial world. Comput. Fraud Secur. **2017**(5), 14–18 (2017)

20. Onolaja, et al.: Conceptual framework for dynamic trust monitoring and prediction. Procedia Comput. Sci. **1**(1), 1241–1250 (2010). https://doi.org/10.1016/j.procs.2010.04.138, iCCS 2010

21. Owens, J.: Blockchain 101 for governments. In: Wilton Park Conference, pp. 27–29 (2017)

22. Saltini, R.: BigFooT: a robust optimal-latency BFT blockchain consensus protocol with dynamic validator membership. Comput. Netw. **204**, 108632 (2022)

23. Suhail, et al.: Blockchain-based digital twins: research trends, issues, and future challenges. ACM Comput. Surv. (2022). https://doi.org/10.1145/3517189

24. Theodoropoulos, G.: Simulation in the era of big data: trends and challenges. In: Proceedings of the 3rd ACM SIGSIM Conference on Principles of Advanced Discrete Simulation. SIGSIM PADS 2015, New York, NY, USA, p. 1. Association for Computing Machinery (2015). https://doi.org/10.1145/2769458.2769484

25. Xu, et al.: Exploration of blockchain-enabled decentralized capability-based access control strategy for space situation awareness. Opt. Eng. **58**(4) (2019)

26. Xu, et al.: Hybrid blockchain- enabled secure microservices fabric for decentralized multi-domain avionics systems. In: Proceedings of Sensors and Systems for Space Applications XIII, vol. 11422 (2020)

27. Zhou, Q., Huang, H., Zheng, Z., Bian, J.: Solutions to scalability of blockchain: a survey. IEEE Access **8**, 16440–16455 (2020)

Adversarial Forecasting Through Adversarial Risk Analysis Within a DDDAS Framework

Tahir Ekin[1]([✉])[iD], Roi Naveiro[3][iD], and Jose Manuel Camacho Rodriguez[2][iD]

[1] Texas State University, San Marcos, TX 78666, USA
tahirekin@txstate.edu
[2] ICMAT, 28049 Madrid, Spain
josemanuel.camacho@icmat.es
[3] CUNEF Universidad, Calle Almansa 101, 28040 Madrid, Spain
roi.naveiro@cunef.edu

Abstract. Forecasting methods typically assume clean and legitimate data streams. However, adversaries' manipulation of digital data streams could alter the performance of forecasting algorithms and impact decision quality. In order to address such challenges, we propose a dynamic data driven application systems (DDDAS) based decision making framework that includes an adversarial forecasting component. Our framework utilizes the adversarial risk analysis principles that allow considering incomplete information and uncertainty. It is demonstrated using a load forecasting example. We solve the adversary's decision problem in which he poisons data to alter an auto regressive forecasting algorithm output, and discuss defender strategies addressing the attack impact.

Keywords: Adversarial forecasting · Adversarial risk analysis · Dynamic data driven applications systems · Load forecasting

1 Introduction

Most integrated decision-making frameworks utilize forecasts as inputs. These forecasting models are typically based on the classical assumption of clean and legitimate data. However, they are not immune to attacks. Adversaries with access to model and/or data inputs may attempt to influence forecasts for their own advantage; which in turn may impact the decision integrity. Examples are often seen in e-commerce, military, and electricity load management. This

This work is supported by Air Force Scientific Office of Research (AFOSR) award FA-9550-21-1-0239 and AFOSR European Office of Aerospace Research and Development award FA8655-21-1-7042. J.M.C. is supported by a fellowship from "la Caixa" Foundation (ID100010434), whose code is LCF/BQ/DI21/11860063. Any opinions, findings, and conclusions or recommendations expressed are those of the authors and do not necessarily reflect the views of the sponsors.

manuscript focuses on load forecasting whose vulnerability could impact energy resilience.

Classical decision theory generally assumes that decisions are made absent of adversaries; uncertainties are controlled by Nature. However, strategic-minded adversaries could have incentives to perturb input data to thwart decision making. Furthermore, most classical game-theoretic models assume some common knowledge and mutual rationality. We present a DDDAS framework rooted in adversarial risk analysis (ARA) that includes an adversarial forecasting component. In so doing, we assimilate ARA within the broader DDDAS framework. DDDAS is a systems-design framework that incorporates statistical methods and computational architectures with data collection and high-dimensional physical model application. Given its flexible and tailorable nature, ARA is a well-suited foundation upon which to build decision making methods robust to adversarially perturbed inputs. ARA contextualizes games as decision-theoretic problems from the expected-utility-maximizing perspective of a given player.

The major contribution of this paper is to adapt the principles of ARA to develop a DDDAS framework for adversarial forecasting. To the best of our knowledge, there is not an existent decision making framework flexible enough to accommodate decision settings characterized by incomplete information, adversarial data, and multi-agent interactions in a dynamic fashion. The proposed framework is illustrated with a load forecasting example by using an adversarial seasonal auto regressive integrated moving average model.

2 Dynamic Data Driven Applications Systems (DDDAS)

DDDAS [13] may be broadly defined as systems that utilize sensor and measurement data collected from a physical environment to dynamically update a computational model of the same environment. Although a broad array of techniques have been utilized under the DDDAS umbrella, there is a general emphasis in the literature on complex simulations of physical environments, and parsimonious statistical models are relatively limited. The Bayesian framework proposed herein is distinct via its adoption of ARA and statistical decision theory as foundational elements. The inherent dynamic feedback loop within DDDAS provides the ability to incorporate operational data into an executing application, and in reverse, the ability of an application to dynamically steer the measurement process [4]. For instance, the sensors can be reconfigured based on the error data through a feedback loop. Similarly, the impact of models and decisions on the measurements are used to improve the overall augmentation of the measurement, forecasting, and decision models. This is valuable within a systems of systems setup wherein the information and data are retrieved from dynamic, multi-modal and heterogeneous sources. Online and offline model integration guides data collection and improves model accuracy within a feedback loop, which could also help address potential privacy and security issues [23]. While they are widely applied to engineering problems [4,14], the DDDAS frameworks are also applicable in decision support applications in adversarial contexts. These include but are not limited to attack detection via anomaly detection modules [12], outlier detection based defenses against data poisoning attacks for classifiers [21] and the iterative dynamic data repair in sensor networks for power network load forecasting models [25].

It is almost impossible to build perfectly secure cyber systems and fully avoid the impact of adversarial attacks [24]. Hence, there has been some emphasis in the DDDAS literature on operating safely in a compromised environment while building defense frameworks that are tolerable to cyberattacks. [15] present a context aware anomaly detection based DDDAS framework that reconfigures the controller based on the differences between the sensor values and system simulation. [24] propose a resilient machine learning ensemble to tolerate adversarial learning attacks that utilizes moving target defense to dynamically change the machine learning models. This prevents adversaries from knowing the active defense and consequently from exploiting its vulnerabilities, thereby increasing resources necessary to perform an attack.

3   Adversarial Forecasting and Adversarial Risk Analysis

Adversarial attacks in forecasting often correspond to the perturbation of a dynamic system to influence either the forecasting output or model parameters; referred to as data-fiddler and structural attacks respectively. While adversarial classification methods are well-studied, there are gaps in classical adversarial forecasting. Stackelberg games are widely leveraged to model adversarial attacks over data generation. [6] use a Stackelberg game between the predictive method and the adversary that identifies the optimal data transformation at minimum cost assuming common knowledge over the adversary's costs and action space. [5] focus on simultaneous games with complete information, and provide conditions for the existence of a unique Nash Equilibrium in specific games. [1] model the poisoning of trained linear autoregressive forecasting models where an attacker manipulates the inputs to drive the latent space towards a region of interest. This is extended to sequential adversarial attack formulations [10]. Solutions for the defender's forecasting problem include the two player non-zero sum Stackelberg game assuming worst case attacker target [2].

Adversarial risk analysis is based on modeling games as decision-theoretic problems from the expected utility maximizing perspective of a given player. The player of interest considers their beliefs of uncertainties through Bayesian models for the goals, capabilities, and strategies of the opponents and by placing subjective distributions on all unknown quantities. This allows to address deficiencies associated with the common knowledge assumption and lack of one-sided decision support in standard game theoretic setups. While ARA has been applied to solve adversarial machine learning problems, its use is limited to specific supervised algorithms and attacks [19]. It is not used with probabilistic forecasting methods including the unsupervised algorithms.

We focus on the case of poisoning batch information release of data for a stochastic time series forecasting model. We assume a limited-knowledge gray-box attack where the attacker has complete access to the data and knows the family of the learning algorithm, but not the exact algorithm and parameters.

4   Illustration: Load Forecasting

Load forecasting algorithms are fundamental to power systems. However, they are also vulnerable to black-box attacks [11]. For example, an attacker may disrupt system operations by manipulating load forecasts in arbitrary directions via the injection of malicious temperature data from online weather forecast APIs. [26] formulate iterative stealthy adversarial attacks on smart grid predictions that maximize prediction loss while minimizing the required perturbations for the worst-case setting. [3] formulate a Stackelberg game wherein the defender selects random sensor measurements to use for an artificial neural network based load forecast, whereas the adversary calculates a bias to inject in a subset of sensors. Auto regressive integrated moving average (ARIMA) models have been widely used for load forecasts [20], hence are chosen for demonstration.

Fig. 1 displays the proposed DDDAS framework which facilitates the dynamic interaction between the data, prediction and decision layers through measurement, application, computational and theoretical foundational blocks. The adversarial attack is modeled through adversarial risk analysis. The adversarial forecast impact the decisions, which in turn affect the measurement through data audits.

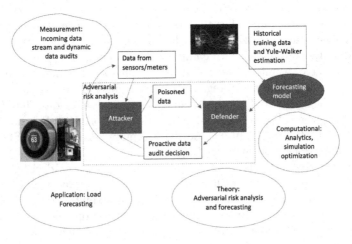

Fig. 1. Proposed DDDAS framework

We utilize public[1] data from an independent system operator, The Electric Reliability Council of Texas (ERCOT). ERCOT oversees the electrical grid of

---

[1] Available at www.ercot.com.

Texas, which is the largest electricity-consuming state in United States. It supplies power to more than 26 million customers corresponding to 90 percent of the state's electric load. There are many methods proposed to model the electricity demand in Texas, for instance see [22]. For demonstration, we focus on the time series of electricity load of Houston region during 2015–2017 summer months due to unique behavior associated with high residential demand [17]. This data could be retrieved from (smart) meters. We filter and transform the data by replacing the outliers with a conservative box-plot maximum of two times inter-quartile range added to third quartile. The hourly data shows high seasonality peaking around 16–17, and bottoming around 4–5. The best training fit was achieved by a seasonal ARIMA (SARIMA) $(5,0,0)x(2,1,0)^{24}$ model using Yule-Walker estimation and data until August 15th, 2017. The parameters are $\boldsymbol{\Phi} = \{2.0154, -1.5330, 0.6392, 0.2037, 0.0530\}$, $\boldsymbol{\Phi}^{S=24} = \{-0.5262, -0.2449\}$, $constant = 0.5955$ with the model having an AIC value of 12.64. The forecast for the next period includes weighted values of the five past observations and 24 h lagged seasonally differenced observations, overall depending on 18 previous observations. Our forecast time of interest is for a 24 h period ($H = 24$) starting with August 16th, 2017. The left panel of Fig. 2 presents the forecast in case of complete information. Next, we explore how poisoning of data and incomplete information could be addressed.

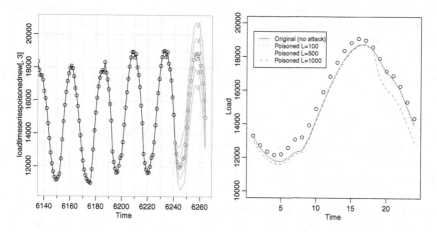

Fig. 2. Historical data (in black) and forecast (in red) under complete information (left) and forecasts with varying levels of attack (right) (Color figure online)

The Attacker poisons the load data, $\boldsymbol{X}$ by adding a perturbation of size, $\boldsymbol{\eta}$, and provides $\boldsymbol{Y} = \boldsymbol{X} + \boldsymbol{\eta}$ in a batch to the Defender. We assume Attacker is interested in decreasing the forecasts, which may have the Defender decrease supply eventually resulting with potential outages. Attacker's utility function consists of the sum of the differences between the original forecasts without the attack, $u(f^h|\boldsymbol{X})$, and under the attack, $u(f^h|\boldsymbol{\eta})$ for each hth forecasting time period. Cost per perturbation size for time period t is referred to as, c_t. The total cost is assumed to be smaller than a budget, B. The individual total perturbation size limit, L, is determined with respect to the sensitivity of the Defender. The Defender may impact $A(\boldsymbol{\eta})$ through data audits which would prevent Attacker from poisoning audited observations. Defender's decision of data auditing, β of time t is denoted as $\beta = t$ which indicates $\eta_t = 0$; updating the Attacker's set of achievable attacks, $A(\boldsymbol{\eta}, \beta)$.

With complete information, Attacker's decision problem could have been solved through the following optimization model where perturbation size, $\boldsymbol{\eta}$ is the decision variable and $A(\boldsymbol{\eta}, \beta)$ is the Attacker's set of achievable attacks.

$$\max_{\eta} \quad \psi_A(\boldsymbol{\eta}, \beta, \Phi) = \sum_{h=1}^{h=H} [u(f^h|\boldsymbol{X}) - u(f^h|\boldsymbol{X}, \boldsymbol{\eta})] \tag{1}$$

$$\text{s.t.} \quad \eta_t <= L \quad \forall t, \sum_{t=1}^{t=T} c_t * \eta_t <= B, \boldsymbol{\eta} \in A(\boldsymbol{\eta}, \beta)$$

Attacker has access to the historical training data, and while he knows the family of the defender's algorithm, i.e., SARIMA; he does not know the exact forecasting model used by the Defender. Therefore, the Attacker's problem is not a deterministic model. ARA allows the Attacker to acknowledge his uncertainty about Defender's forecasting model, $p_D(\Phi)$ as well as the data audit decision, β affecting $A(\boldsymbol{\eta}, \beta)$; and their impact on $\psi_A(\boldsymbol{\eta}, \beta)$. Attacker's expected utility is

$$\psi_A(\boldsymbol{\eta}) = \int \psi_A(\boldsymbol{\eta}, \beta) \, p_A(\beta|\boldsymbol{\eta}) d\beta = \int \left[\int u_A(\boldsymbol{\eta}, \beta, \Phi) \, p_A(\Phi|\boldsymbol{\eta}, \beta) d\Phi \right] p_A(\beta|\boldsymbol{\eta}) d\beta$$

To find $p_A(\beta|\boldsymbol{\eta})$, he would induce a distribution over the Defender's expected utility $\psi_D(\boldsymbol{\eta})$, from which random optimal alternative, $\beta^*(\boldsymbol{\eta})$ is computed for each $\boldsymbol{\eta}$ decision alternative. This could be substituted into the Attacker's expected utility maximization problem to find the optimal decision, as in $\boldsymbol{\eta}^*_{ARA} = argmax_{\boldsymbol{\eta} \in A(\eta, \beta)} \psi_A(\boldsymbol{\eta})$.

After Defender retrieves the data (which could be poisoned) and makes a forecast, that is compared to a predetermined threshold within a simple decision support tool. If the load forecast is less than the predetermined threshold for h^{th} time period, PT^h, the system decision makers may make supply adjustment decisions. PT^h could be set as the mean historical load. If the adjustments are often needed, then the Defender may want to reevaluate the choice of model, and may want to retrain the model. As a proactive mechanism, the Defender conducts a data audit for the next time forecasting period of interest. Defender aims to minimize the maximum potential attack impact assuming a worst case (optimal) attack. She selects one observation (at time t) to be audited, i.e., $\beta = t$, and update observed data of $\boldsymbol{Y}$ in order to create the largest difference in forecast corresponding to optimal attack. This, in turn would eliminate prospective adversarial activities for the audited observation. Since Attacker observes

Defender's decisions, it would be optimal for him not to poison audited data in the next time period. In particular, this would impact the Attacker's feasible set of prospective perturbation, $A(\eta)$. It should be noted that Defender has access to the model, and hence can compute $u(f^h|Y, \beta)$. The Defender's problem is written as

$$\min_{\beta} \max \sum_{h=1}^{h=H} [PT^h - u(f^h|Y, \beta)] \tag{2}$$

In this implementation, we assumed the Attacker to have positive probability for three models: $(5, 0, 0)$ x $(2, 1, 0)^{24}$ with 80 percent probability, and $(4, 0, 0)$ x $(2, 1, 0)^{24}$, $(5, 0, 0)$ x $(1, 1, 0)^{24}$ each with probability of 0.1. Attacker's decision problem is solved with different L values for a large enough B. The right panel of Fig. 2 presents the comparison of forecasts with original and poisoned data alternatives (L=100, 500 and 1000). For the first forecasting period in our illustration for the optimal attack alternative, β^* is found to be $t - 1$. This could be explained by the parameter coefficient values. The sensitivity analysis demonstrating the impact of data audits (for various β values) on the attack could provide additional insights, but is not shared given space limitations.

5   Discussion

Herein, we illustrate the utility of ARA-based DDDAS framework on load forecasting under incomplete information, adversarial attacks and with two decision makers. We demonstrate the optimal attack, and present a Defender strategy against data poisoning over time. Such proactive and dynamic approaches can prove to be vital for energy resilience. They can enable the Defender to understand the nature of attacks and adapt their preparations to the reality that they are inevitable. These proactive adjustments could help with load availability that provides for mission assurance and readiness.

Our ongoing work includes a detailed overview [16] in addition to a comprehensive computational illustration and demonstration of the proposed method in scenarios that could be relevant to the needs of Air Force Joint All-Domain Command and Control decision making as described by [7, 8]. In doing so, we are exploring the feasibility of using adversarial Hidden Markov Models [9]. In addition, other relevant defense and security application areas could include defending airports against 21st century terrorist threats. That may require proactive defenses that allow interaction between multiple sensors and adversarial forecasting. ARA models entail integration and optimization procedures that can be computationally challenging. While this manuscript focuses on small instances that can be solved through using Monte Carlo simulation, efficient computational methods are required for real world instances. We are exploring the feasibility of using augmented probability simulation methods [18] in addition to approximate Bayesian inference techniques.

References

1. Alfeld, S., Zhu, X., Barford, P.: Data poisoning attacks against autoregressive models. In: Proceedings of the AAAI Conference on Artificial Intelligence, vol. 30 (2016)
2. Alfeld, S., Zhu, X., Barford, P.: Explicit defense actions against test-set attacks. In: Thirty-First AAAI Conference on Artificial Intelligence (2017)
3. Barreto, C., Koutsoukos, X.: Design of load forecast systems resilient against cyber-attacks. In: Alpcan, T., Vorobeychik, Y., Baras, J., Dan, G. (eds.) Decision and Game Theory for Security. Lecture Notes in Computer Science(), vol. 11836, pp. 1–20. Springer, Cham (2019). https://doi.org/10.1007/978-3-030-32430-8_1
4. Blasch, E., Ravela, S., Aved, A.: Handbook of Dynamic Data Driven Applications Systems. Springer, Cham (2018)
5. Brückner, M., Kanzow, C., Scheffer, T.: Static prediction games for adversarial learning problems. J. Mach. Learn. Res. **13**(1), 2617–2654 (2012)
6. Brückner, M., Scheffer, T.: Stackelberg games for adversarial prediction problems. In: Proceedings of the 17th ACM SIGKDD international conference on Knowledge discovery and data mining, pp. 547–555 (2011)
7. Caballero, W.N., Friend, M., Blasch, E.: Adversarial machine learning and adversarial risk analysis in multi-source command and control. Sig. Proc., Sens./Inf. Fus. Target Recogn. XXX **11756**, 98–108 (2021)
8. Caballero, W.N., Lunday, B.J., Deckro, R.F., Pachter, M.N.: Informing national security policy by modeling adversarial inducement and its governance. Socioecon. Plann. Sci. **69**, 100709 (2020)
9. Caballero, W.N., Camacho, J.M., Ekin, T., Naveiro, R.: Manipulating hidden-Markov-model inferences by corrupting batch data. Comput. Oper. Res. 162, 106478 (2024)
10. Chen, Y., Zhu, X.: Optimal attack against autoregressive models by manipulating the environment. In: Proceedings of the AAAI Conference on Artificial Intelligence, vol. 34, pp. 3545–3552 (2020)
11. Chen, Y., Tan, Y., Zhang, B.: Exploiting vulnerabilities of load forecasting through adversarial attacks. In: Proceedings of the Tenth ACM International Conference on Future Energy Systems, pp. 1–11 (2019)
12. Combita, L.F., Giraldo, J.A., Cardenas, A.A., Quijano, N.: DDDAS for attack detection and isolation of control systems. In: Blasch, E., Ravela, S., Aved, A. (eds.) Handbook of Dynamic Data Driven Applications Systems, pp. 407–422. Springer, Cham (2018). https://doi.org/10.1007/978-3-319-95504-9_17
13. Darema, F.: Dynamic data driven applications systems: a new paradigm for application simulations and measurements. In: Bubak, M., van Albada, G.D., Sloot, P.M.A., Dongarra, J. (eds.) Computational Science - ICCS 2004. Lecture Notes in Computer Science, vol. 3038, pp. 662–669. Springer, Berlin (2004). https://doi.org/10.1007/978-3-540-24688-6_86
14. Darema, F., Blasch, E., Ravela, S., Aved, A.: Dynamic Data Driven Applications Systems: Third International Conference, DDDAS 2020, Boston, MA, USA, October 2–4, 2020, Proceedings, vol. 12312. Springer Nature, Cham (2020). https://doi.org/10.1007/978-3-030-61725-7
15. Dsouza, G., Hariri, S., Al-Nashif, Y., Rodriguez, G.: Resilient dynamic data driven application systems (rDDDAS). Procedia Comput. Sci. **18**, 1929–1938 (2013)
16. Ekin, T., Cabellaro, W.N., Camacho, J.M., Naveiro, R.: Adversarial forecasting: a decision theoretic approach (2022)

17. Ekin, T., Damien, P., Zarnikau, J.: Estimating marginal effects of key factors that influence wholesale electricity demand and price distributions in Texas via quantile variable selection methods. J. Energy Markets **13**(1), 1–29 (2020)

18. Ekin, T., Naveiro, R., Insua, D.R., Torres-Barrán, A.: Augmented probability simulation methods for sequential games. Eur. J. Oper. Res. (2022). https://doi.org/ 10.1016/j.ejor.2022.06.042

19. Insua, D.R., Naveiro, R., Gallego, V., Poulos, J.: Adversarial machine learning: perspectives from adversarial risk analysis. arXiv preprint: arXiv:2003.03546 (2020)

20. Lee, C.M., Ko, C.N.: Short-term load forecasting using lifting scheme and ARIMA models. Expert Syst. Appl. **38**(5), 5902–5911 (2011)

21. Li, X., Miller, D.J., Xiang, Z., Kesidis, G.: A scalable mixture model based defense against data poisoning attacks on classifiers. In: Darema, F., Blasch, E., Ravela, S., Aved, A. (eds.) Dynamic Data Driven Applications Systems. Lecture Notes in Computer Science(), vol. 12312, pp. 262–273. Springer, Cham (2020). https://doi. org/10.1007/978-3-030-61725-7_31

22. Nguyen, H., Hansen, C.K.: Short-term electricity load forecasting with time series analysis. In: 2017 IEEE International Conference on Prognostics and Health Management (ICPHM), pp. 214–221. IEEE (2017)

23. Xiong, L., Sunderam, V., Fan, L., Goryczka, S., Pournajaf, L.: Privacy and security issues in DDDAS systems. In: Blasch, E.P., Darema, F., Ravela, S., Aved, A.J. (eds.) Handbook of Dynamic Data Driven Applications Systems, pp. 615–630. Springer, Cham (2018). https://doi.org/10.1007/978-3-030-74568-4_27

24. Yao, L., Tunc, C., Satam, P., Hariri, S.: Resilient machine learning (rML) ensemble against adversarial machine learning attacks. In: Darema, F., Blasch, E., Ravela, S., Aved, A. (eds.) Dynamic Data Driven Applications Systems. Lecture Notes in Computer Science(), vol. 12312, pp. 274–282. Springer, Cham (2020). https://doi. org/10.1007/978-3-030-61725-7_32

25. Zhou, X., Canady, R., Li, Y., Koutsoukos, X., Gokhale, A.: Overcoming stealthy adversarial attacks on power grid load predictions through dynamic data repair. In: Darema, F., Blasch, E., Ravela, S., Aved, A. (eds.) Dynamic Data Driven Applications Systems. Lecture Notes in Computer Science(), vol. 12312, pp. 102–109. Springer, Cham (2020). https://doi.org/10.1007/978-3-030-61725-7_14

26. Zhou, X., et al.: Evaluating resilience of grid load predictions under stealthy adversarial attacks. In: 2019 Resilience Week (RWS), vol. 1, pp. 206–212. IEEE (2019)

Main-Track Plenary Presentations - Distributed Systems

Power Grid Resilience: Data Gaps for Data-Driven Disruption Analysis

Maureen S. Golan[1]([✉]), Javad Mohammadi[1], Erika Ardiles Cruz[2],
David Ferris[2], and Philip Morrone[2]

[1] Cockrell School of Engineering, The University of Texas at Austin, Austin, USA
mgolan@utexas.edu
[2] Air Force Research Laboratory, Rome, NY, USA

Abstract. A resilient and reliable power grid is crucial for energy security, sustainability, and reducing service restoration time and economic burdens to society. Electric utility companies and power system regulatory bodies rely on metrics based on tracked data to assess the power grid's reliability and plan for future needs. While reliability and resilience are often used interchangeably, their distinction should be recognized. Reliability assumes the power grid is operating under standard system conditions; resilience requires a disruption to occur to be evaluated and measured. Historically, reliability standards have been tracked with standardized metrics and enforced with great oversight to plan for predicted contingencies. However, power grid resilience has not been tracked and metrics remain elusive for utilities and regulatory authorities needing to ensure system performance in the changing context of power grid operations and development. In this paper, we evaluate existing power grid data sources (the Department of Energy's Electric Emergency Incident and Disturbance Events form, DOE-417, a mandatory form for electric utilities) and identify data gaps that adversely impact data-driven resilience analysis. We consider different event types and their system-wide implications to evaluate how the missing data can impact power grid resilience assessments.

Keywords: Power grid · Resilience · Reliability

1 Introduction

As deep decarbonization gains momentum, an increased burden will be placed on the electric grid as a source of cleaner energy for traditionally large users of fossil fuels (e.g., transportation, fertilizer and cement manufacture, etc.), and also increasing electric demand from new sources, such as big data management and Industry 4.0 sectors (e.g., data centers, digital currency mining, etc.) [20]. The increased burden is coupled by not only the changing grid edge and aging infrastructure, but also other unprecedented disruptions such as climate change,

This research is supported by the Visiting Faculty Research Program of the Air Force Labs, Rome, New York as well as the Block Center for Technology and Society of Carnegie Mellon University.

supply chain and geopolitical disturbances [13]. Resilience, a concept comple-
mentary to reliability, must be matured in order to meet society's energy needs.
Finding ways to use existing reporting and oversight mechanisms to gain a systems view of power grid disruptions is essential for maintaining grid reliability
and resilience, and optimal performance.

1.1 Reliability and Resilience

Reliability is centered on risk-based metrics, which use expected vulnerabilities
and expected consequences of a given hazard to weigh the cost and benefit of
hardening the system [9]. Reliability can also be interpreted as the N-1 or N-X
contingency analysis, which addresses grid operations under predicted scenarios
[15], and the standardized metrics SAIDI, SAIFI, and CAIDI [9]. Resilience,
on the other hand, focuses on the recovery and adaptation of the system post-
disruption, assuming that the system cannot be hardened against all hazards,
focusing instead on the ability of the system to recover its critical function as
quickly as possible and with the least amount of lost function (see Fig. 1) [5,15].
Further complementing reliability, resilience emphasizes outage consequence, or
the impacts that an outage would have on individuals and society: "the concept
of reliability must be augmented with a resilience approach - one that looks at
the grid not strictly as a flow of electrons but as a grid that services, interfaces
with, and impacts people and societies" [22].

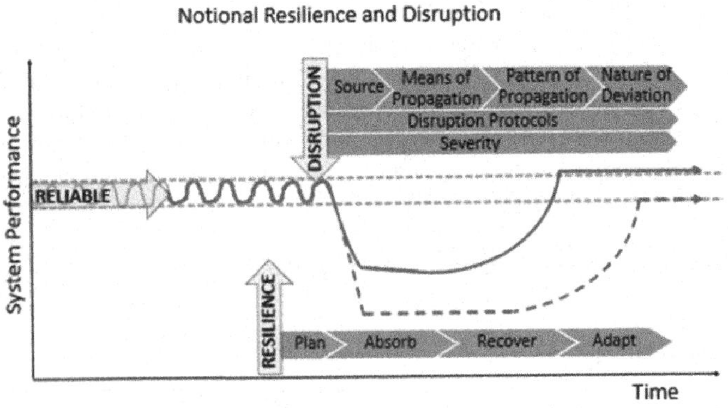

Fig. 1. A reliable system maintains system performance within minimum and maximum thresholds (dashed grey) despite some disturbances; upon a system disruption, a
more resilient system (solid red) regains minimal performance faster and with less lost
performance than a less resilient system (dashed red), and better adapts to changing
contexts. Disruption and resilience stages align over the resilience curve. (After [5,7]).
(Color figure online)

Reliability allows utilities and regulators to plan and operate the system
under normal (undisrupted) conditions, while resilience allows for operation and

planning during and post disruption. Hence, both resilience and reliability are essential in grid operations, but resilience is behind reliability in model maturity.

1.2 Modeling Resilience

Reviews of the power grid resilience literature indicate that although it is a burgeoning field, quantitatively and comprehensively measuring resilience for wide application is not yet possible [14, 19, 27]. Most research is based on natural causes of disruption or single sources of disruption [10, 11, 26] and develop metrics that have a degree or more of separation from resilience, such as value of avoided interruption, risk profiles, grid capacity, economics, operating cost, load shed, physical damage [1, 4, 8, 12, 24]. These "proxies" do not provide a resilience metric, but rather insight into a related system measurement [6, 23]. This is largely in part because the underlying models lack connection to the larger system and interconnected networks, are confined to preconceived scenarios and mitigation strategies, or oversimplify the resilience curve. Further, myopic focus on natural causes of disruption has been shown to miss large sources of bulk electric system disruption [17]. Many existing studies also limit themselves to synthetic data sets, simulations, or hard to gather data [8, 12, 24].

In order to build on existing literature and avoid the identified gaps in power grid resilience modeling, we evaluate resilience from the point that differentiates it from reliability: disruption itself. By addressing the entire disruption timeline, and providing a data-driven analysis that incorporates system interactions and domains - cyber, physical, human - we take a step towards enabling resilience quantification and development of a metric that addresses the entire resilience curve. We use publicly available data that utilities are required to collect and report on, and analyze it for feasibility in disruption mapping and for data gaps.

1.3 DOE-417 Data Set

The Department of Energy (DOE) maintains the Electric Emergency Incident and Disturbance Report form, DOE-417, a mandatory national security filing under Public Law 93-275 [18]. The Report allows the DOE to fulfill national security and emergency management responsibilities and is approved by the North American Electric Reliability Corporation (NERC) [18]. The form is reviewed and updated about every three years [21]. The publicly available annual summaries include: date/time event began, date/time of restoration, area affected, NERC region, alert criteria, event type, demand loss, and number of customers affected. "Event Types" provide a high-level categorization of the incidents, while "Alert Criteria" provide more disruption source detail. Alert Criteria were only added in 2015.

For perspective, between 2012 and 2021, there were 2,286 total reported incidents. Table 1 gives an overview of the available disruption data. Note that the date reflects the starting point of the disruption. For example, a five-day outage starting on December 17, 2021 and ending on January 1, 2022 will be recorded under December in the 2021 Annual Summary Report. Also, note that all summaries include "unknown" classifications in the "Demand Loss" and "Number of Customers Affected" categories and is counted towards the event count as it does not rule out a disruption. However, it is not considered in the average calculations; only events with demand loss are considered for average calculations. All years also include "unknown" in outage start and stop times, and these entries are not included in the average calculation. The average outage time was calculated for events greater than 0 min and less than 2 weeks. On their own, the number of incidents do not describe power grid resilience, but need to be associated with consequences, such as demand loss and customers affected.

Table 1. DOE-417 Reportable events from 2012–2021

Year	No. events	No. events w/demand loss	No. events w/ customers affected	Avg. demand loss per event (MW)	Avg. no. customers affected per event	Avg. outage time (min)
2021	387	223	212	692	156,956	1,392
2020	383	219	214	1,003	151,022	1,045
2019	278	135	130	2,745	101,978	820
2018	220	137	134	1,903	168,105	1,312
2017	150	101	98	402	195,329	1,663
2016	141	84	82	688	143,362	1,122
2015	143	122	116	1,362	97,012	1,021
2014	214	196	196	1,215	210,637	1,268
2013	174	138	138	575	113,184	1,185
2012	196	91	107	872	252,005	2,224
Total	2,286	1,446	1,427	1,123	158,846	1,275

2   Disruption Mapping: Definitions and Methods

2.1   Defining Network Disruption

The power grid is a complex interconnected network, where system-wide failures can result from component-level disruptions. This is similar to supply chains, where individual components - nodes and links - can be impacted by any number of factors [2, 6, 25]. We therefore take a systems approach to map disruptions from the source to the ultimate consequence, as defined below.

– **Disruption Source:** the origination of an adverse deviation from normal operations, i.e., the unforeseen triggering event that results in anomalies across the network. The disruption sources are broadly defined as *human* and *nature*. Under these sources, the scope of the disruption is classified as *local* (nodes/links subject to isolated disturbances), *regional* (nodes/links dispersed over an area subject to regional disturbances), and *global* (nodes/links dispersed globally subject to global disturbances). The source's direction is defined as *supply-side/upstream* (nodes/links associated with generation through distribution; event impacting a system from within the utility), *demand-side/downstream* (nodes/links associated with electricity use), and *catastrophic* (force majeure; extraordinary event impacting a system from all directions).

- **Means of Disruption Propagation:** the domain through which the violation from normal operations spreads from origination. The three domains are *physical* (physical infrastructure and natural environment, including structures, processes and designs), *cyber* (information and data flow), and *human* (organizational structure, communication, and societal context of decision-making).
- **Pattern of Propagation:** characteristics of how the negative deviation from normal operations disseminates throughout the network.
- **Nature of Deviation:** how normal operations have been negatively impacted.

2.2 Disruption Mapping Methods

Mapping was done twice for the 1,048 reported incidents in the 2019–2021 annual summaries: once using Event Types and once using Alert Criteria for the largest three event categories. First the Event Types were grouped into 46 event categories that enabled similar or identical events to be processed together. These event categories were then assessed for probabilities (0–1) for each step in the disruption timeline. This means that each event category has a total of 1 across each: Disruption Source - Scope (human/nature: local, regional, global), Disruption Source - Upstream/Downstream (human/nature: supply-side, demand-side, catastrophic), and Means of Disruption Propagation (physical, cyber, human). These probabilities were then applied to the DOE-417 data to analyze frequencies of incidents, incidents with nonzero demand loss and incidents with nonzero customers affected.[1] Due to data limitations in DOE-417, mapping the Pattern of Disruption Propagation and Nature of Deviation was not possible because more detailed timeline and impact information is necessary (e.g., data on the demand loss curve, rather than cumulative demand loss over the entire disruption).

This process was also completed for the Alert Criteria, using the largest three event categories. Combined, severe weather/natural disaster, physical attack/vandalism/sabotage, and system operations account for over three quarters of all reported events, events with nonzero demand loss, and events with nonzero customers affected. For the 806 incidents that comprise the largest three events, there were 33 unique Alert Criteria. The Alert Criteria mapping was then compared to the Event Criteria mapping for the largest three events to further detail data gaps and areas for improved reporting.

---

[1] "Unknown" is occasionally entered in the "Demand Loss" and "Number of Customers Affected" categories, which is counted towards events affecting customers/demand because it does not rule out a disruption.

3 Data Gaps: Results and Discussion

3.1 Disruption Mapping

Starting at the source stage of a disruption, humans cause a slight majority of the total number of events, while the reported events with demand and customer impacts are more often attributed to natural sources. Human initiated events may be more visible than natural events, which may only be known to a utility if an outage occurs, and may skew the data towards overreporting of human events and underreporting of natural events. Better resilience strategies may also be in place for human-caused events, which may plan for and absorb the impacts prior to any system-wide disturbances. Strategies for nature-caused incidents may bypass the first stages of resilience, with utilities focusing on recovery stage strategies - after the disruption is felt and reported (see Fig. 1). Additional data on when the utility responds to the incidence would help fill this data gap.

Further, Fig. 2 shows that under scope, nature-caused events tend to be evenly split between regional and local sources of disruption, while with human-caused, events tend to be more localized. This intuitively makes sense because physical attacks/vandalism/sabotage make up a majority of human-caused events and are more likely to occur at single substations or other specific locations. And under upstream/downstream, upstream/supply-side sources make up over two thirds of events. However, about one fifth of total events that do not report zero customer impact are mapped to demand-side sources, indicating that under certain conditions, system operators may favor controlled load shedding to minimize the depth of the resilience curve or the consequence of the disruption, rather than uncontrolled system failure. Again, data on customer categories of impact (e.g., commercial, residential, etc.) and the distribution of demand loss over the timespan of the disruption would allow for improved resilience analysis.

Under the disruption propagation category, mapping shows that once a disruption begins, it has similar probabilities of causing power grid impacts via all three domains. This highlights the fact that cyber, physical, and human considerations must be engrained in disruption response and resilience strategies. As the power grid becomes more interconnected and nonlinear, in order to provide maximum security and maintain agility in system optimization, better records on domains and associated disruption impacts may be warranted by utilities and regulatory authorities.

3.2 Alert Criteria Analysis

Mapping of the largest three event categories by both Event Types and Alert Criteria highlight where a lack of detailed reporting in Event Types may lead to a misunderstanding of the disruption process (see Fig. 3). Overall, the largest inconsistency was within the System Operations category, with an overestimate of upstream-human source of disruption in the Event Type mapping. This is due largely to two Alert Criteria within that Event Type that are mapped to downstream-human: firm load shedding of 100 MW or more implemented under

Fig. 2. DOE-417 Event Types for all reported events from 2019–2021 mapped to human (H) and natural (N) sources of local, regional, global, supply-side, demand-side, or catastrophic disruption, and to propagation domains.

emergency operational policy, and public appeal to reduce the use of electricity for purposes of maintaining the continuity of the Bulk Electric System. With the exception of one public appeal event, all incidents in both these Alert Criteria reported demand loss and affected customers. As System Operations comprises a large percentage of reported incidents, greater DOE-417 specificity would provide more efficient disruption analyses and more effective resilience quantification.

Also significant was the underestimation of cyber propagation in the Physical Attack category. This is not overwhelmingly due to any one Alert Criteria, but rather the open-ended nature of what the attack/vandalism/sabotage targets, resulting in an assumption that in most cases, all three domains will be sources of disruption propagation. System Operations also had local-human disruption source underestimated, but this only significantly impacted the number of events reported rather than those leading to demand loss or affecting customers. This is largely due to the two largest Alert Criteria under the System Operations category: complete loss of monitoring or control capability at its staffed Bulk Electric System control center for 30 continuous minutes or more, and unplanned evacuation from its Bulk Electric System control center facility for 30 continuous minutes or more, both of which have few reportable events resulting in demand loss or customer impact, and are local events that were underrepresented in the assumption that System Operations would likely be impacted more so by regional disruptions due to their nature. Regardless, greater information on the target of the Physical Attack would allow for better propagation measurements and resilience models.

In general, the Event Criteria provide a fairly accurate representation of the disruption process as compared to the Alert Criteria. However, there are specific areas where more detailed reporting can advance system-wide resilience analytics. In this manner, targeting gaps reduces burdens on utilities and regulatory authorities while providing necessary information for reliability and resilience.

| | | Disruption Source | | | | | | | | | | | | Domains | | |
| | | Local | | Regional | | Global | | Upstream | | Downstream | | Catastrophic | | | | |
		H	N	H	N	H	N	H	N	H	N	H	N	P	C	H
Severe Weather/ Natural Disaster	Total Events	0.0%	-8.1%	0.0%	8.1%	0.0%	0.0%	0.0%	-13.0%	18.7%	-3.9%	0.0%	-1.7%	-6.6%	-1.6%	8.2%
	≠0 demand loss	0.0%	-7.9%	0.0%	7.9%	0.0%	0.0%	0.0%	-13.0%	18.4%	-3.7%	0.0%	-1.7%	-6.5%	-1.3%	7.7%
	≠0 customers	0.0%	-7.8%	0.0%	7.8%	0.0%	0.0%	0.0%	-13.1%	18.2%	-3.4%	0.0%	-1.7%	-6.4%	-1.3%	7.6%
Physical Attack/ Vandalism/ Sabotage	Total Events	9.9%	0.0%	-9.9%	0.0%	0.0%	0.0%	7.7%	0.0%	-10.0%	0.0%	2.3%	0.0%	-37.3%	37.8%	-0.5%
	≠0 demand loss	10.0%	0.0%	-10.0%	0.0%	0.0%	0.0%	7.7%	0.0%	-10.0%	0.0%	2.3%	0.0%	-37.1%	37.9%	-0.9%
	≠0 customers	10.0%	0.0%	-10.0%	0.0%	0.0%	0.0%	7.4%	0.0%	-10.0%	0.0%	2.6%	0.0%	-37.7%	37.3%	0.4%
System Operations	Total Events	34.4%	0.0%	-34.4%	0.0%	0.0%	0.0%	-5.5%	0.0%	-0.8%	0.0%	6.3%	0.0%	30.8%	-17.3%	-13.5%
	≠0 demand loss	6.3%	0.0%	-6.3%	0.0%	0.0%	0.0%	-35.0%	0.0%	29.1%	0.0%	5.9%	0.0%	23.2%	-20.3%	-2.9%
	≠0 customers	-0.5%	0.0%	0.5%	0.0%	0.0%	0.0%	-43.0%	0.0%	37.4%	0.0%	5.7%	0.0%	20.8%	-20.6%	-0.2%

Fig. 3. The mapping using the Event Criteria subtracted from Alert Criteria mapping, showing data gaps from using broad categories over specific event categorization. (H = human; N = nature; P = physical; C = cyber)

4   Conclusions

Executing power grid resilience models in real time is essential for meeting the needs of regulatory authorities, utilities, and the public. Dynamically incorporating existing data with additional select detailed data into disruption modeling, will enable a data-driven systems approach to quantifying power grid resilience in real time. Specifically, we find that there are less data gaps in how utilities report severe weather/natural disaster events as there is more nuance in the reporting documents. However, data gaps may simultaneously exist in natural events that occur, but do not result in noticeable impacts, providing non-reliable information on the plan and absorb stages of a disruption. System operations and physical attack/vandalism/sabotage also present major data gaps in understanding disruption propagation, as cyber, physical and human information is not indicated in the event description, with noticeable gaps in the upstream/ downstream system operations category as well. Although the DOE-417 form cannot possibly include all event nuances, developing a form that can better distinguish the disruption process, elaborating on certain events that are more convoluted can provide better insight into the resilience curve and facilitate effective quantification of resilience. Our study is limited by the fact that although we aim to show data gaps, the mapping we conduct is still confined to the assumptions derived from the DOE definitions. We are also limited by the assumption that utilities consistently report all events. Future works will further explore Dynamic Data Driven Applications Systems (DDDAS) paradigms to mitigate data gaps through methods such as scenario definitions in [3], dynamic incorporation of field expertise in [28], and attack detection in cyber physical systems in [16].

Acknowledgements. Thanks to Drs. Erik Blasch and Alexander Aved for review and feedback. The views and conclusions contained herein are those of the authors and should not be interpreted as necessarily representing the official policies or endorsements, either expressed or implied, of the AFRL or the U.S. Government.

References

1. Ahrens, M., Kern, F., Schmeck, H.: Strategies for an adaptive control system to improve power grid resilience with smart buildings. Energies **14**(15), 4472 (2021)
2. Bugert, N., Lasch, R.: Supply chain disruption models: a critical review. Logist. Res. **11**(1), 1–35 (2018)
3. Darville, J., Curia, J., Celik, N.: Microgrid operational planning using a hybrid neural network with resource-aware scenario selection. Simul. Model. Pract. Theory **119**, 102583 (2022)
4. Diahovchenko, I.M., Kandaperumal, G., Srivastava, A.K., Maslova, Z.I., Lebedka, S.M.: Resiliency-driven strategies for power distribution system development. Electric Power Syst. Res. **197**, 107327 (2021)
5. Galaitsi, S.E., Keisler, J.M., Trump, B.D., Linkov, I.: The need to reconcile concepts that characterize systems facing threats. Risk Anal. **41**(1), 3–15 (2021)
6. Golan, M.S., Jernegan, L.H., Linkov, I.: Trends and applications of resilience analytics in supply chain modeling: systematic literature review in the context of the Covid-19 pandemic. Environ. Syst. Decisions **40**(2), 222–243 (2020)
7. Golan, M.S., Trump, B.D., Cegan, J.C., Linkov, I.: The vaccine supply chain: a call for resilience analytics to support COVID-19 vaccine production and distribution. In: Linkov, I., Keenan, J.M., Trump, B.D. (eds.) COVID-19: Systemic Risk and Resilience. RSD, pp. 389–437. Springer, Cham (2021). https://doi.org/10.1007/978-3-030-71587-8_22
8. Huang, G., Wang, J., Chen, C., Qi, J., Guo, C.: Integration of preventive and emergency responses for power grid resilience enhancement. IEEE Trans. Power Syst. **32**(6), 4451–4463 (2017)
9. Jin, A., Trump, B., Golan, M., Hynes, W., Young, M., Linkov, I.: Building resilience will require compromise on efficiency. Nat. Energy **6**(11), 997–999 (2021)
10. Jufri, F.H., Widiputra, V., Jung, J.: State-of-the-art review on power grid resilience to extreme weather events. Appl. Energy **239**, 1049–1065 (2019)
11. Mar, A., Pereira, P., Martins, J.F.: A survey on power grid faults and their origins: a contribution to improving power grid resilience. Energies **12**(24), 4667 (2019)
12. McGrath, J.: Will updated electricity infrastructure security protect the grid? A case study modeling electrical substation attacks. Infrastructures **3**(4), 53 (2018)
13. NERC: 2021 long-term reliability assessment (2021). https://www.nerc.com/pa/RAPA/ra/Reliability%20Assessments%20DL/NERC_LTRA_2021.pdf
14. NERC: 2021 state of reliability: an assessment of 2020 bulk power system performance (2021). https://www.nerc.com/pa/RAPA/PA/Performance%20Analysis%20DL/NERC_SOR_2021.pdf
15. NERC: Reliability Issues Steering Committee: Report on Resilience. Atlanta, Georgia (2018). https://www.nerc.com/comm/RISC/Related%20Files%20DL/RISC%20Resilience%20Report_Approved_RISC_Committee_November_8_2018_Board_Accepted.pdf
16. Pantopoulou, S., Lagari, P.L., Townsend, C.H., Tsoukalas, L.H.. In: Dynamic Data Driven Applications Systems, pp. 283–290. Springer, Cham (2020)
17. Papic, M., Ekisheva, S., Robinson, J., Cummings, B.: Multiple outage challenges to transmission grid resilience. In: 2019 IEEE Power & Energy Society General Meeting (PESGM), pp. 1–5 (2019). https://doi.org/10.1109/PESGM40551.2019.8973606
18. Department of Energy: Electric disturbance events (DOE-417). https://www.oe.netl.doe.gov/oe417.aspx

19. Department of Energy: North American energy resilience model. Office of Electricity. Washington, D.C (2019)
20. National Academies of Sciences Engineering and Medicine: The Future of Electric Power in the United States. https://doi.org/10.17226/25968
21. OE-417 Help Desk: email communication (2022)
22. Sandia National Laboratories: Grid resilience (2022). https://energy.sandia.gov/programs/electric-grid/resilient-electric-infrastructures/
23. Rickerson, W., Gillis, J., Bulkeley, M.: The value of resilience for distributed energy resources: an overview of current analytical practices. Report prepared for the National Association of Regulatory Utility Commissioners (2019)
24. Shao, C., Shahidehpour, M., Wang, X., Wang, X., Wang, B.: Integrated planning of electricity and natural gas transportation systems for enhancing the power grid resilience. IEEE Trans. Power Syst. **32**(6), 4418–4429 (2017)
25. Wagner, S.M., Bode, C.: An empirical investigation into supply chain vulnerability. J. Purch. Supply Manag. **12**(6), 301–312 (2006)
26. Waseem, M., Manshadi, S.D.: Electricity grid resilience amid various natural disasters: challenges and solutions. Electricity J. **33**(10), 106864 (2020)
27. Willis, H.H., Loa, K.: Measuring the Resilience of Energy Distribution Systems, Santa Monica, CA (2015)
28. Yavuz, A., et al.: Advancing self-healing capabilities in interconnected microgrids via dynamic data driven applications system with relational database management. In: 2020 Winter Simulation Conference (WSC), pp. 2030–2041 (2020)

Attack-Resilient Cyber-Physical System State Estimation for Smart Grid Digital Twin Design

M. Rana[2(✉)], S. Shetty[1], Alex Aved[2], Erika Ardiles Cruz[2], David Ferris[2], and Philip Morrone[2]

[1] Old Dominion University, Norfolk, VA, USA
[2] AFRL, Rome, NY, USA
md_r71@yahoo.com

Abstract. Before implementing the microgrid testbed and SCADA electricity monitoring systems, computer aided tools can be used to design and validate technical specifications and performance. In this way, the system and product can be implemented digitally reducing cost, time, efforts, and visualizing expected quality. In real-time, designing and implementing the smart grid incorporating renewable microgrids is also a critical and challenging task due to random generation patterns of foreseeable green energy. In order to solve this impending problem, the microgrid digital twin incorporating renewable distributed energy resources is designed using physical and governing laws such as Kirchhoff's laws, and input-output relationships. After modeling the distribution grid into a set of first-order differential equations, the microgrid digital framework is transformed into a compact state-space representation. Using a set of IoT sensors, the measurements are collected from the distribution grid at common coupling points. Indeed, the increased rate of cyber-attacks on the smart grid communication network requires for innovative solutions to ensure its resiliency and operations. When the IoT sensing information is under cyber attacks, designing the optimal smart grid state estimation algorithm that can tolerate false data injection attacks is a crucial task for energy management systems. To address aforementioned issue, this article had proposed a physics-informed based optimal grid state estimation. The simulation results have to be demonstrated the improved performance in grid state estimation accuracy, and computational efficiency compared to the traditional method. The availability of smart grid digital twin model can assist in monitoring the grid status which is precursor for controller design to regulate grid voltage at common coupling points.

Keywords: Controller · cyber attacks · digital twin · loss function · system state estimation · microgrids

1 Introduction

The U.S. Air Force has laid out a vision for energy science and technology needs over the next 15 years, which includes plans for secure grid systems, space-based uninterrupted power stations, and integration of environmental-friendly

E. Blasch et al. (Eds.): DDDAS 2022, LNCS 13984, pp. 315–324, 2024.
https://doi.org/10.1007/978-3-031-52670-1_31

renewable microgrids into the electric grid. To achieve these key objectives, we need secure integration of disparate information and communication technologies. Optimal state estimation models can play an important role for effectively integrating, monitoring, improving reliability and efficiency of the mission critical smart grid [1]. However, the following challenges will need to be resolved (a) avoid intermittent power delivery due to the loading effects from electric vehicles and microgrids and (b) lack of resilience in smart grid communication network to malicious intruders attempt at rendering power grid unstable and vulnerable [2]. Cyber-attacks on the smart grid result in erroneous decisions by the control center. The key consequences of these decisions could be transmission congestion or effects like cascading failures which causes catastrophic blackouts [3].

1.1   Related Works

Due to bidirectional way of smart grid implementation, the sensory information, communication network and measurements are vulnerable to cyber attacks such as false data injection attack (FDIA). It can be seen that such attacks can significantly impact the grid operations, planning, threat the security of state estimation and monitoring process. Consequently, the smart grid security and privacy has become a key challenge. For example, in Ukraine, due to cyber-attacks on the smart grid, over 225,000 customers were affected by power outages for several hours [4]. The detrimental effects of such attacks can propagate across the entire smart infrastructure due to the tight integration of smart devices, distributed resources, and network systems [5]. Generally, attackers can break down the temporal correlation among state dynamics lead to abnormal changes. Apart from intermittent energy sources such as solar cells, the grid states can change along with the variation of energy loads. In order to achieve a reliable and stable grid operation, limited research work exists on grid attack detection in the renewable energy generation domain. Considering these clean energy sources, it is a challenging task to represent them into a state-space framework.

In order to reduce energy loss, electricity cost and environmental concerns, the renewable energy sources such as wind turbine, fuel cells and photovoltaic arrays are widely used in today world. They can supply energy to the main grid after satisfying own demand or can act as a islanded mode. For this interfacing, the power electronic devices such as inverters can play an important role to improve the transient characteristics. Using a set of differential equations of inverter, a state-space model is developed for controller design [6]. Moreover, the Kalman filter based optimal control strategy for microgrid state tracking is proposed in [7,8]. In [9], a least square regression based suboptimal state estimation algorithm is designed. Moreover, a hybrid-learning method for online state estimation in multi-machine power systems is proposed in [10]. In addition, a microgrid state-space model and controller is designed based on the small-signal stability [11]. Furthermore, the linear parameter varying robust controller is deigned for microgrid [12]. Clearly, the designed state-space framework is very simple and they cannot use it real-time.

From grid security point of view, a blockchain-based energy charging coordination algorithm is proposed [13]. The charging request is uniquely signed which can prevent cyber attackers from sending valid charging requests [14]. A blockchain is a distributed electronic database where blocks are growing and linked through a string of hash values to the front block [15]. The hash value is generated based on cryptographic process such as secure hash algorithm-256 [16]. The use of cryptographic hash is to ensure that if the previous block is changed, then all subsequent blocks must also be changed. Therefore, the blockchain can create more efficient, secure and reliable grid systems. In order to trade energy, a multi-agent based blockchain algorithm is proposed [17]. It reduces transaction costs and improves grid security [18]. In order to identify malicious PMU nodes, a blockchain scheme is proposed in [19].

Generally, modeling the electricity network incorporating intermittent renewable energy is a challenging task. In the considered network, the observation information is transmitted to the control center over an unreliable communication channel that is subject to attacks. Developing an effective attack-resilient state estimation is crucial to recover the actual grid states within a short period of time. This work bridges the gap between modeling grid integrating renewable energy and robust state estimation communities.

1.2 Key Contributions

The specific contributions of the paper are summarized as follows [20]:

- In order to know the operating condition of power systems, a new grid state estimation algorithm is proposed. Using the property of semi-definite matrix and Lyapunov composition function, the proposed state estimation algorithm is derived.
- Numerical simulation results considering renewable microgrid show that the developed algorithm provides significant performance improvement compared with the existing method.

2 Digital Twin Design: Smart Grid Incorporating Microgrid with DERs

The smart grid incorporating renewable microgrid provides higher efficiency, reliability, and consumer-centricity in an environment of growing power demand [21]. The state-space representation of power networks is obtained on the basis of a set of differential equations of DERs, power networks and uncertainties. Using Kirchhoff's laws, a set of differential equations are written and after simplifying them, the state-space compact form is obtained.

Generally speaking, the distributed energy resources (DERs) such as solar cells and wind turbines are connected to the power network. The connecting point are point common coupling (PCC) voltages. The PCC voltages and DER voltages are denoted by $\mathbf{V}_b = [V_{b1}, V_{b2}, \cdots, V_{bn}]'$ and $\mathbf{V}_s = [V_{s1}, V_{s2}, \cdots, V_{sn}]'$,

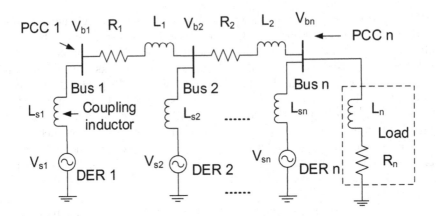

Fig. 1. The n-bus system connected to DERs [22, 23].

where V_{bi} and V_{si} are the i-th PCC voltages and DER voltages, respectively [22, 23] (Fig. 1).

After applying Kirchhoff's voltage law at each bus with simplifications, the digital grid can be written as [20, 22, 23]:

$$\dot{\mathbf{x}} = \mathbf{A}_c\mathbf{x} + \mathbf{B}_c\mathbf{u}. \tag{1}$$

Here, $\mathbf{x} = \Delta\mathbf{V}_b - \mathbf{L}\Delta\mathbf{V}_s$ is the PCC voltage deviation from the reference value, $\mathbf{A}_c = \mathbf{A}^c$ for notional consistency, $\mathbf{B}_c = \mathbf{A}^c\mathbf{L} + \mathbf{B}^c$ and $\mathbf{u} = \Delta\mathbf{V}_s$ is the DER input voltage.

The continuous time digital grid can be represented into discrete-time state-space framework as follows:

$$\mathbf{x}_{k+1} = \mathbf{A}_d\mathbf{x}_k + \mathbf{B}_d\mathbf{u}_k + \mathbf{w}_k. \tag{2}$$

Based on the step size parameter μ, the continuous-time system is discretise to $\mathbf{A}_d = \mathbf{I} + \mu\mathbf{A}_c$ and $\mathbf{B}_d = \mu\mathbf{B}_c$. Here, $\mathbf{w}_t \backsim N(\mathbf{0}, \mathbf{R})$ is the sensing noise which can follow the Gaussian distribution with mean zero and covariance is $\mathbf{R}$.

The sensing measurements from the digital grid are obtained by a set of smart sensors. These phasor measurements sensors are placed into the point common coupling point of the grid. Mathematical, the sensing grid information can be written as follows:

$$\mathbf{y}_k = \mathbf{C}\mathbf{x}_k + \mathbf{v}_k. \tag{3}$$

Here, $\mathbf{y}_k \in \mathbb{R}^p$ is the sensing measurement, $\mathbf{v}_k \backsim N(\mathbf{0}, \mathbf{Q})$ is the noise, and $\mathbf{C}$ is the sensing matrix.

The sensor locally processes raw measurements, and the measurement innovation sequence $\mathbf{z}_k = \mathbf{y}_k - \mathbf{C}\hat{\mathbf{x}}_k^-$ ($\hat{\mathbf{x}}^-$ is the priori state information) is transmitted through communication channel that is subject to attacks as shown in Fig. 2. The

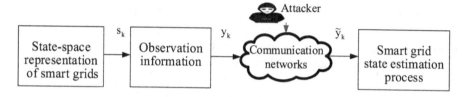

Fig. 2. Design a smart control center under cyber attacks.

manipulated innovation is $\tilde{\mathbf{z}}_k = \mathbf{T}_k \mathbf{z}_k + \mathbf{a}_k$, where $\mathbf{T}_k$ is the attacker matrix, and $\mathbf{a}_k$ is the channel noise. An attacker can hack into a communication network, access and modify the operator and user information [24]. Technically, attackers choose the form and the placement of the attack based on their own purpose and ability [25].

3   Problem Formulation: Grid Observer Design

Today's expeditionary military forces require steady, and reliable green energy sources to power worldwide missions. Diverse field environments and a move towards cost-effective, resilient and agile energy supplies are driving a new look to the way of Defense Department powers a mission, and the Air Force Research Laboratory's advanced power technology office leads the innovation from the front. This paper is concerned with the attack-resilient smart grid state estimation and stabilization for robust observer design. When the sensing information is under cyber attacks what is the optimal smart grid state estimation algorithm that can tolerate cyber attacks. The attackers inject the malicious information into the targeted network to mislead the observer. In light of the aforementioned discussion, the grid state estimation problem can be mathematically written as:

$$\tilde{\mathbf{x}}_k = \tilde{\mathbf{x}}_k^- + \mathbf{K}\tilde{\mathbf{z}}_k. \tag{4}$$

Here, $\tilde{\mathbf{x}}_k^-$ is the prior state estimation, $\tilde{\mathbf{x}}_k$ is the desired estimation, and $\mathbf{K}$ is the gain which needs to be designed. Technically, the estimation gain can reinforce to minimize estimation error dynamics leads to accurate estimated states. After perfectly estimating the grid states, designing and developing the state feedback controller is the second problem. This work addresses the aforementioned issues after developing the optimal state estimation and feedback controller techniques.

4   Proposed Smart Grid State Estimation Method

Identification of system states are essential for optimal power flow, contingency analysis and economic dispatch [26]. Otherwise, the outputs of state estimator can deviate from its actual value and mislead energy management to make unexpected decisions which can lead to serious impact on safety, control, threats, outage, and economic dispatch [27,28]. Using the property of semi-definite matrix and Lyapunov composition function, an attack-resilient estimation algorithm is developed.

Theorem: Based on the mean squared error principle, the optimal grid state estimation algorithm is derived. The optimal estimation gain is derived as follows [29]:

$$\mathbf{K}_k = \tilde{\mathbf{P}}_k \mathbf{C}'(\mathbf{C}\tilde{\mathbf{P}}\mathbf{C}' + \mathbf{R})^{-1}. \tag{5}$$

The state prediction and estimation are computed by:

$$\tilde{\mathbf{x}}_k^- = \mathbf{A}_d \tilde{\mathbf{x}}_{k-1}, \qquad \tilde{\mathbf{x}}_k = \tilde{\mathbf{x}}_k^- + \mathbf{K}\tilde{\mathbf{z}}_k. \tag{6}$$

The predicted and updated error covariance matrices are:

$$\tilde{\mathbf{P}}_k^- = \mathbf{A}_d \tilde{\mathbf{P}}_{k-1} \mathbf{A}_d' + \mathbf{Q}, \quad \tilde{\mathbf{P}}_k = \tilde{\mathbf{P}}_k^- + \bar{\mathbf{P}}\mathbf{C}'(\check{\mathbf{P}} - \mathbf{T}_k'\check{\mathbf{P}} - \check{\mathbf{P}}\mathbf{T}_k)\mathbf{C}\bar{\mathbf{P}}.$$

The gain corrects and minimizes the residual error lead to an accurate states over time. The optimal attack strategy is given by the solution of the optimization problem:

$$\min_{\mathbf{T}_k} tr[\mathbf{C}\bar{\mathbf{P}}\bar{\mathbf{P}}\mathbf{C}'(\mathbf{C}\bar{\mathbf{P}}\mathbf{C}' + \mathbf{R})^{-1}\mathbf{T}_k] \tag{7a}$$

$$\text{subject to } \begin{bmatrix} \rho & \mathbf{T}_k \\ \mathbf{T}_k' & \rho^{-1} \end{bmatrix} \geq \mathbf{0}. \tag{7b}$$

Here, $\tilde{\mathbf{P}}_0^- = E[(\mathbf{x}_0 - \tilde{\mathbf{x}}_0^-)(\mathbf{x}_0 - \tilde{\mathbf{x}}_0^-)'] = \bar{\mathbf{P}}$. The derivation of the algorithm is illustrated in [29], [20].

5   Simulation Results and Discussions

This paper focuses on grid state estimation for robust observer design. Figure 3 shows the main steps of the proposed method. After getting the system and measurements by (6) and (7), the state estimation algorithm runs with (10). Finally, the prior and posterior error covariances are obtained. The algorithm runs until exists.

The simulation parameters are described in Table 1. It assumes that the attacker is added FDIA into measurement during 0.05 to 0.25 s. The attack pattern follows the Gaussian distribution with mean and covariance are 0.9 and $3 \times \mathbf{I}$, respectively [20, 30, 31]. Based on the simulation results, it is easy to conclude as follows (Fig. 5):

- The proposed scheme outperforms the existing baseline approach in [32] as the designed method computes the optimal gain and estimation error covariance.
- The proposed grid estimation process can able to properly recover grid states within 30 iterations. For example, Fig. 4 demonstrates the true PCC voltage dynamic at bus-1 and estimated one. It can be seen that the estimated state is fluctuated during iteration k=10 to 20. Interestingly, the proposed algorithm can able to perfectly estimate the system state within 30 iterations (i.e., *iterations* × *step size* = 0.3 sec) while existing approach requires 150 iterations (i.e., 1.5 sec).

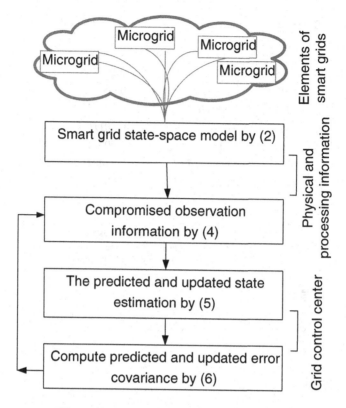

Fig. 3. Main steps of the proposed method.

6   Conclusion and Future Work

This paper addresses the design problem of robust observer, which is composed of grid state estimator. For doing this, the grid digital twin framework is firstly introduced, and it can be used for analysis and investigation. We have formulated and solved the grid state estimation problem based on the mean squared error principle. The superiority and practicability of the obtained theoretical results are illustrated by numerical examples such as IEEE-4 bus systems. It shows the proposed algorithm can be able to recover the PCC voltages within a short period of time. Hopefully, this framework is able to help utility operators to design, verify and investigate online observer for smart grid incorporating microgrids with renewable distributed energy resources.

Table 1. Considered design parameters with Matlab.

Symbols	Values	Symbols	Values
R_1	0.175Ω	R_2	0.1667Ω
R_3	0.2187Ω	L_1	$0.0005H$
L_2	$0.0004H$	L_3	$0.0006H$
R_4	0.0148Ω	L_{sn}	$0.001H$
μ	$0.01sec$	$\mathbf{Q}$	$0.00001 \times \mathbf{I}$
$\mathbf{R}$	$0.0001 \times \mathbf{I}$	L_4	$0.0148H$
ρ	$0.01 \times \mathbf{I}$	$\mathbf{P}_0$	$0.05 \times \mathbf{I}$

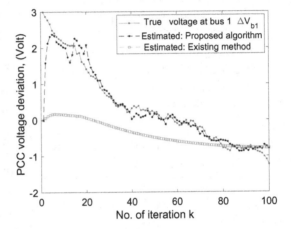

Fig. 4. PCC voltage ΔV_{b1} and its estimation without sensor faults.

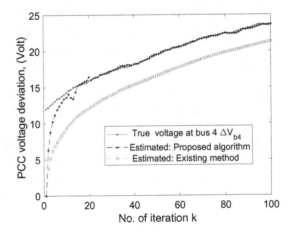

Fig. 5. PCC voltage ΔV_{b4} and its estimation without sensor faults.

Acknowledgement. This work is supported by the U.S. Air Force Office of Scientific Research (AFOSR) FA8750-19-3-1000 Program via grant PIA FA8750-20-3-1003. The views and conclusions contained herein are those of the authors and should not be interpreted as necessarily representing the official policies or endorsements, either expressed or implied, of the U. S. Air Force.

References

1. Rana, M.M., Li, L., Su, S.W.: Cyber attack protection and control of microgrids. IEEE/CAA J. Automatica Sinica **5**(2), 602–609 (2017)
2. Rana, M.M., Bo, R., Abdelhadi, A.: Distributed grid state estimation under cyber attacks using optimal filter and Bayesian approach. IEEE Syst. J. **15**(2), 1970–1978 (2021)
3. Che, L., Liu, X., Shuai, Z., Li, Z., Wen, Y.: Cyber cascades screening considering the impacts of false data injection attacks. IEEE Trans. Power Syst. **33**(6), 6545–6556 (2018)
4. Li, Y., Huo, W., Qiu, R., Zeng, J.: Efficient detection of false data injection attack with invertible automatic encoder and long-short-term memory. IET Cyber-Phys. Syst. Theory Appl. **5**(1), 110–118 (2020)
5. Mohammadi, M., Al-Fuqaha, A., Sorour, S., Guizani, M.: Deep learning for IoT big data and streaming analytics: a survey. IEEE Commun. Surv. Tutor. **20**(4), 2923–2960 (2018)
6. Paduani, V., Kabalan, M., Singh, P.: Small-signal stability of islanded-microgrids with DC side dynamics of inverters and saturation of current controllers. In: Power & Energy Society General Meeting, pp. 1–5 (2019)
7. Feng, Y., Yang, D.: Kalman filter-based centralized controller design for smart microgrid. In: 2019 Chinese Automation Congress, pp. 2185–2190 (2019)
8. Damgacioglu, H., Celik, N.: A two-stage decomposition method for integrated optimization of islanded ac grid operation scheduling and network reconfiguration. Int. J. Electr. Power Energy Syst. **136**, 107647 (2022)
9. Karimipour, H., Dinavahi, V.: Robust massively parallel dynamic state estimation of power systems against cyber-attack. IEEE Access **6**(18), 2984–2995 (2018)
10. Tian, G., Zhou, Q., Birari, R., Qi, J., Qu, Z.: A hybrid-learning algorithm for online dynamic state estimation in multimachine power systems. IEEE Trans. Neural Netw. Learn. Syst. **31**(12), 5497–5508 (2020)
11. Unnikrishnan, B.K., Johnson, M.S., Cheriyan, E.P.: Small signal stability improvement of a microgrid by the optimised dynamic droop control method. IET Renew. Power Gener. **14**(5), 822–833 (2019)
12. Otoofi, F., Asemani, M.H., Vafamand, N.: Polytopic-LPV robust control of power systems connected to renewable energy sources. In: International Conference on Control, Instrumentation and Automation, pp. 1–6 (2019)
13. Baza, M., Nabil, M., Ismail, M., Mahmoud, M., Serpedin, E., Rahman, M.A.: Blockchain-based charging coordination mechanism for smart grid energy storage units. In: International Conference on Blockchain, pp. 504–509 (2019)
14. Wang, J., Wu, L., Choo, K.K.R., He, D.: Blockchain based anonymous authentication with key management for smart grid edge computing infrastructure. IEEE Trans. Ind. Inform. **16**(3), 1984–1992 (2020)
15. Wu, X., Duan, B., Yan, Y., Zhong, Y.: M2M blockchain: the case of demand side management of smart grid. In: International Conference on Parallel and Distributed Systems, pp. 810–813 (2017)

16. Upreti, A., Cardell, J., Thiebaut, D.: Data privacy in the smart grid: a decentralized approach. In: Proceedings of the 52nd Hawaii International Conference on System Sciences (2019)

17. Mezquita, Y., Gazafroudi, A.S., Corchado, J.M., Shafie-Khah, M., Laaksonen, H., Kamišalić, A.: Multi-agent architecture for peer-to-peer electricity trading based on blockchain technology. In: International Conference on Information, Communication and Automation Technologies, pp. 1–6 (2019)

18. Darville, J., Curia, J., Celik, N.: Microgrid operational planning using a hybrid neural network with resource-aware scenario selection. Simul. Model. Pract. Theory **119**, 102583 (2022)

19. Iyer, S., Thakur, S., Dixit, M., Agrawal, A., Katkam, R., Kazi, F.: Blockchain based distributed consensus for byzantine fault tolerance in PMU network. In: International Conference on Computing, Communication and Networking Technologies, pp. 1–7 (2019)

20. Rana, M.M., Abdelhadi, A.: Attack-resilient smart grid dynamic state estimation algorithm. In: IEEE International Symposium on Systems Engineering, pp. 1–5 (2020)

21. He, Y., Mendis, G.J., Wei, J.: Real-time detection of false data injection attacks in smart grid: a deep learning-based intelligent mechanism. IEEE Trans. Smart Grid **8**(5), 2505–2516 (2017)

22. Mishra, S.R., Korukonda, M.P., Behera, L., Shukla, A.: Enabling cyber physical demand response in smart grids via conjoint communication and controller design. IET Cyber-Phys. Syst. Theory Appl. **4**(4), 291–303 (2019)

23. Li, H., Lai, L., Poor, H.V.: Multicast routing for decentralized control of cyber physical systems with an application in smart grid. IEEE J. Sel. Areas Commun. **30**(6), 1097–1107 (2012)

24. Singh, S.K., Khanna, K., Bose, R., Panigrahi, B.K., Joshi, A.: Joint-transformation-based detection of false data injection attacks in smart grid. IEEE Trans. Industr. Inf. **14**(1), 89–97 (2018)

25. Manandhar, K., Cao, X., Hu, F., Liu, Y.: Detection of faults and attacks including false data injection attack in smart grid using Kalman filter. IEEE Trans. Control Netw. Syst. **1**(4), 370–379 (2014)

26. Dou, C., Wu, D., Yue, D., Jin, B., Xu, X.: A hybrid method for false data injection attack detection in smart grid based on variational mode decomposition and OS-ELM. IEEE Trans. Power Syst. **8**(6), 1697–1707 (2021)

27. Xie, L., Mo, Y., Sinopoli, B.: Integrity data attacks in power market operations. IEEE Trans. Smart Grid **2**(4), 659–666 (2011)

28. Yuan, Y., Li, Z., Ren, K.: Modeling load redistribution attacks in power systems. IEEE Trans. Smart Grid **2**(2), 382–390 (2011)

29. Guo, Z., Shi, D., Johansson, K.H., Shi, L.: Optimal linear cyber-attack on remote state estimation. IEEE Trans. Control Netw. Syst. **4**(1), 4–13 (2017)

30. Kurt, M.N., Ogundijo, O., Li, C., Wang, X.: Online cyber-attack detection in smart grid: a reinforcement learning approach. IEEE Trans. Smart Grid **10**(5), 5174–5185 (2018)

31. Sanjab, A., Saad, W.: Data injection attacks on smart grids with multiple adversaries: a game-theoretic perspective. IEEE Trans. Smart Grid **7**(4), 2038–2049 (2016)

32. Rana, M.M.: Least mean square fourth based microgrid state estimation algorithm using the internet of things technology. PLoS ONE **12**(5), e0176099 (2017)

Applying DDDAS Principles for Realizing Optimized and Robust Deep Learning Models at the Edge

Robert Canady, Xingyu Zhou, Yogesh Barve, Daniel Balasubramanian, and Aniruddha Gokhale(✉)

Department of Computer Science, Vanderbilt University, Nashville, TN 37203, USA
a.gokhale@Vanderbilt.Edu

Abstract. Edge computing is an attractive avenue to support low-latency applications including those that leverage deep learning (DL)-based model inferencing. Due to constraints on compute, storage and power at the edge, however, these DL models must be quantized to reduce their footprint while minimizing loss of accuracy. However, DL models and their quantized equivalents are often prone to adversarial attacks requiring them to be made robust against such attacks. The resource constraints at the edge, however, preclude any quantization and robustness design operations directly at the edge. Moreover, the changing dynamics of edge-based computations and resulting concept drifts in the models require an iterative approach to meet the needs of robust DL models at the edge. To address these challenges, this paper presents initial results on an iterative procedure involving a DDDAS feedback loop. DDDAS is used to dynamically instrument the edge-deployed, quantized DL models for data on the effectiveness of their quantization and robustness abilities, which in turn is used to drive an automated, cloud-based process that uses tools, such as Apache TVM, to generate quantized, optimized and robust DL models suitable for the edge. These models subsequently are automatically deployed at the edge using orchestration tools. Preliminary studies using this approach have shown its effectiveness in image classification and object detection applications.

Keywords: Deep Learning · Dynamic Data-driven System · Adversarial Machine Learning · Edge Computing · Model quantization

1 Introduction

Deep learning (DL) can be used to detect and segment distinct objects in an image, translate speech to text, detect fraud, etc., which has led to a number of DL-based cloud-hosted services. For several reasons, such as low latency response, conservation of bandwidth, security, privacy, environmental concerns, etc., however, applications are being designed to shift most of their computations away from the cloud and closer to the edge. Consequently, there is a need to deploy these traditional DL models on a diverse range of edge hardware from

© The Author(s), under exclusive license to Springer Nature Switzerland AG 2024
E. Blasch et al. (Eds.): DDDAS 2022, LNCS 13984, pp. 325–339, 2024.
https://doi.org/10.1007/978-3-031-52670-1_32

FPGAs to GPUs to ASICs [36]. Moreover, due to capacity and power constraints of these edge resources, the footprint of these models for edge resources must be reduced – a process known as quantization. Thus, a change in model execution performance and accuracy across these different devices can be expected since these models are typically optimized for the smaller footprint and resource-constrained devices. There have been several efforts, such as Apache TVM [7], Once-for-all (OFA) [4], and knowledge distillation (KD) [14] that attempt to optimize DL models for heterogeneous platform deployment especially at the edge.

Unfortunately, current machine learning models are generally vulnerable to adversarial machine learning (AML) attacks, which are bound-limited perturbations unnoticeable to the human eye but that cause the model to misunderstand the data. While frameworks like TVM, OFA and KD work effectively under normal, non-adversarial conditions, prior studies on determining whether or not the compiler-generated smaller DL models are vulnerable to the same adversarial machine learning (AML) attacks as traditional DL models are generally lacking. Moreover, defense mechanisms against such attacks, e.g., an effective defense strategy such as adversarial training, take much longer to deploy than traditionally trained DL models.

These prohibitive costs make it very difficult to use existing AML defense techniques directly on the edge. To address this problem, we propose a novel application of the DDDAS paradigm [9] to generate, evaluate and deploy robust and optimized edge-based DL models. In our approach, the DDDAS feedback loop manifests between a powerful cloud or fog server and multiple edge-based devices. The end result is a robust and reduced-size DL model that will be deployed on the edge devices. Our novel application of the DDDAS paradigm operates as follows: An initial reduced size and robust DL model is deployed at the edge; then this edge device will periodically stream dynamically instrumented data concerning attack robustness as well as accuracy of model predictions back to the server where several larger models will check the performance and robustness of the edge-based model. Based on these performance results, adaptive retraining of the edge-based model will occur at the cloud server. Subsequently, this new model will be deployed on the edge-device after adversarial retraining and optimization.

In this paper, we lay out the general idea behind our approach and present preliminary results using this approach. In Sect. 2 we provide the necessary background information. This will be followed by Sects. 3 and 4 where we describe our approach. We present some of our initial results in Sect. 5. We then conclude the paper and discuss future directions in Sect. 6.

2   Background and Related Work

To make this paper self-contained, we provide background information on the use of deep learning in computer vision focusing primarily on adversarial machine learning, and on model quantization/optimization. We also describe related efforts and compare them to our proposed ideas.

2.1 Edge-Based Computer Vision Applications

Edge Computing is the idea of moving the computation from the cloud closer to the sensors or Internet of Things (IoT). One of the benefits of this approach is that it eliminates the need to send data back to the cloud, which can be very costly depending on the type of data.

There have been several recent works, such as OpenDataCam [28], Coral-Pie [34] and DeepLite [15], that demonstrate different object detection applications at the edge and the need for edge accelerators. In Coral-Pie, the application is vehicle tracking using two Raspberry Pi's connected to a Coral USB. The authors did not use the full YOLOv3 [27] for object detection because it was too computationally expensive for the CORAL USB. OpenDataCam is an open source tool for monitoring and tracking moving objects in a live video stream. This application uses YOLOv3 on a desktop machine and recommends using YOLOv3-tiny for edge devices like Jetson Nano.

2.2 Adversarial Machine Learning and Defenses

Adversarial machine learning attacks on deep learning is a relatively new field with its start in machine learning models [3]. The work on adversarial evasion attacks [2] led to the seminal work on adversarial work on deep learning models [30]. The idea behind the attacks is that the image that is to be classified is perturbed enough to make the ML model misclassify but not so much that a human observer would notice. Several more efforts followed, such as FGSM [13], PGD [23] and DeepFool [25].

There have been efforts to defend against such attacks. One of the most successful defenses is a proactive method called Adversarial Training [23]. The idea is to augment the training with adversarial examples so that the model will hopefully learn smoothened decision boundaries taking into account the adversarial perturbations, and later can correctly classify the adversarial samples. There have been other defense works that utilize data augmentation [22] and/or preprocessing [26,33] where the idea is to remove the perturbations from the image, or at least to mitigate the impact of the perturbation. A combination of these data transformations with adversarial training [1,32] have also been proposed as a set of defense strategies.

Generally speaking, adversarial machine learning attacks and defenses have gained much research interest. A roadmap on improving and evaluating adversarial examples was given in [6], where the authors outline an approach to test the robustness of models and describe some of the usual pitfalls that can occur with defenses.

2.3 Deep Learning Computer Vision Attacks and Defenses

Since we focus on computer vision DL models with the goal of making them robust for the edge, we provide some background in this area. Computer vision is a very large area including many kinds of tasks like detection, classification and segmentation. Among them, object detection is a quite fundamental one. Object detection in computer vision can be broken into two categories: Single shot and Two-stage. Two-stage object detectors like R-CNN [12] have a region proposal network in the first stage that narrows down the number of Region of Interests (ROIs), and in the second stage completes the classification and refines the bounding box. Two-stage models achieve good performances but training and inference are both expensive. Single shot detectors like SSD models [20] predict the boundary box and the class at the same time. Single shot detectors often trade accuracy (on objects too close or far away) for the inference speed.

Due to the rise in the number of applications using deep learning-based object detectors, research has focused on attacks against object detectors in a direction guided by the previous adversarial machine learning works. The purpose of these attacks can be different, ranging from causing many false objects to be detected like adding an adversarial patch [21] to not detecting any objects at all as in TOG [8].

Another recent extended attack on a tracking application was presented in [17] in which the authors try to fool the Multiple Object Tracking (MOT). They present an early work in autonomous driving to explore an attack on the complete computer vision pipeline. Since adversarial ML attacks on object detectors is still a relatively new field, the research on defense techniques is still scarce. Efforts, such as [16], have used a two-stage adversarial training algorithm to improve the robustness in safety-critical scenarios. In [35] the authors present a similar adversarial training approach with a model trained on PGD attacked data. They however only use one type of attack to augment their training data.

2.4   Summary of Prior Efforts and Unresolved Challenges

Much work has been done to address the challenges of adversarial machine learning and deploying models on edge-based devices, but little work has been done on evaluating the robustness of edge-based models before and after optimization and as models incur concept drift. We posit that the adversarial robustness challenges, particularly on edge hardware deployments, requires further investigations into the following questions, which formulates the need to apply the DDDAS paradigm as discussed in this paper:

1. Although research on model optimization with adversarial robustness exists, these are mostly input-dependent solutions raising the question whether such settings are too ideal without considering realistic resource limitations?
2. Past research has focused heavily on the single computation device node adversarial vulnerability raising the question whether more adaptive system-level attack evaluations can be designed?
3. The high transferability of the adversary across model settings raises the question whether models can be deployed at the edge in a resilient way to mitigate potential adversarial risks for new edge-based data-driven applications?

Given these unresolved challenges, our objective is to explore solutions from a system-level perspective. To make this system robust and resilient, we are adopting DDDAS principles because the DDDAS's dynamic and adaptive feedback loop can ensure that the ML models never start to decrease in performance nor lose their robustness.

3 Methodology

In this section, we present our work. First, we describe the overall system model and then discuss the approach highlighting the DDDAS loop and its components that are distributed across the edge and cloud.

3.1 System Model

The system consists of one or more edge devices and a cloud server. In this scenario, the edge device is assumed to run an application such as surveillance of a parking lot or aiding an augmented reality (AR) device that is giving real-time guidance to a user. It will be directly connected to a camera or any other type of sensor being used.

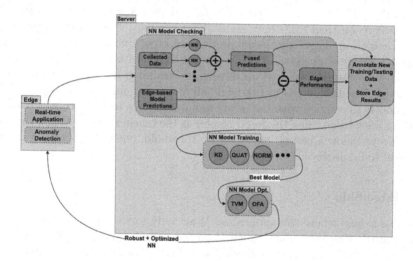

Fig. 1. System Model with the DDDAS Feedback Loop

3.2 Server Side

As seen in Fig. 1, the server is used to train the original model that is to be deployed to an edge device. This is also where multiple larger networks are deployed that have been trained using different architectures, defense methods, etc. The goal is to cover enough 'adversarial' ground with multiple networks that a fusion of all of the models cannot be fooled by whatever perturbations are crafted.

There have been many research works on machine learning model optimizations for specific hardware, and in particular smaller edge-based devices [29]. Among them we specifically choose three critical techniques of Once-for-all (OFA) [4], Apache TVM [7] and Knowledge Distillation (KD) [14]. These three techniques involve three model phases of adversarial training, defensive model optimization and robustness-preservation model structure simplification. OFA [4] trains one network, and then uses a generalized pruning approach to obtain many smaller networks that have reduced dimensions in depth, width, kernel size, and resolution. Apache TVM [7] is a deep learning optimizer and compiler, that takes in models trained using frameworks such as Tensorflow, PyTorch, MXNet, etc. and generates code optimized to run the models on diverse hardware backends. KD [14] utilizes a teacher-student approach, where the teacher is a large model and the student is a smaller model. In our framework, the candidate models go through Apache TVM or OFA where they are optimized for the given device.

We emphasize the *dynamic data-driven* aspect by continuously checking prediction results on incoming data from the edge side in a dynamic way. The server takes in data from the edge device and passes it through multiple networks, where we then fuse those results and compare them to the edge-device model prediction. From the fused results, we create new ground truth training and validation images. These images can then either be used directly, or they can be attacked and used for adversarial training. Using terminology from Knowledge Distillation (KD) [14], while all of this is going on, an edge model is continuously fine-tuned with the newly annotated data while a student model is distilling knowledge using the fused results as the teacher model. When the edge device drops below a certain performance threshold, one of the updated models is chosen to be deployed.

3.3 Edge Side

The edge side is where the application is actually executing. It is continuously collecting the streaming data which is passed through the quantized and optimized network. These results are saved and checked for any potential anomalies. If there are any, then this triggers a certain amount of data and predictions to be sent back to the server. Periodically, data and predictions are sent back for model checking, where a trigger causes more data to be sent back.

We emphasize the 'dynamic data-driven' view from two aspects. First of all, the periodic prediction result checking and calibration with the server side enables the 'dynamic' model updating to guarantee consistent adversarial robustness. Secondly, the feedback information from the server side should also enable potential threat type detection and estimation(for example which L_p norm attack), leading to 'dynamic' selection and execution of robust candidate models.

3.4   Expected Use Cases

In ongoing work we are applying these ideas to augmented reality edge applications used to provide interactive maintenance support. We expect this framework to allow users to deploy applications to the edge that can then be dynamically adapted during deployment to perform most optimally when faced with clean or adversarial data. We combine multiple ideas to obtain the most robust, optimized edge-based DL models.

4   Experimental Setup

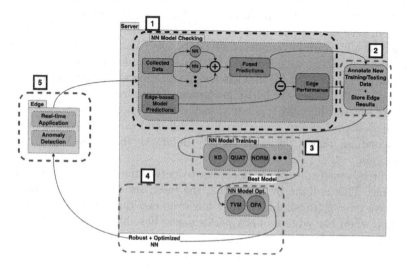

Fig. 2. System Model with the DDDAS Feedback Loop

In this section, we will explore each part of the DDDAS feedback loop and how we plan to implement and evaluate each component. To more easily visualize the distinct parts of the loop, we have highlighted the parts in Fig. 2.

4.1   Feedback Loop Components

To begin testing the proposed DDDAS feedback loop from Fig. 1, there were several experimental logistical issues that we needed to resolve. The first was determining how can we use pre-existing datasets to test out our feedback loop with 'new' data. Our initial idea was to take datasets and reduce the number of training images, and use the extra training images for the feedback loop. These could then be combined with the validation and test sets to simulate new data being presented. This data would then be used later on in the retraining phase of the loop.

We want to split the data so that there are examples from each class in the new data. We also needed to separate some data for the anomaly detection. For example, we would want an anomaly to be detected for object detection if there are too few or too many detected objects. We discovered a tool [24] that would aid us in doing this. This tool allows the user to more easily visualize and rearrange datasets for their needs.

NN Model Checking. The first component we will discuss is how we evaluate the edge model's performance. There has been some research done on fusion and its impact on robustness, which has mostly shown that fusion does improve robustness [31]. We want to utilize this fact to evaluate the edge model.

To setup this component, we use pretrained networks, which are adversarially trained networks as well as a variety of different neural networks. The idea is to get a diverse enough ensemble of models that when fused together will not be fooled by adversarial examples, OOD data, etc. When we say fused together, we mean by utilizing decision-level fusion. Decision-level fusion is where the output probabilities from ML models trained on different modalities of data are combined in order to achieve better performance than from just one type of data. There are several different types of decision-level fusion: average ranks, naive-bayes, highest probability, generalized chernoff, and sandia probabilistic fusion. Each algorithm, makes decisions slightly differently based on the probability distributions.

We directly use the robust decisions to determine how our edge model is performing. This is done by comparing the fused results with the edge model results. We have a threshold set, that has been determined through testing, to determine when the edge model's performance has dipped enough to warrant retraining.

Preparation of New Data and History Tracking. We discussed briefly how we plan to approach sending 'new' data through a network for evaluation and retraining. A benefit of splitting up the data more than just train and test splits, is that we have already annotated data. This way we can check how well our annotation techniques work without having to hand annotate new data.

In this component, we use the predictions from the NN model check to help us annotate the new images. We also store the results of the models to be able to compare performance going forward.

NN Model Training/Retraining. At this point, new data has been annotated and there is a need to retrain the edge model. We utilize several different approaches to determine the best and most robust model. Several of the different training techniques we are using are our own QUAT [5] approach, normal training, knowledge distillation training, regular adversarial training, etc. We then compare each of these models to each other and send the best one to the next step of the feedback loop.

NN Model Optimization. In this component, we optimize the trained model to be deployed on a specified edge device using preexisting frameworks such as Apache TVM. This step will optimize the model to be deployed on whatever hardware is being used. There has been some previous work showing that model quantization can actually improve adversarial robustness [37].

Real-Time Edge Application. This component is where the real-time edge application will be deployed. At the moment we are focusing solely on computer vision tasks like image classification, object detection, semantic segmentation. With one of our motivating applications being AI-assisted AR for smart maintenance, we would like to include other tasks such as natural language processing in further work.

While the application is being run, there is periodic offloading of some of the collected data as well as the model's predictions. There also is an anomaly detector running that checks the model's outputs at runtime and checks to see if there is any anomalous behavior. An example of this for object detection would be detecting many more objects than is normal or not detecting any objects for an extended period of time.

To aid in decision-making further on in the loop, we also collect resource usage data. We see this mainly being used to help in the optimizations of the models.

4.2 Datasets

We want to evaluate our framework using several computer vision tasks; image classification, object detection, and semantic segmentation. To carry out the best and most informative evaluation, we wanted to select several datasets for each task.

In Table 1, there is a dataset that we collected ourselves, Car Engines. The goal for this dataset was to show a proof of concept for engine maintenance guidance using ML. This dataset was collected with help from collaborators in our research group. The dataset was then self-annotated for semantic segmentation. An example image and segmentation map from this dataset can be seen in Fig. 3.

4.3 Models

The models we are using and their deployment level are each outlined in Table 2. We also present the model size to illustrate the differences between cloud and edge models.

Table 1. Description of CV Tasks and Corresponding Datasets

CV Task	Dataset	Train/Val/Test Images	Classes
Image Classification	CIFAR10 [18]	60,000	10
	CIFAR100	60,000	100/20
	Imagenet [10]	14,197,122	1000
Object Detection	Pascal VOC [11]	21,493	20
	MS COCO [19]	123,287	80
	VisDrone [38]	8632	10
Semantic Segmentation	Pascal VOC	123,287	20
	Car Engines (Ours)	58	14

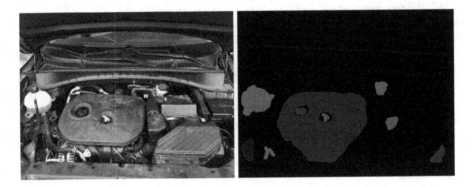

Fig. 3. Example Engine Images (L: RGB Image, R: Segmentation Map)

4.4 Devices

In Table 3, we list out the devices we have used/plan to use for evaluation of our feedback loop. We have also included selected resource information such as CPU cores, RAM, GPU, and typical power consumption when idle and busy.

5 Preliminary Studies and Challenges

Typically accuracy of the model is the only metric for image classification. Object detection and semantic segmentation, have slightly more complex metrics.

To determine how well models perform on object detection/semantic segmentation datasets like PASCAL-VOC [11], they are judged on their inference time and their mean average precision (mAP). The mAP is calculated using a metric called *Intersection over Union (IoU)*. The higher the mAP the better, but its semantics for object detection are different compared to image classification accuracy, where the classification is either correct or incorrect. In contrast, the goal of object detection is to draw bounding boxes around objects and then correctly classify the object(s). To calculate the mAP, the analyst needs the

Table 2. Description of CV Models

CV Task	Model	Backbone	Model Size (MB)	Device Type
Image Classification	Resnet(18/50)	N/A	45/98	Cloud
	Densenet121	N/A	33	Cloud
	Mobilenet v3	N/A	16	Edge
Object Detection	YOLO v3	Darknet	237	Cloud
	Tiny YOLO v3	Darknet	34	Edge
	FasterRCNN	Resnet50	160	Cloud
	FasterRCNN	Mobilenet v3	74	Edge
Semantic Segmentation	Unet	Resnet50	164	Cloud
	Unet	Mobilenet v3	43	Edge

Table 3. Description of Devices Used

Deployment Level	Device Name	CPU	RAM	GPU	Power
Cloud	Desktop	12 Core	32 GB	NVIDIA RTX 2060	70 W/175 W 175 W/500 W
Edge	Raspberry Pi	4 Core	1 GB	N/A	1.9 W/5 W
	Jetson Nano	4 Core	4 GB shared	NVIDIA Maxwell	5 W/10 W
	Jetson TX2	2 + 4 Core	8 GB shared	NVIDIA Pascal	7.5 W/15 W

ground-truth and predicted bounding box coordinates, which can then be used to calculate the IoU. The IoU is calculated as the amount the predicted bounding box overlaps with the ground-truth bounding box divided by the total area of the union of both boxes.

To determine the efficacy of the model, the analyst sets a threshold percentage for the overlap. The threshold is usually set at 0.5 per convention and because of the fact that humans can barely tell the difference between 0.3 and 0.5 IoU. For some different datasets or competitions, a different confidence threshold is used. The mAP is then calculated by drawing precision-recall curves with the IoU set at different thresholds. This is done for each class, and at this point it is just the average precision (AP). The average AP across all classes is then the mAP.

5.1 Image Classification

We have done extensive evaluation of image classification models. Most of this is still waiting to be publicly released, but that will all be included in the final framework. We have looked at different techniques for adversarial training as well as potential preprocessing techniques for improving adversarial robustness.

5.2   Object Detection

In our earlier work, we evaluated our QUAT [5] algorithm using object detection models. Through experimentation, our approach appeared to be a promising solution, but there is still room for improvement as well as more evaluation which we plan to do throughout developing this framework.

5.3   Semantic Segmentation

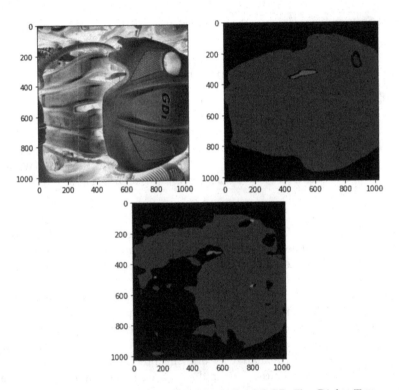

Fig. 4. Example Engine Images (Top left: Normalized RGB, Top Right: Test seg, Bottom: Predicted seg) (mIoU = 30.7%, Accuracy = 79.3%)

Some of our initial results on the Car Engine dataset are shown in Fig. 4. While the mIoU might seem slightly low at 30.7%, it is actually not bad for a semantic segmentation task. The model was able to locate close to 3/4 of the engine as well as the dip stick. Also, the validation accuracy of 79.3% is promising showing that the model was able to classify the predicted objects reasonably well.

We believe these results can be further improved by training the model more as well as add more engine images.

6   Conclusions

This paper presented preliminary ideas and results on applying DDDAS principles to address the challenges of realizing adversarially robust deep learning models that are suitable for edge devices. Presently, this research has shown the effectiveness of using a DDDAS feedback loop to keep a real-time application from decreasing in performance and improve in robustness to unseen circumstances. There is still much work to be done to further explore this area of research. We plan to explore this problem by utilizing applications such as semantic segmentation and object detection that will be evaluated on a range of edge device types and application use cases, such as augmented reality-based maintenance.

References

1. Addepalli, S., Vivek, B.S., Baburaj, A., Sriramanan, G., Babu, R.V.: Towards achieving adversarial robustness by enforcing feature consistency across bit planes. In: Proceedings of the IEEE/CVF Conference on Computer Vision and Pattern Recognition (CVPR), June 2020
2. Biggio, B., et al.: Evasion attacks against machine learning at test time. CoRR abs/1708.06131 (2017). http://arxiv.org/abs/1708.06131
3. Biggio, B., Roli, F.: Wild patterns: ten years after the rise of adversarial machine learning. CoRR abs/1712.03141 (2017). http://arxiv.org/abs/1712.03141
4. Cai, H., Gan, C., Wang, T., Zhang, Z., Han, S.: Once for all: train one network and specialize it for efficient deployment. In: International Conference on Learning Representations (2020). https://arxiv.org/pdf/1908.09791.pdf
5. Canady, R., Zhou, X., Barve, Y., Balasubramanian, D., Gokhale, A.: Adversarially robust edge-based object detection for assuredly autonomous systems. In: 2022 IEEE International Conference on Assured Autonomy (ICAA), pp. 97–106 (2022). https://doi.org/10.1109/ICAA52185.2022.00021
6. Carlini, N., et al.: On evaluating adversarial robustness. CoRR abs/1902.06705 (2019). http://arxiv.org/abs/1902.06705
7. Chen, T., et al.: TVM: an automated End-to-End optimizing compiler for deep learning. In: 13th USENIX Symposium on Operating Systems Design and Implementation (OSDI 2018), pp. 578–594. USENIX Association, Carlsbad, CA, October 2018. https://www.usenix.org/conference/osdi18/presentation/chen
8. Chow, K.H., et al.: Adversarial objectness gradient attacks in real-time object detection systems. In: IEEE International Conference on Trust, Privacy and Security in Intelligent Systems, and Applications, pp. 263–272. IEEE (2020)
9. Darema, F., Blasch, E., Ravela, S., Aved, A. (eds.): Dynamic Data Driven Applications Systems: Third International Conference, DDDAS 2020, Boston, MA, USA, 2–4 October 2020, Proceedings, vol. 12312. Springer, Cham (2020). https://doi.org/10.1007/978-3-030-61725-7
10. Deng, J., Dong, W., Socher, R., Li, L.J., Li, K., Fei-Fei, L.: ImageNet: a large-scale hierarchical image database. In: 2009 IEEE Conference on Computer Vision and Pattern Recognition, pp. 248–255 (2009). https://doi.org/10.1109/CVPR.2009.5206848

11. Everingham, M., Van Gool, L., Williams, C.K.I., Winn, J., Zisserman, A.: The pascal visual object classes (VOC) challenge. Int. J. Comput. Vision **88**(2), 303–338 (2010)
12. Girshick, R.B., Donahue, J., Darrell, T., Malik, J.: Rich feature hierarchies for accurate object detection and semantic segmentation. CoRR abs/1311.2524 (2013). http://arxiv.org/abs/1311.2524
13. Goodfellow, I.J., Shlens, J., Szegedy, C.: Explaining and harnessing adversarial examples (2015)
14. Hinton, G., Vinyals, O., Dean, J.: Distilling the knowledge in a neural network. In: NIPS Deep Learning and Representation Learning Workshop (2015). http://arxiv.org/abs/1503.02531
15. Ho, D.H., Marri, R., Rella, S., Lee, Y.: DeepLite: real-time deep learning framework for neighborhood analysis. In: 2019 IEEE International Conference on Big Data (Big Data), pp. 5673–5678 (2019). https://doi.org/10.1109/BigData47090.2019.9005651
16. Hu, Z., Zhong, Z.: Towards practical robustness improvement for object detection in safety-critical scenarios. In: Wang, G., Ciptadi, A., Ahmadzadeh, A. (eds.) MLHat 2020. CCIS, vol. 1271, pp. 66–83. Springer, Cham (2020). https://doi.org/10.1007/978-3-030-59621-7_4
17. Jia, Y., et al.: Fooling detection alone is not enough: adversarial attack against multiple object tracking. In: International Conference on Learning Representations (2020). https://openreview.net/forum?id=rJl31TNYPr
18. Krizhevsky, A., Hinton, G., et al.: Learning multiple layers of features from tiny images (2009)
19. Lin, T., et al.: Microsoft COCO: common objects in context. CoRR abs/1405.0312 (2014). http://arxiv.org/abs/1405.0312
20. Liu, W., et al.: SSD: single shot multibox detector. CoRR abs/1512.02325 (2015). http://arxiv.org/abs/1512.02325
21. Liu, X., Yang, H., Song, L., Li, H., Chen, Y.: DPatch: attacking object detectors with adversarial patches. CoRR abs/1806.02299 (2018). http://arxiv.org/abs/1806.02299
22. Lopes, R.G., Yin, D., Poole, B., Gilmer, J., Cubuk, E.D.: Improving robustness without sacrificing accuracy with patch gaussian augmentation. CoRR abs/1906.02611 (2019). http://arxiv.org/abs/1906.02611
23. Madry, A., Makelov, A., Schmidt, L., Tsipras, D., Vladu, A.: Towards deep learning models resistant to adversarial attacks (2019)
24. Moore, B.E., Corso, J.J.: Fiftyone. GitHub. Note (2020). https://github.com/voxel51/fiftyone
25. Moosavi-Dezfooli, S., Fawzi, A., Frossard, P.: DeepFool: a simple and accurate method to fool deep neural networks. CoRR abs/1511.04599 (2015). http://arxiv.org/abs/1511.04599
26. Raff, E., Sylvester, J., Forsyth, S., McLean, M.: Barrage of random transforms for adversarially robust defense. In: Proceedings of the IEEE/CVF Conference on Computer Vision and Pattern Recognition (CVPR), June 2019
27. Redmon, J., Farhadi, A.: YOLOv3: an incremental improvement. CoRR abs/1804.02767 (2018). http://arxiv.org/abs/1804.02767
28. Sawadski, V.: Opendatacam. https://github.com/opendatacam/opendatacam
29. Sze, V., Chen, Y.H., Yang, T.J., Emer, J.S.: Efficient processing of deep neural networks: a tutorial and survey. Proc. IEEE **105**(12), 2295–2329 (2017)
30. Szegedy, C., et al.: Intriguing properties of neural networks. arXiv preprint arXiv:1312.6199 (2013)

31. Wang, S., Wu, T., Chakrabarti, A., Vorobeychik, Y.: Adversarial robustness of deep sensor fusion models. In: Proceedings of the IEEE/CVF Winter Conference on Applications of Computer Vision (WACV), pp. 2387–2396, January 2022
32. Xie, C., Wu, Y., van der Maaten, L., Yuille, A.L., He, K.: Feature denoising for improving adversarial robustness. In: 2019 IEEE/CVF Conference on Computer Vision and Pattern Recognition (CVPR), pp. 501–509 (2019). https://doi.org/10.1109/CVPR.2019.00059
33. Xie, C., Wang, J., Zhang, Z., Ren, Z., Yuille, A.L.: Mitigating adversarial effects through randomization. CoRR abs/1711.01991 (2017). http://arxiv.org/abs/1711.01991
34. Xu, Z., Shah, H.S., Ramachandran, U.: Coral-Pie: a geo-distributed edge-compute solution for space-time vehicle tracking. Association for Computing Machinery, New York, NY, USA (2020). https://doi.org/10.1145/3423211.3425686
35. Zhang, H., Wang, J.: Towards adversarially robust object detection. In: Proceedings of the IEEE/CVF International Conference on Computer Vision (ICCV), October 2019
36. Zhou, X., Canady, R., Bao, S., Gokhale, A.: Cost-effective hardware accelerator recommendation for edge computing. In: 3rd {USENIX} Workshop on Hot Topics in Edge Computing (HotEdge 2020) (2020)
37. Zhou, X., et al.: Guarding against universal adversarial perturbations in data-driven cloud/edge services. In: 2022 IEEE International Conference on Cloud Engineering (IC2E), pp. 233–244 (2022). https://doi.org/10.1109/IC2E55432.2022.00032
38. Zhu, P., et al.: Detection and tracking meet drones challenge. IEEE Trans. Pattern Anal. Mach. Intell. 44(11), 7380–7399 (2021). https://doi.org/10.1109/TPAMI.2021.3119563

Main-Track: Keynotes

DDDAS2022 Keynotes - Overview

Frederica Darema[1]([envelope]) and Erik Blasch[2]

[1] InfoSymbiotic Systems Society, Bethesda, MD, USA
fredericadarema@hotmail.com
[2] MOVEJ Analytics, Fairborn, OH, USA

Abstract. The DDDAS2022 Conference featured five keynote presentations, and an invited talk, which addressed important science and technology topics, and provided examples of advances in capabilities enabled or supported by DDDAS-based methods. The presentations covered a ange of areas such as: aerospace systems, cyber-security, bio-infomatics and genomics, and adverse environmental events. Together with the present overview, papers contributed by the keynote speakers are included in these proceedings. In addition, the keynotes' slides are available at www.1dddas.org.

Keywords: Aerospace · Mechanics · Dynamics · Systems Analytics · Autonomous Vehicles · UAVs · Robotics · Performance-Resource Tradeoffs · Network Security · Intrusion Detection · Continual Domain Adaptation · Unsupervised Adaptation · Computer Vision · AI Genomics · Proteomics · Epigenetics · Cybersecurity · Distributed Spacecraft Constellations · Environmental Monitoring · Soil Moisture · Urban Floods: Hurricanes · Cyclones

1 Introduction

An overview of the keynotes and an invited talk are provided here, followed by additional information on each of these presentations, as provided by each of the speakers: Professors Yuri Bazilevs, Sertac Karaman, Manolis Kellis, Nathaniel Bastian, and Andreas Savakis, and Dr. Sreeja Nag. Overall, the keynotes and the invited talk anchored presentations and papers in the main track sessions which are part of the present proceedings, but also in their scope relate to the broader and considerable body of work that has been conducted under the rubric of DDDAS, for example [1–4]. Following the present overview, papers contributed by each of the keynote speakers are included in this part of the proceedings, and in the Keynotes Appendix.

2 Overview of the Keynotes' Presentations

In the area of aerospace systems, the keynote by Professor Yuri Bazilevs of Brown University, titled: *DDDAS for Systems Analytics in Applied Mechanics*, discussed DDDAS-based methods for maximizing the predictive power of models to support systems-analytics in structural health monitoring of aerospace structures and take actions to mitigate damage and provide enhanced operational decision support. One of the aspects of important emergent directions emplasized in Prof. Bazilevs was the aspect of multi-scale modeling enabled through the DDDAS paradigm at several spatial and temporal scales involved in the modeling and deployment of composite structural systems (Fig. 1). The paper contributed by Prof. Bazilevs on his talk in this Conference (and included in this section of the Proceedings) expands on such and prior DDDAS efforts, [5–9] and provides further details and references on the body of work towards the above stated objectives, and more broadly on aspects of design of aircraft wings and wind turbines, composite structures, as well as robotic systems. Additional information is provided in the presentation slides, posted in www.1dddas.org (DDDAS2022Conference Agenda).

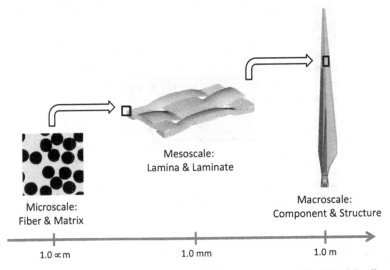

Fig. 1. Emegent Siection: Multiscale DDDAS (Figure, courtesy of Prof. Yuri Bazilevs).

Also in the area of aerospace systems, the keynote by Professor Sertac Karaman of MIT, titled: *Computing for Emerging Aerospace Autonomous Vehicles*, discussed the Low-Energy Autonomous Navigation (LEAN) project which employs DDDAS-based methods for autonomous operation and coordination of new generations of aerial vehi-cles, "ranging from the most miniature and long-endurance to the fastest and the most agile". The methods discussed in this work are aimed to enable motion planning for com-plex dynamics, including aerodynamics, integrated robotics algorithms, which enable energy efficiencies and support performance-resource tradeoffs with provable guarratees of correctness, optimality, and robustness. These are novel approaches and create new and interesting applications such as low-energy aerospace robotics applications – for miniature aerial vehicles, lighter than air vehicles, micro-unmaned gliders, and minia-ture satellites. The DDDAS-based approach in this work for computing hardware and algorithms for low-energy aerospace robotics, is shown in Fig. 2. Additional informa-tion is provided in [10–14], and the abstract and representative references contributed by Prof. Karaman, and included in the Keynotes-Appendix Section of the proceedings, as well as in the presentation slides posted in www.1dddas.org (DDDAS2022Conference Agenda).

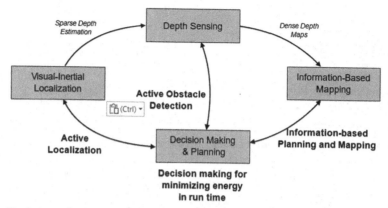

Fig. 2. Depiction of DDDAS feedback-loop low-energy autonomous-navigation aerospace robotics capabilities (Figure, courtesy of Prof. Sertac Karaman).

In the area of bioinformatics, the keynote by MIT Professor Manolis Kellis, titled: *From genomics to therapeutics: Single-cell dissection and manipulation of disease circuitry*, addressed novel, multidisciplinary approaches, at the confluence of computer sciences, biology, controls methods, genetics, genomics, epigenetics, genes coding, noncoding Ribonucleic Acid (RNAs), and other, for understanding how gene-regulatory circuitry impacts human disease, such as Alzheimer's, Schizophrenia, Obesity, Cardiac Disorders, Cancer, and Immune System and multiple other disorders. The human body is a collection of dynamic control networks, interoperarating in a dynamic reconfigurable and as systems-of-systems fashion; and moreover on needs to also consider human health as an end-to-end systems circuity, as depicted in Fig. 3. In addition to the abstract on the talk contributed by Prof. Kellis, together with representative publications examples (e.g., [15]), additional information can be found in the abstract with references, included in the Keynotes-Appendix Section of the proceedings, and in the presentation slides and videotaped presentation, posted in www.1dddas.org (DDDAS2022Conference Agenda).

The keynote by Nathaniel Bastian, PhD and Lieutenant Colonel in the U.S. Army and Academy Professor at the United States Military Academy (USMA) at West Point, titled: *Data Augmentation to Improve Adversarial Robustness of AI-Based Network Security Monitoring*, addressed the area of cyber-security, an area is increasingly important, in the commercial, civilian, and national defense sectors, with the growth of the world-wide inter-connectedness. To counter cyberthreats, involves continually collecting features of Internet Protocol (IP) communications at tap points within a network and continually analyzing them, as shown in Fig. 4. The talk addressed how Artificial Intelligence (AI) technologies, DDDAS-related, and the ensuing Network intrusion detection systems (NIDS), en able the collection of synchronized, real-time capabilities to discover, define, analyze, and mitigate cyber threats and vulnerabilities with limited human intervention, and are effective in supporting network security monitoring for

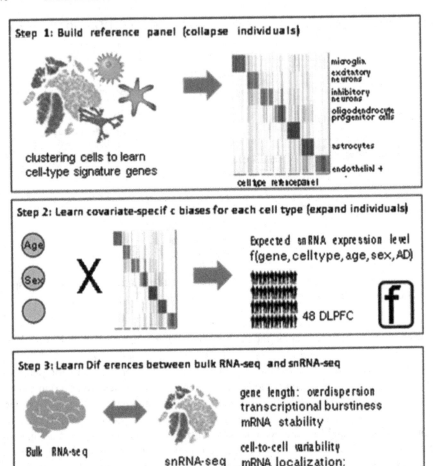

Fig. 3. Human Health as a DDDAS feedback loop (Figure, courtesy of Prof. Manolis Kellis).

cyber security operations at enterprise, operational and tactical environments, as per Dr. Bastian's related work [16]. The talk abstract with references is included in the Keynotes-Appendix Section of the proceedings, and the presentation slides posted in www.1dddas.org (DDDAS2022Conference Agenda).

The keynote by Sreeja Nag, PhD, covered methods for adaptive and cognizant coordination of multimodal instrumentation for DDDAS-based capabilities in the environmental applications area, and specifically in the area of wildfires, which in the Summer of 2022 exhibited unprecedent extremes in the number of incidents as well as numbers and destructive effects – a "hot topic" (no pun intended) that has drawn wide attention in multiple continents. The keynote by Dr. Sreeja Nag, Senior Research Scientist at NASA Ames Research Center, titled: *Improving Predictive Models for Environmental*

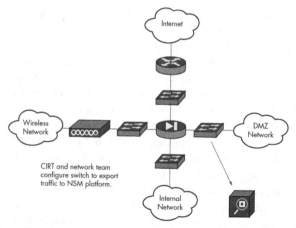

Fig. 4. Schematic of Network security monitoring components (Figure, courtesy of Prof. Nathaniel Bastian).

Monitoring using Distributed Spacecraft Autonomy, addressed advanced instrumentation methods, DDDAS-based for adaptive management of heterogeneous constellations of spacecraft (UAVs and satellites), as depicted in Fig. 5.

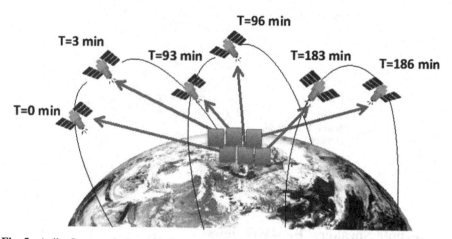

Fig. 5. Agile Spacecraft Constellations Maximizing Coverage and Revisit through DDDAS (Figure, courtesy of Dr. Sreeja Nag).

These constellations ae inter-connected and co-ordinated, along DDDAS-based modeling and instrumentation approaches, for a range of environmental concerns areas, spanning many aspects and spatial scales, from urban floods, hurricanes and cyclones events, to soil moisture and vegetation monitoring scenarios, [17, 18]. In addition to the abstract on the talk with references provided by Dr. Nag, and and included in the Keynotes-Appendix Section of the proceedings, other supplementary information can be found in the presentation slides and videotaped presentation, posted in www.1dddas.org (DDDAS2022Conference Agenda).

The invited talk by Andreas Savakis, PhD, presented work that introduces a new method, for more efficiently and effectively discerning relevant features for image classification in computer vision. The method, Continual Domain Adaptation (ConDA), creates capabilities overcoming limitations in the Domain Adaptation (DA) methods. The traditional DA methods, when given a training dataset, then assume information for the entire domain of reference to be available at once in performing adaptation for the actual (target) dataset. However such an approach is not practically effective. The new methodology, DDDAS-based, allows continuous adaptation, as each partial set of data becomes available, as depicted in Fig. 6. In the ensuing unsupervised adaptation framework, the improvements or refinements are performes as new batches of data become available. The presentation showed results that demonstrate the accuracy of the proposed method over other methods (e.g., in a Table in the paper contributed by Prof. Savakis, and included in this section of the proceedings. Additional information is providedin the presentation slides, posted in www.1dddas.org (DDDAS2022Conference Agenda).

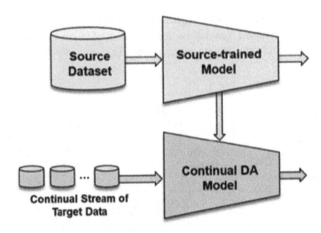

Fig. 6. Continual Domain Adaptation (Figure, courtesy of Andreas Savakis).

3   Keynote Speakers' Bio-Overviews

3.1   Prof. Yuri Bazilevs

Yuri Bazilevs, PhD, is the E. Paul Sorensen Professor of Engineering. Bazilevs' research interests lie in the field of computational science and engineering with an emphasis on computational mechanics. Prior to that he was Vice Chair and Professor of Engineering in the Structural Engineering Department at the University of California, San Diego. His work addresses complex problems in the areas of biomedicine; renewable energy, and protecting infrastructures against disasters. He is the original developer of a computational technology called Isogeometric Analysis (IGA), which has had a significant impact in computational mechanics, and remains a prevailing research direction in the field today, and is a method that has advanced capabilities in the field of computational solid and structural mechanics. In addition, his work on fluid-structure interaction (FSI) analysis and the development of a fully-integrated FSI framework - methods and software, are employed in simulations for a range of industrial-scale applications, ranging from large civilian infrastructures to military aerospace systems.

Prior to joining the faculty at UCSD, he earned his Ph.D. in Computational and Applied Mathematics from the University of Texas-Austin in 2006, and his M.S. and B.S. in Mechanical Engineering from Rensselaer Polytechnic Institute in 2001 and 2000, respectively, and was J.T. Oden Postdoctoral Fellow at the Institute for Computational Engineering and Sciences (2006–2008) and a lecturer in the Department of Aerospace Engineering and Engineering Mechanics (2007–2008), both at the University of Texas-Austin. More detailed CV in:Yuri Bazilevs Joins School of Engineering at Brown | Engineering | Brown University.

3.2 Prof. Sertac Karaman

Sertac Karaman,PhD, is a faculty member at the Massachusetts Institute of Technology, and the Director of the Laboratory for Information and Decision Systems (LIDS) -- an interdepartmental research laboratory home to more than 30 faculty members and hundreds of researchers working on the next-generation automated inference and decision making systems that closely interact with the existing physical and social systems around us. Sertac Karaman's research focuses on computing for autonomous vehicles. It spans a large breadth of topics, ranging from the foundational aspects in algorithms, computer architecture, integrated circuits, control and estimation theory and robotics to its applications in driverless cars, next-generation drones and autonomous space vehicles. He is the director of the Autonomy and Embedded Robotics Accelerated (AERA) research group, which focuses on fast and agile autonomous vehicles (https://aera.mit.edu), and the co-director of the Low-Energy Autonomy and Navigation (LEAN) research group, which focuses on miniature or long-endurance autonomous vehicles, particularly focusing on the co-design of algorithms and computing hardware (https://lean.mit.edu). He is a co-founder of Optimus Ride, where he acted as President and Chief Scientist. Optimus Ride focused on safe, efficient and sustainable autonomous mobility for all. It was acquired by the automotive-supplier-giant Magna in 2020.

3.3 Prof. Manolis Kellis

Manolis Kellis,PhD, is a professor of computer science at MIT, a member of the Broad Institute of MIT and Harvard, a principal investigator of the Computer Science and Artificial Intelligence Lab at MIT, and head of the MIT Computational Biology Group (compbio.mit.edu). His research includes disease circuitry, genetics, genomics, epigenomics, coding genes, non-coding RNAs, regulatory genomics, and comparative genomics, applied to Alzheimer's Disease, Obesity, Schizophrenia, Cardiac Disorders, Cancer, and Immune Disorders, and multiple other disorders. He has led several large-scale genomics projects, including the Roadmap Epigenomics project, the ENCODE project, the Genotype Tissue-Expression (GTEx) project, and comparative genomics projects in mammals, flies, and yeasts. He received the US Presidential Early Career Award in Science and Engineering (PECASE) by US President Barack Obama, the Mendel Medal for Outstanding Achievements in Science, the NIH Director's Transformative Research Award, the Boston Patent Law Association award, the NSF CAREER award, the Alfred P. Sloan Fellowship, the Technology Review TR35 recognition, the AIT Niki Award, and the Sprowls award for the best Ph.D. thesis in computer science at MIT. He has authored over 245 journal publications cited more than 125,000 times. He has obtained more than 20 multi-year grants from the NIH, and his trainees hold faculty positions at Stanford, Harvard, CMU, McGill, Johns Hopkins, UCLA, and other top universities. He lived in Greece and France before moving to the US, and he studied and conducted research at MIT, the Xerox Palo Alto Research Center, and the Cold Spring Harbor Lab. Additional information in: compbio.mit.edu.

3.4 LtCol/Prof. Nathaniel Bastian

Nathaniel D. Bastian, PhD, is a Lieutenant Colonel in the U.S. Army, where he serves as Academy Professor at the United States Military Academy (USMA) at West Point. As an expert analytics and innovation professional, he balances responsibilities as a science, technology, engineering and mathematics (STEM) leader, researcher and educator, as well as a technical program manager. At USMA, Nate is Chief Data Scientist and Senior Research Scientist at the Army Cyber Institute (ACI), as well as Assistant Professor of Operations Research and Data Science with a dual faculty appointment in the Department of Systems Engineering and the Department of Mathematical Sciences. At the ACI, Nate leads the Data and Decision Sciences research team while also serving as Director of the Intelligent Cyber-Systems and Analytics Research Lab, overseeing the ACI's computing capabilities, resources and services while leading a $2M + externally-funded research portfolio as a Principal Investigator in support of Army, other Services, Department of Defense (DoD), and Intelligence Community stakeholders. His prior military assignments include serving as Chief Artificial Intelligence Architect at the DoD Joint Artificial Intelligence Center, Operations Research Scientist at the ACI/USMA, Analytics Officer at the U.S. Army Human Resources Command, and Aeromedical Evacuation Officer and UH-60 Black Hawk Aviator at the 25th Combat Aviation Brigade. Nate earned his Ph.D. in Industrial Engineering and Operations Research from the Pennsylvania State University (PSU), M.Eng. in Industrial Engineering from PSU, M.S. in Econometrics and Operations Research from Maastricht University, and B.S. in Engineering Management (Electrical Engineering) with Honors from USMA. He is an active professional member of INFORMS, MORS, ACM, IEEE, and AAAI.

3.5 Dr. Sreeja Nag

Sreeja Nag,PhD, is a Senior Research Scientist at NASA Ames Research Center, contracted by BAER Institute, where she serves as the PI on the D-SHIELD project. She also leads Software Systems Engineering at Nuro, a Silicon Valley start- up that is building and deploying safe, self-driving robotic fleets for public roads. Sreeja completed her PhD from the Department of Aeronautics and Astronautics at Massachusetts Institute of

Technology, Cambridge, USA. Her research interests include distributed space systems, space robotics for Earth observation, space traffic management, and vehicular robotics validation.

3.6 Prof. Andreas Savakis

Andreas Savakis, PhD, is Professor of Computer Engineering at Rochester Institute of Technology (RIT) and Director of the Center for Human-aware Artificial Intelligence (CHAI). He received the B.S. and M.S. degrees in Electrical Engineering from Old Dominion University in Virginia, and the Ph.D. in Electrical and Computer Engineering with Mathematics Minor from North Carolina State University. Prior to joining RIT He was Senior Research Scientist with the Kodak Research Labs. At RIT, he has served as department head of Computer Engineering for 10 years and founded the Vision and Image Processing Lab. His research interests include computer vision, deep learning, domain adaptation, human pose estimation, robust and efficient learning, visual tracking, and scene analysis. Prof. Savakis has co-authored over 120 publications and holds 12 U.S. patents. He became Fellow of the American Council on Education (ACE) and received the NYSTAR Technology Transfer Award for Economic Impact, the IEEE Region 1 Award for Outstanding Teaching and the RIT Trustees Scholarship Award. Additional information in: https://www.rit.edu/chai/people.

References

1. Darema, F., Blasch, E., Ravela, S., Aved, A. (eds.) Dynamic Data Driven Applications Systems: Third International Conference on DDDAS 2020, Boston, MA, USA, 2–4 October (2020)
2. Blasch, E., Ravela, S., Aved, A. (eds.): Handbook of Dynamic Data Driven Applications Systems. Springer, Cham (2018). https://doi.org/10.1007/978-3-319-95504-9
3. Blasch, E.P., Darema, F., Ravela, S., Aved, A.J. (eds.) Handbook of Dynamic Data Driven Applications Systems. Springer, Cham (2022). https://doi.org/10.1007/978-3-319-95504-9
4. Darema, F., Blasch, E.P., Ravela, S., Aved, A.J. (eds.) Handbook of Dynamic Data Driven Applications Systems. Springer, Cham (2023). https://doi.org/10.1007/978-3-030-74568-4
5. Korobenko, A., Hsu, M.-C., Bazilevs, Y.: A computational steering framework for large-scale composite structures: part i—parametric-based design and analysis. In: Blasch, E.P., Darema, F., Ravela, S., Aved, A.J. (eds.) Handbook of Dynamic Data Driven Applications Systems, vol. 1, pp. 163–180. Springer International Publishing, Cham (2022). https://doi.org/10.1007/978-3-030-74568-4_8
6. Korobenko, A., Pigazzini, M., Deng, X., Bazilevs, Y.: Multiscale DDDAS framework for damage prediction in aerospace composite structures. In: Blasch, E., Ravela, S., Aved, A. (eds.) Handbook of Dynamic Data Driven Applications Systems, pp. 677–696. Springer, Cham (2018). https://doi.org/10.1007/978-3-319-95504-9_30
7. Bazilevs, Y., Deng, X., Korobenko, A., Lanza di Scalea, F., Todd, M.D., Taylor, S.G.: Isogeometric fatigue damage prediction in large-scale composite structures driven by dynamic sensor sata. J. Appl. Mech. 82(9) (2015)
8. Korobenko, A., et al.: Dynamic-data-driven damage prediction in aerospace composite structures. In: AIAA/ISSMO Multidisciplinary Analysis and Optimization Conference, 2016
9. Y. Bazilevs, J. Yan, X. Deng, X. et al. Computer Modeling of Wind Turbines: 2. Free-Surface FSI and Fatigue-Damage. Arch Computat Methods Eng 26, 1101–1115, 2019

10. Carlone, L., Axelrod, A., Karaman, S., Chowdhary, G.: Aided optimal search: data-driven target pursuit from on-demand delayed binary observations. In: Blasch, E.P., Darema, F., Ravela, S., Aved, A.J. (eds.) Handbook of Dynamic Data Driven Applications Systems: Volume 1, pp. 303–343. Springer International Publishing, Cham (2022). https://doi.org/10.1007/978-3-030-74568-4_14

11. Carlone, L., Axelrod, A., Karaman, S., Chowdhary, G.: Data-driven prediction of confidence for EVAR in time-varying datasets. In: Blasch, E.P., Darema, F., Ravela, S., Aved, A.J. (eds.), Handbook of Dynamic Data Driven Applications Systems, vol. 1, 2nd ed. pp 389–412. Springer (2022). https://doi.org/10.1007/978-3-030-74568-4_16

12. Ryou, G., Tal, E., Karaman, S.: Multi-fidelity black-box optimization for time-optimal quadrotor maneuvers. Inter. J. Robotics Res. **40**(12–14), 1352–1369 (2021)

13. Tal, E., Karaman, S.: Accurate tracking of aggressive quadrotor trajectories using incremental nonlinear dynamic inversion and differential flatness. IEEE Trans. Control Syst. Technol. **29**(3), 1203–1218 (2020)

14. Tal, E., Ryou, G., Karaman, S.: Aerobatic trajectory generation for a vtol fixed-wing aircraft using differential flatness. arXiv preprint arXiv:2207.03524, (2022)

15. Ruzicka, W.B.: Single-cell dissection of schizophrenia reveals neurodevelopmental-synaptic axis and transcriptional resilience. 2020. Single-cell dissection of schizophrenia reveals neurodevelopmental-synaptic axis and transcriptional resilience | medRxiv

16. Alhajjar, E., Maxwell, P., Bastian, N.: Adversarial machine learning in network intrusion detection systems. Expert Syst. Appli. **186** (2021)

17. Levinson, R., Niemoeller, S., Nag, S., Ravindra, V.: Planning satellite swarm measurements for earth science models: comparing constraint processing and MILP methods. In: Proceedings of the International Conference on Automated Planning and Scheduling, vol. 32, pp. 471–479 (2022)

18. Melebari, A., Nag, S., Ravindra, V., Moghaddam, M.: Soil moisture retrieval from multi-instrument and multi-frequency simulated measurements in support of future earth observing systems. In: International Geoscience and Remote Sensing Symposium, pp. 5594–5597 (2022)

DDDAS for Systems Analytics in Applied Mechanics

A. Korobenko[2], S. Niu[1], X. Deng[3], E. Zhang[1], V. Srivastava[1], and Y. Bazilevs[1(✉)]

[1] School of Engineering, Brown University, Providence, USA
yuri_bazilevs@brown.edu
[2] Department of Mechanical and Manufacturing Engineering, University of Calgary, Calgary, Canada
[3] Department of Civil Engineering, University of Hong Kong, Hong Kong, China

Abstract. This contribution is comprised of two parts. In the first part we provide an overview of the Dynamically Data-Driven Applications Systems (DDDAS) concept, with particular emphasis on the analytics of systems coming from the field of Applied Mechanics and focusing on the applications to aerospace structures. Aerospace composite materials and structures exhibit a strong multiscale behavior, which necessitates the development of a multiscale DDDAS framework wherein measurements and models interact at all the relevant spatial scales of the system of interest to maximize the resulting predictive power. We present a large-scale structural system example where the combination of dynamic data and advanced models are needed to be truly predictive. In the second part we examine the Neural Network (NN)-based data-driven approaches for systems analytics in applied mechanics, in particular, the Physics-Informed Neural Networks (PINNs) framework. The main idea of PINNs is to compensate for the lack of sufficient volume of measured data by forcing the system to obey the laws of physics expressed in the form of boundary-value problems (BVPs) based on partial differential equations (PDEs). A distinguishing feature of PINNs is that the discretization of a BVP does not make use of traditional methods, but rather NNs themselves. We focus on the ability of the approaches, incorporating NNs (as a tool) into DDDAS, to model large-deformation elastoplastic behavior of solids and structures so that they can be seamlessly integrated into structural systems analytics and beyond.

Keywords: DDDAS · Applied Mechanics · Composite Structures · PINNs

E. Blasch et al. (Eds.): DDDAS 2022, LNCS 13984, pp. 353–361, 2024.
https://doi.org/10.1007/978-3-031-52670-1_34

1   Multiscale DDDAS for Composite Structures

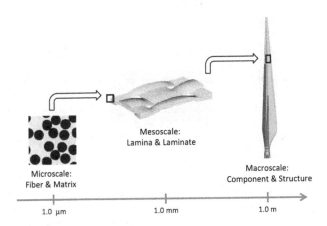

Fig. 1. Multiple spatial scales involved in the modeling of the behavior of composite materials and structures, necessitating the development of a DDDAS framework that reflects this strong multiscale behavior.

In recent years, there has been a significant increase in the use of Unmanned Aerial Vehicles (UAV) by the US military. UAVs are expected to fly a large number of long (48 or more hours) missions, and operate without failure. Furthermore, in order to increase the durability of these vehicles and to decrease weight, composite materials are currently experiencing a widespread adoption in applications related both to military and civilian aerospace structures. As a result, in order to decrease costs associated with the operation, maintenance, and, in some cases, loss of these vehicles, it is desirable to have a Dynamically Data-Driven Applications Systems (DDDAS) [5] framework that can reliably predict the onset and progressions of structural damage in geometrically and materially complex aerospace composite structures operating in the environments typical of UAVs. DDDAS is a general framework in which sensor and measurement data collected for a given physical system are used to dynamically update a computational model of that system. Using measurement data, the computational model geometry, boundary conditions, external forces, and material parameters may be updated to better represent physical reality. At the same time, the updated computational model can produce higher-fidelity outputs for the quantities of interest for which measurements are not readily available, and provide feedback to a measurement system.

Here we present a multiscale DDDAS Interactive Structure Composite Element Relation Network (DISCERN) framework. Build upon an early version of the DISCERN framework, the multiscale DISCERN framework reflects the multiscale nature of laminated composites by applying the DDDAS concept at all spatial scales involved in the modeling of composite materials damage (see Fig. 1):

- At the microscale level, the Representative Volume Element (RVE) computations are often employed to obtain material properties such as directional elastic moduli or failure stresses. X-ray digital micro-tomography may be employed concurrently with the RVE simulations for precise strain measurements intended to calibrate the RVE model parameters.

- At the mesoscale level (i.e. the "coupon" level) smaller-scale experiments may be performed concurrently with the simulation of simple geometry specimens (rectangular and, possibly, notched) to extract the parameters for the damage model and to assess the sensitivity of the damage model with respect to these parameters. The optimal set of damage-model parameters can be obtained, for instance, by minimizing a misfit functional between the experimental and computational results.

- The full richness and power of DDDAS can be exercised at the macroscale level (i.e., the structural component level) where the accelerometer and strain-gauge data can be used to adjust the external forces, boundary conditions and other structural-model input data to better represent physical reality and predict damage onset and growth. The location(s) of the damage-zone formation predicted by the steered computational model may be, in turn, used to make decisions about future sensor placement. This represents a true feedback loop between the actual structure and its computational model.

1.1   Fatigue Damage in Wind-Turbine Blades

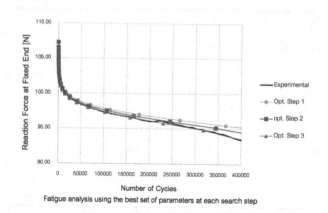

Fatigue analysis using the best set of parameters at each search step

Fig. 2. Reaction force at the clamped support plotted as a function of cycle number. Convergence of the resulting force-vs-cycle curve to the experimental result is clearly seen in the plot.

Fig. 3. Stress distribution in the composite plate specimen after 400,000 cycles of applied non-reversible loading. Left: View of the surface under compression; Right: View of the surface under tension.

The multiscale DISCERN framework is applied to the simulation of fatigue damage of an actual wind-turbine blade subjected to realistic wind loading. The results for this case may be found in [2,3].

In the first step, we make use of the coupon-level experimental tests to calibrate the fatigue-damage model from [6,9]. We make use of the methods based on surrogate management framework (SMF) [4] and minimize the objective function that makes use of the error between the experimental and computational results. The plate is clamped on one end and driven by a prescribed vertical displacement on the other end. Figure 2 shows the reaction force versus cycle number. Convergence of the predicted force-cycle history to the experimental curve is evident from the figure, and the results obtained using the optimized fatigue-model parameters are remarkably close to the experimental data. Figure 3 shows the stress distribution after 400,000 cycles in the composite specimen.

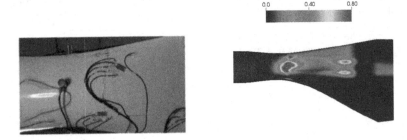

Fig. 4. Visual comparison of the fatigue-test and simulation results. Location and shape of the damage zone in a DBM layer near the root are in very good agreement with the location and orientation of the crack observed in the fatigue test.

As the next step, we employ the calibrated fatigue model in the simulation of the actual composite wind-turbine blade. The wind turbine blade employed is CX-100 with the details of the structural model (i.e., geometry, material data, and lamination sequence) provided in [2]. We carry out further fatigue model calibration using the experimental data from [11]. The dynamic sensor data are employed to simultaneously calibrate the magnitude of the applied displacement loading, as well as to further improve the input parameters of the fatigue-damage model. To this end, we devise two DDDAS loops - the *inner loop* responsible for displacement forcing amplitude calibration, and the *outer loop* responsible for simulation of damage growth and calibration of the associated material constants. Closer to 8 M cycles a part of the root section is fully damaged, and the damage location is in excellent agreement with that of the crack observed on the blade surface during the fatigue test. See Fig. 4 for a visual comparison of the fatigue-test and simulation results.

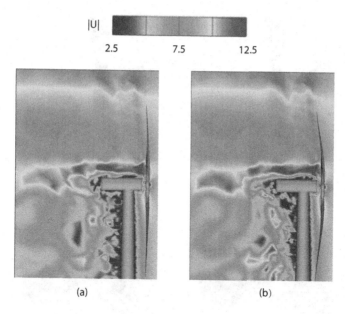

Fig. 5. Isocontours of air speed (in m/s) on a plane cut after 100,000 (left) and 150,000,000 (right) cycles. The right graphic corresponds to the cycle right before the blade failure. Large bending deformation is due to significant loss of blade stiffness.

As the final step, the structural model of the CX-100 composite wind-turbine blade calibrated for fatigue response is introduced into a fully-coupled fluid–structure interaction (FSI) framework for wind turbine simulation [3]. The CX-100 bladed is mounted on a Micon 65/13M [12] turbine and simulated under realistic wind conditions. The wind-turbine rotor spins at 55 rpm, and is subjected to a wind speed of 10.5 m/s. After 150 M cycles, damage propagates through the blade, eventually leading to loss of stiffness and large bending deformation (see Fig. 5.) Fig. 6 shows a contour plot of the fiber damage index in the DBM-1708 ply at the outer surface of the leading edge zone. Note that the manner in which the damage initiates and progresses during the wind-turbine operation is quite different from what is observed in the laboratory fatigue test (see Fig. 4 for comparison).

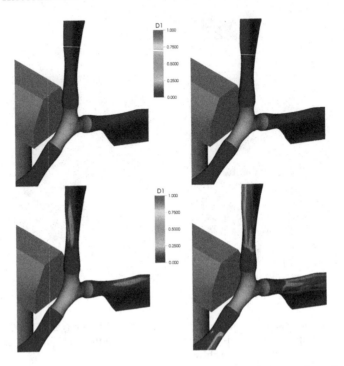

Fig. 6. Fiber damage index in the DBM-1708 layer near the blade aerodynamic zone after 10,000,000, 40,000,000, 100,000,000, and 150,000,000 cycles (left to right, top to bottom).

2 PINNs for Large-Deformation Elasto-Plasticity

While machine learning has had great success for purely data-driven applications, and is thus a popular direction in the DDDAS research and development, many applications in engineering and science are not sufficiently rich in data to take advantage of these methods. To overcome the data sparsity issue, physics-informed neural networks (PINNs) were introduced in [10] as a novel approach to encode the known BVPs explicitly into data-driven models. As such, PINNs are may be viewed as an instantiation of the DDDAS concept that effectively blends data and physics-based models to achieve superior predictability. Significant research effort has been exerted to use PINNs as a deep learning technique to solve PDEs, largely enabled by the built-in ability of the neural networks to perform automatic differentiation [1]. For many BVPs, PINNs may be implemented as a forward solver in a straightforward manner. It is able to solve the governing PDEs without invoking the training data due to the fact that a neural network deep enough can be regarded as a universal function approximator [7].

On the modeling and simulation front, solid mechanics, which often deals with BVPs that are posed in a weak form over complex geometry and are governed by mechanical equilibrium involving sophisticated kinematics and constitutive relations, is dominated by Finite Element Methods (FEMs). However, compared to FEM that requires sophisticated mesh generation, PINN is "mesh-free" in that it only makes use of individual residual points to represent the problem domain, an attribute that is shared with Meshfree methods and that makes the domain discretization simpler. Modeling the elasto-plastic response of materials and structures using PINNs was previously limited to problems with small strains and rotations. However, finite-strain plasticity is important in practice and was only recently addressed using PINNs in [8] and we summarize the methodology and key findings in what follows.

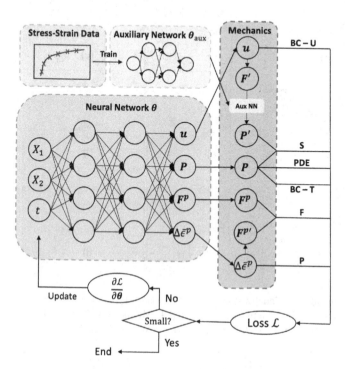

Fig. 7. Architecture of the PINN for finite-strain elasto-plasticity. The setup is based on the plane-strain formulation with multiple (pseudo)-time steps. The loss function is formulated based on the PDE, boundary conditions and constitutive relations. An auxiliary network is embedded in the architecture and used to model the material response by learning from discrete stress-strain data prior to the training of PINN.

The PINN architecture for finite-strain plasticity is shown in Fig. 7. The network input is comprised of the reference-configuration spatial locations, referred to as the residual points, and time levels. (Here, the formulation is quasi-static so pseudo-time is employed.) The network output is the displacement, stress, plastic deformation, and effective plastic strain at the residual points. A mixed form of the BVP is employed where the network output is considered as the problem unknowns. The mixed-form approach necessitates the computation of the first-order space derivatives, which makes the formulation simpler to implement and more efficient than using a displacement-only approach that is typical in FEM. The Mechanics module shown in Fig. 7 evaluates the objective function that simultaneously penalizes the PDE, constitutive-model, and boundary-condition residual. An auxiliary neural network is employed to encode a constitutive law for the hardening behavior with the input coming from experimental measurements.

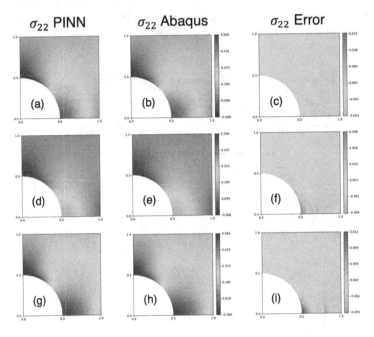

Fig. 8. 22-component of Cauchy stress from PINN and FEM (using Abaqus software) and the error for the three loading steps. Top row: Step 1. Middle row: Step 2. Bottom row: Step 3. (a)(d)(g): PINN, (b)(e)(h) FEM, (c)(f)(i) Pointwise error.

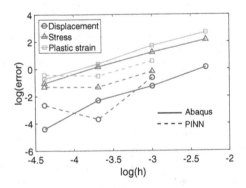

Fig. 9. Comparison of the errors between PINN and FEM for the residual-point mesh densities of 10×10, 20×20, 40×40 and 80×80. Errors are plotted on a log-log scale.

Figure 8 benchmarks the PINN solution with that of the FEM using linear quads and a similar mesh of nodal points (equivalent to the residual points in PINN). The problem corresponds to a classical plate-with-a-hole geometry subjected to uniaxial tension in the 2-direction. The PINN solution for the 22-component of the Cauchy stress distribution shows excellent accuracy compared to the FEM results for a complex loading scenario. Figure 9 shows the accuracy of the PINN solution compared to that of the FEM for a sequence of refined meshes. In this case, on the coarsest grid, PINN fails to capture a meaningful solution, while the error saturates on the finest grid. On the intermediate grids, however, PINN is able to deliver solutions that are at least as accurate as the FEM solutions. These result demonstrate that while PINNs present a powerful and promising approach to incorporate physics into data-driven models, further research and development is needed to overcome some of the robustness and accuracy shortcomings of the present methodology.

Acknowledgment. This work was supported by the AFOSR Award FA9550-12-1-0005, AFOSR Award FA9550-16-1-0131, and ONR Award N00014-21-1-267. The authors greatly acknowledge this support.

References

1. Baydin, A.G., Pearlmutter, B.A., Radul, A.A., Siskind, J.M.: Automatic differentiation in machine learning: a survey (2015)
2. Bazilevs, Y., Deng, X., Korobenko, A., Lanza di Scalea, F., Todd, M.D., Taylor, S.G.: Isogeometric fatigue damage prediction in large-scale composite structures driven by dynamic sensor data. J. Appl. Mech. **82**, 091008 (2015)
3. Bazilevs, Y., Korobenko, A., Deng, X., Yan, J.: FSI modeling for fatigue-damage prediction in full-scale wind-turbine blades. J. Appl. Mech. **83**(6), 061010 (2016)
4. Booker, A.J., Dennis, J.E., Jr., Frank, P.D., Serafini, D.B., Torczon, V., Trosset, M.W.: A rigorous framework for optimization of expensive functions by surrogates. Struct. Optim. **17**, 1–13 (1999)
5. Darema, F.: Dynamic data driven applications systems: a new paradigm for application simulations and measurements. In: Proceedings of ICCS 2004–4th International Conference on Computational Science, pp. 662–669 (2004)
6. Degrieck, J., Paepegem, W.V.: Fatigue damage modeling of fiber-reinforced composite materials: review. Appl. Mech. Rev. **54**(4), 279–300 (2001)
7. Hornik, K., Stinchcombe, M., White, H.: Multilayer feedforward networks are universal approximators. Neural Netw. **2**(5), 359–366 (1989)
8. Niu, S., Zhang, E., Bazilevs, Y., Srivastava, V.: Modeling finite-strain plasticity using physics informed neural network and assessment of the network performance. J. Mech. Phys. Solids **172**, 105177 (2023)
9. Paepegem, W.V., Degrieck, J.: Simulating in-plane fatigue damage in woven glass fibre-reinforced composites subject to fully reversed cyclic loading. Fatigue Fract. Eng. Mater. Struct. **27**, 1197–1208 (2004)
10. Raissi, M., Perdikaris, P., Karniadakis, G.E.: Physics-informed neural networks: a deep learning framework for solving forward and inverse problems involving nonlinear partial differential equations. J. Comput. Phys. **378**, 686–707 (2019)
11. Taylor, S.G., Park, G., Farinholt, K.M., Todd, M.D.: Fatigue crack detection performance comparison in a composite wind turbine rotor blade. Struct. Health Monit. **12**, 252–262 (2013)
12. Zayas, J.R., Johnson, W.D.: 3X-100 blade field test. Wind Energy Technology Department, Sandia National Laboratories, page Report (2008)

Towards Continual Unsupervised Data Driven Adaptive Learning

Andeas Savakis$^{(\boxtimes)}$

Rochester Institute of Technology, Rochester, NY 14623, USA
andreas.savakis@rit.edu

Abstract. Domain Adaptation (DA) techniques are important for overcoming the domain shift between the training dataset (called source domain) and the testing dataset (called target domain). Standard DA methods assume that the entire target domain is available during adaptation, but this assumption is often violated in practice. We consider DA in a data constrained scenario, where target data become available in small batches over time, and adaptation takes place continually. Hence, continual DA is a framework to instantiate the Dynamic Data Driven Applications Systems (DDDAS) paradigm, wherein a model is developed from the data available to discern the relevant features, and subsequently when the model is deployed, it needs to be adapted (i.e., through a learning process) from the new real-world data. We discuss a novel source-free method for Continual Domain Adaptation (ConDA) that utilizes a buffer for selective replay of previously seen samples. In our unsupervised adaptation framework, we selectively mix samples from incoming batches with data stored in a buffer and use them to adapt our model as new batches are received. Our results using ConDA demonstrate the benefits of our framework when operating in data constrained environments.

Keywords: Dynamic Data Driven Applications Systems · continual domain adaptation · unsupervised adaptation · computer vision · AI

1 Introduction

Recent advances in Artificial Intelligence (AI) include deep learning (e.g., for detection, classification and segmentation), reinforcement learning (e.g., controls), and transfer learning (e.g., multi-domain adaptation). In computer vision, the terms "source domain" and "target domain" refer to the training and testing datasets respectively. Domain Adaptation (DA) techniques aim to overcome the distribution shift between the source domain used for training and the target domain where testing takes place [1]. However, current DA methods assume that the entire target domain dataset is available during adaptation, which is generally not realistic in practice. Hence, DA is a method to instantiate the Dynamic Data Driven Applications Systems (DDDAS) paradigm wherein a model is developed from the data available to discern the relevant features, and subsequently when the model is deployed, it needs to be adapted (i.e., a learning process) from the new real-world data. To utilize new data to update a model of salient features to classify an object

E. Blasch et al. (Eds.): DDDAS 2022, LNCS 13984, pp. 362–366, 2024.
https://doi.org/10.1007/978-3-031-52670-1_35

in an image, we introduce a data-constrained DA paradigm, illustrated in Fig. 1, where unlabeled target samples are received in batches and adaptation is performed continually [2]. The data-constrained DA methods is a novel source free approach for continual unsupervised domain adaptation (ConDA) [3] that utilizes a buffer for selective replay of previously seen samples. In our ConDA framework, the system selectively mixes samples from incoming batches with data stored in a buffer to incrementally update the classifier model.

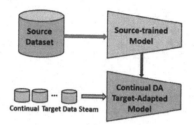

Fig. 1. Continual DA paradigm. Initial training is performed with labeled data in the source domain and the trained model is deployed in the target domain. During deployment, unlabelled target domain data are received in streaming batches and the model is continuously adapted with each new batch of target data.

2   Methodology

Our ConDA framework is extending the Source Hypothesis Transfer (SHOT) method [4] in a continual setting. The source model consists of a feature extractor, with bottleneck and batch normalization layers, and a source hypothesis model, consisting of a fully connected layer and a weight normalization layer (Fig. 2). The feature extractor backbone in ConDA is either ResNet-101, that was also used by SHOT, or HRNet to obtain a higher resolution representations. We train the source model in a supervised manner with label smoothing, using the source domain labels. Then we transfer the source hypothesis model to the target hypothesis model, which remains unchanged during adaptation.

The source feature generation model is transferred to the target feature generation model and is adapted with the incoming batches of target data. Since the target data do not have labels, our ConDA system generates pseudo-labels based on deep clustering of the source and target data. The buffer manager selects only the highest confidence target samples to place in the buffer for replay. The buffer placement strategy enforces equal number of samples from each class to promote balanced training. Our model only requires access to the samples stored in the buffer for subsequent adaptation along with new target batches that are received incrementally.

For our objective function, we consider the information maximization (IM) loss, which is a combination of the entropy loss and the equal diversity loss, along with the mixup cross-entropy loss, combined using hyperparameters. We train using an stochastic gradient descent (SGD) optimizer with momentum.

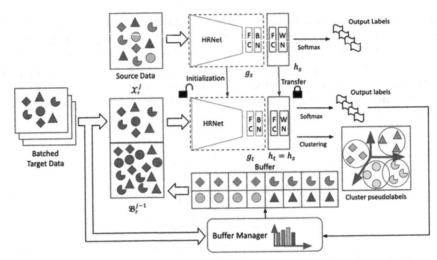

Fig. 2. Proposed ConDA framework [3] adapting on target domain data that arrive in small batches, where a sample subset already seen by the network are stored in a buffer for replay with the incoming batches. The buffer manager selects the samples that populate the buffer. The incoming target samples are mixed with the current buffer samples and sent to the network for adaptation.

3 Results and Conclusions

To test ConDA, three datasets were used: Office-31, Office-Home and VisDA-C. Office-31 [5] is a popular dataset that has 3 domains, Amazon (A), DSLR (D), and Webcam (W), containing 31 object classes of items found in an office environment in each of the domains. Office-home [6] is a medium scale dataset with 4 domains, Art (Ar), Clip Art (Cl), Product (Pr), Real-World (Rw). The dataset has 65 classes of items found in everyday office and home environments. VisDA-C [7] is a large scale dataset with 2 domains, Synthetic (S) and Real (R). The dataset has 12 classes. The synthetic samples are generated using 3D rendering, and the real samples are taken from MS COCO dataset [8] (Fig. 3).

Table 1 shows mean percent accuracy results after adaptation on Office-31, Office-Home and VisDA-C datasets, using ConDA or SHOT with different backbones, adapting using the full target dataset or in a continual setting with small batches. Training took place over multiple epochs depending on the dataset. Additional results and comparisons with other methods are found in [2, 3, 9]. We note that using the HRNet backbone with SHOT (HR-SHOT) increases accuracy significantly. We further observe that ConDA improves upon continual SHOT (HR-SHOT-C), illustrating the benefit of continual domain adaptation.

Future directions on continual adaptation are planned using the DDDAS paradigm for physics-based learning, control of the data collection process, and leveraging multimodal data sets (e.g., synthetic aperature radar and electro-optical imagery) and multi-domain datasets (e.g., data collected from spaceborne and airborne sensors).

Fig. 3. Example images from the four domains in Office-Home. From top to bottom: Art, Clip Art, Product, Real-World. Image source: https://www.hemanthdv.org/officeHomeDataset.html

Table 1. Results using ConDA and comparison with SHOT.

Method	Target Data	Backbone	Office-31 Mean Acc	Office-Home Mean Acc	VisDA-C Mean Acc
SHOT[4]	Full	ResNet	88.6	71.8	82.9
HR-SHOT	Full	HRNet	92.4	82.8	86.4
HR-SHOT-C	Continual	HRNet	88.9	77.8	84.3
ConDA	Continual	HRNet	90.7	79.3	84.6

Acknowledgments. The author acknowledges the valuable contributions by his students Abu Taufique and Chowdhury Sadman Jahan, and would like to thank Dr. Erik Blasch for participating in the research conceptualization. The research was partly supported by the Air Force Office of Scientific Research (AFOSR) grant FA9550–20–1–0039.

References

1. Nagananda, N., et al.: Benchmarking domain adaptation methods on aerial datasets. Sensors (2021)
2. Taufique, A., Jahan, C.S., Savakis, A.: Unsupervised continual learning for gradually varying domains. In: Computer Vision and Pattern Recognition (CVPR) Workshop on Continual Learning in Computer Vision (CLVision) (2022)
3. Taufique, A., Jahan, C.S., Savakis, A.: Continual unsupervised domain adaptation in data-constrained environments. IEEE Trans. Artifi. Intell. (2023)
4. Liang, J., Hu, D., Feng, J.: Do we really need to access the source data? Source hypothesis transfer for unsupervised domain adaptation. In: International Conference on Machine Learning (ICML), pp. 6028–6039. PMLR (2020)
5. Saenko, K., Kulis, B., Fritz, M., Darrell, T.: Adapting visual category models to new domains. In: Daniilidis, K., Maragos, P., Paragios, N. (eds.) Computer Vision – ECCV 2010: 11th European Conference on Computer Vision, Heraklion, Crete, Greece, September 5-11, 2010, Proceedings, Part IV, pp. 213–226. Springer Berlin Heidelberg, Berlin, Heidelberg (2010). https://doi.org/10.1007/978-3-642-15561-1_16

6. Venkateswara, H., Eusebio, J., Chakraborty, S., Panchanathan, S.: Deep hashing network for unsupervised domain adaptation. In: Proceedings of the IEEE Conference on Computer Vision and Pattern Recognition, pp. 5018– 5027 (2017)
7. Peng, X., Usman, B., Kaushik, N., Hoffman, J., Wang, D., Saenko, K.: VisDA: the visual domain adaptation challenge. arXiv preprint arXiv:1710.06924, (2017)
8. Lin, T.-Y., Maire, M., Belongie, S., Hays, J., Perona, P., Ramanan, D., Piotr Dollár, C., Zitnick, L.: Microsoft COCO: common objects in context. In: Fleet, D., Pajdla, T., Schiele, B., Tuytelaars, T. (eds.) Computer Vision – ECCV 2014: 13th European Conference, Zurich, Switzerland, September 6-12, 2014, Proceedings, Part V, pp. 740–755. Springer International Publishing, Cham (2014). https://doi.org/10.1007/978-3-319-10602-1_48
9. Taufique, A., Jahan, C.S., Savakis, A.: ConDA: Continual Unsupervised Domain Adaptation, arXiv preprint arXiv:2103.11056, (2021)

Main-Track: Wildfires Panel

Overview of DDDAS2022 Panel on Wildfires

Frederica Darema[✉]

InfoSymbiotic Systems Society, Bethesda, MD 20817, USA
`fredericadarema@hotmail.com`

Abstract. The DDDAS2022 Conference featured a Panel on Wildfires, which addressed DDDAS-based modeling and instrumentation methods for the detection, prediction of the onset and propagation of wildfires, smoke generation and spread, as well as utilizing such information for containment of the wildfire; the DDDAS-based wildfire methods also support other emergency response actions, infrastructure safety and evacuation of humans out of the fire and smoke harm's way. Together with the present overview, papers contributed by the panelists are included in this part of the proceedings. In addition, the keynotes' slides and presentations recordings are available at www.1dddas.org.

Keywords: Dynamic Data Driven Applications Systems · DDDAS · wildfire weather-fire coupled physics models · radiative heat flux · bidirectional feedback · turbulence · multisource data · continuous data assimilation · physics-aware machine learning · multi-UAVs (Unmanned Aerial Vehicles) coordination; satellite constellations measurements · visible and infrared imaging radiometric suite (VIIRS) imagery · synthetic aperture radar (SAR) data · high resolution spatiotemporal data · aerosol backscatter · clouds · moisture · smoke-pollution health effects

1 Introduction

The DDDAS2022 Conference featured a Panel on Wildfires, comprising of seven experts from academe and Research Laboratories, namely Profs. Ilkay Altintas, Janice Coen, Fatemeh Afghah, Mrinal Kumar, Milton Halem, Thomas Huang, and Kamran Mohseni. The panel convened in-tandem with the keynote which preceded it, by Dr. Sreeja Nag, on space-based instrumentation, through satellite constellations, for environmental observations including wildfire-related observations. The panel was co-chaired by Drs. Frederica Darema and Sreeja Nag.

The panelists addressed a range of DDDAS-based modeling and instrumentation methods and infrastructures supporting DDDAS-methods and environments, for detecting and predicting the onset and propagation of wildfires, taking into account a number of relevant environmental factors, such as: fuel, moisture, weather, topography. Additional focus was on dynamic factors, such as interaction of the radiative heat transfer process with the air turbulence, caused by prevalent winds as well as that which is induced by the heat gradients in the ambient air within the boundary of the fire and the adjacent regions.

E. Blasch et al. (Eds.): DDDAS 2022, LNCS 13984, pp. 369–381, 2024.
https://doi.org/10.1007/978-3-031-52670-1_36

Other objectives of the advanced DDDAS-based wildfires methods presented are: utilizing such information for containment of the wildfire and other emergency response actions, such as safety and evacuation of humans out of harm's way. Notably, the effect of wildfires is not limited to the damage caused locally or regionally; in addition, wildfire events affect weather and increase atmospheric pollution due the wildfire-generated smoke, emitted and carried in more extended areas, and that in-turn also impacts human health (respiratory ailments). These and other end-to-end aspects were addressed in the DDDAS wildfires panel presentations.

Increasingly over the years and especially this past summer (Summer 2022), wildfires became an issue of major concern and at a worldwide scale, not only in terms of increased numbers, extent, and duration of the wildfires in the usual geographical areas, such as California in the US and the southern Mediterranean, but also the increase of the incidence and extent, and over a broader set of regions, such as up and down in the US Western and mid-Western States, as well as in Europe, with wildfires reaching countries in central Europe and the UK.

According to the National Interagency Fire Center (NIFC) in the US, its statistics show that as of late October 2022 (the time of the writing of the present summary), in 2022, 56,660 fires have already burned a total of 6,946,391 acres; which is about three times the 10-year average of 17,742 fires; with the average acreage of 6,705,736. In California alone, the 2022 wildfire season has seen an ongoing series of wildfires, burning throughout the state, and as of 21 September 2022, a total of 6,473 fires were recorded, totaling approximately 365,140 acres (147,770 hectares) across the state. In other states, in the September/October 2022 timeframe, there were over 226 fires, a number of them in states traditionally not plagued by such a deluge of wildfires (New Mexico. 1 fire. 341,735 acres; Oregon. 16 fires. 132,570 acres; Alaska. 8 fires; 39,044 acres; Washington. 53 fires. 34,398 acres; Montana. 42 fires. 20,303 acres; Idaho. 24 fires. 8,344 acres; Oklahoma. 71 fires. 966 acres; Texas. 14 fires. 479 acres). Up-to-date information and statistics on the above are provided in [1].

In Europe, not only the Mediterranean countries (e.g., Greece, Spain, Italy, Portugal) saw an unprecedented frequency and number of wildfires, but also countries in mid- and northern Europe (e.g., UK, France, Germany, Czech Republic, Romania, Slovenia), with a total of 2,297 fires (as of August 13, 2022), around 4–5 times the prior 15-year average [2]. Unprecedented wildfires incidence and frequenscy happened also elsewhere in the world, such as in Asia, North Africa [3].

Furthermore, it is increasingly becoming evident that the smoke generated by the wildfires, can transport over far distances and affect the quality of air; for example on March 27, 2023, as reported by the Washington Post, burning smell in the Washington DC area (DC-VA-MD) was attributed to wildfire in North Carolina [4]. The wildfire-induced smoke can have lasting effects on human health air transport and commercial businesss. The wildfire produced ash creates challenges for agriculture water resources and vegetation. Determining these multi-scale effects requires modeling forecasting and prepartion methods to respond to the growing numbers of wildfires and their secondary effects on a wide scope of environmental aspects ranging from lost forests to air pollution to animal habitats and migration.

Hence, there is a need to develop more advanced methods of containing the wildfires and stemming-off individual fires' spread and their broader adverse effects. The DDDAS2022 panel addressed prediction of the onset, spread and other effects of wildfires, including impact to environment and human health. Along this scope, the panelists discussed an array of DDDAS-based methods and infrastructures for advanced modeling and instrumentation capabilities, aimed at improving the accuracy for detecting and predicting the onset and the spread of the fire and ambient smoke, and creating real-time decision support tools for effectively containing and mitigating the damage caused by wildfires.

2 Overview of Panelists' Presentations

Prof. Ilkay Altintas, (UCSD – University of California, San Diego), in her talk: *Using Dynamic Data Driven Cyberinfrastructure for Next Generation Disaster Intelligence*, spoke about the DDDAS-based WIFIRE Cyberinfrastrure efforts she established and is leading at UCSD, where end-to-end methods span wildfire modeling, multisource data collection, machine learning and advanced computing which integrates edge-, cloud-, and high-performance computing, to assess and predict the extent and dynamics of wildfires and their impacts. WIFIRE technology aims to provide real-time and long-term data-driven knowledge for a wide range of public and private sector users, ranging from first responders to municipal entities, as well as use of such capabilities for scientific and educational objctives. A schematic of the DDDAS-based fire-modeling framework, aimed to characterize the dynamic fire environment (that is, the variation of winds, smoke, moisture, vegetation-fuels, fire-perimeter, etc.), the detection of fire ignitions, decision support management, and prediction of potential fire ignitions, and propagation (spread) of fire, is shown in Fig. 1. A paper by Prof. Altintas, is included in this part of the proceedings (Part 3); and additional, supplementary information can be found in the presentation slides and videotaped presentation, posted in www.1dddas.org.

Dr. Janice Coen, (NCAR – National Center for Atmospheric Research and University of San Francisco), in her talk titled: *Simulating large wildland and urban (WUI) fires with a physics-based weather-fire behavior model: Understanding, prediction, and data-shaped products*, she addressed methods of: (1) coupled physics models - those representing the factors relating to wildfire (including radiative heat transfer, rate of fire spread, fuel consumption, related environmental effects such as vegetation and terrain), and (2) computational fluid dynamic models, those representing atmospheric turbulence, not only by ambient winds, but also air turbulence created by the heat flux feedbacks from the fire itself and plume-driven fires, as depicted in Fig. 2.

Such plume- and wind-dominated factors are important to consider, as they affect the fire spread - directions and intensity, and therefore become crucial in accurately predicting the wildfire spread and take mitigating actions to stem-off the fire propagation. Additionally, the results support evacuation decisions, and protect first responders from harm. The complex "Coupled Fire-Weather Pediction and Fire-Behavior Modeling Framework – the Coupled Atmosphere-Wildland Fire-Environment (CAWFE) Modeling Systems is depicted in Fig. 3.

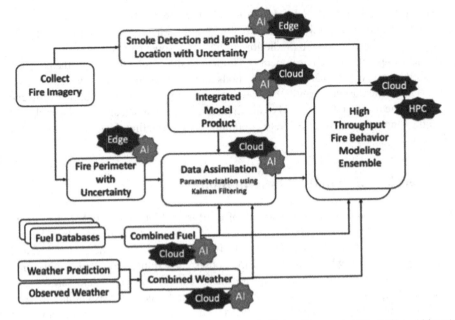

Fig. 1. Dynamic Data Driven Fire modeling framework (Figure Courtessy of Prof. Ilkay Altintas)

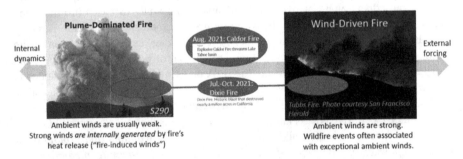

Fig. 2. Fire propagation driven by ambient winds and by fire-induced winds (Figure Courtessy of Dr. Janice Coen)

The above referenced CAWFE models also are dynamically updated and enhanced with instrumentation data, ranging from UAVs (Unmanned Aerial Vehicles) and satellite-constellations' measurements, visible infrared imaging radiometric suite (VIIRS) imagery, synthetic aperture radar (SAR) data, and other airborne observations as well as terrain and vegetation data. The abstract of Dr. Coen's talk together with a representative set of references, is provided in the Wildfires-Appendix of the Proceedings; and additional, supplementary information can be found in the presentation slides and videotaped presentation, posted in www.1dddas.org..

CAWFE® Modeling System

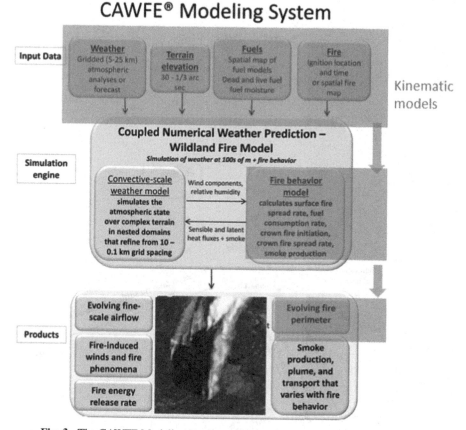

Fig. 3. The CAWFE Modeling Framework (Figure Courtessy of Dr. Janice Coen)

Prof. Fatemeh Afghah (Clemson University), who collaborates with Dr. Coen, presented her work on: *Autonomous Unmanned Aerial Vehicle systems in Wildfire Detection and Management-Challenges and Opportunities.* The talk focused on methods for adaptive, autonomic coordination of collections of UAVs for aerial imaging of wildfires conditions. The talk included discussion on Leader-Follower coalition formation of the UAV's constellation and a DDDAS-based framework (Fig. 5b) for such hierachical dynamic and adaptive coordination (based on observations and bidirectional communications across leader-follower clusters (Fig. 5b) for ochestated data collection, as shown in Fig. 5 (a and b).

Examples were presented where on coordination of the UAVs in a cluster and across clusters, for distributed task search and allocation, timely completion of the tasks, efficient allocation of network resources. The UAVs' wildfire monitoring and imaging data collected are subsequently used for wildfire-spread prediction, such as those that are part of the instrumentation data utilized by Dr. Coen in the work she presented, for determining the location, extent, and movement of the fire. More details on Prof. Afghah's work is

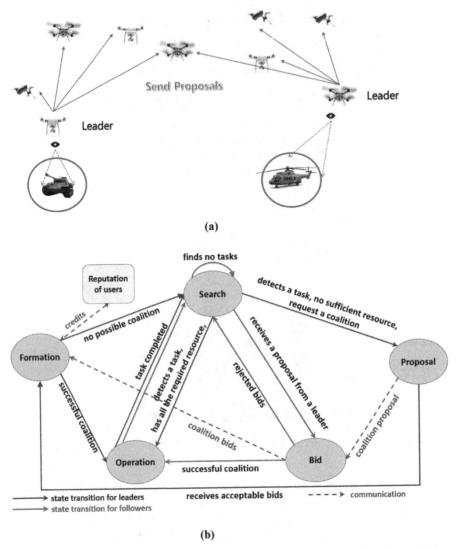

(a)

(b)

Fig. 4. Leader-Follower coalition formation and and adaptive coodination for UAVs constellations (Figure Courtessy of Prof. Fatemeh Afghah)

presented in the paper by her, in this section of the proceedings; and additional, supplementary information can be found in the presentation slides and videotaped presentation, posted inwww.1dddas.org.

Prof. Mrinal Kumar (Ohio State University), in his talk on the*: Role of Autonomous Unmanned Aerial Systemsin Prescribed Burn Projects*, addressed the use and autonomic management of UASs (Unmanned Aerial Systems). The application for the methods discussed in this wok are in the context of *controlled burn*. The autonomic coordination of UASs to adaptively monitor relevant aspects of a fire (for example, advance of the fire-fronts), is shown in Fig. 5.

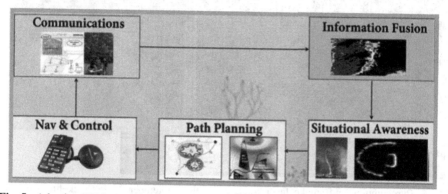

Fig. 5. Adaptive UAS coordination and dispatching for fire situational awareness (Figure Courtessy of Prof. Mrinal Kumar)

The situational awareness capabilities developed under this work (on controlled burning; that is, controlled fires), are also applicable for the case of wildfires' events. Namely, the understanding and methods derived from controlled fires environments, with respect to the bidirectional feedback between UAS-based sensing and computational components at multiple scales, are also applicable to wildfires detection, and for command and control of first-responder movements. More details of these methods are presented in the paper by Prof. Kumar, in this part of the proceedings (Part 3); and additional supplementary information can be found in the presentation slides and videotaped presentation, posted in www.1dddas.org.

Prof. Milton Halem (University of Maryland, Baltimore County), in his talk on: *Towards a Dynamic Data Driven Wildfire Digital Twin (WDT): Impact on Deforestation, Air Quality and Cardiopulmonary Disease,* he presented work on the development of a real-time "Digital Twin" for wildfires, to model smoke generation (from wildfires), and propagation and impact of the smoke-pollution in distant locales, as illustrated in Fig. 6.

In addition, Prof. Halem presented impact scenarios and studies parametrized by season, location, intensity, and time-dependent and varying atmospheric conditions. The methods presented involve multi-level and multi-modal modeling including microphysics, cloud radiative processes, surface fuel chemistry transport, near-time continuous data assimilation schemes for radar data, physics-aware machine learning for aerosol backscatter, coupled with corresponding instrumentation from satellite-borne Lidar for aerosol-backscatter data, and infrared-range measurements for detection of elements on Earth's surface or in atmosphere, such as trees, water, clouds, moisture or smoke. The use DDDAS-based methods coupled with machine-learning methods speed-up the modeling, and enable for example 1–2 h forecasts obtained in seconds. More details of these methods are presented in the paper by Prof. Halem later in this part of the Proceedings (Part 3); additional, supplementary information can be found in the presentation slides and videotaped presentation, posted in www.1dddas.org.

Fig. 6. Pyrocumulus and smoke layer forecast for the September 2020 Creek fire event in California (Figures Courtessy of Prof. Milt Halem)

Dr. Thomas Huang (NASA JPL), in his talk on: *Earth System Digital Twin for Air Quality*, presented DDDAS-based methods for effectively and efficiently collecting the diverse set of data involved in environmental observations, including wildfires, and managing the diverse collections of multi-discipline, high resolution spatiotemporal data involved. Figure 7, shows an illustration of the multi-source and diverse instrumentation and such data, in the context of an integrated information system that, for example, enables continuous assessment of impact from naturally occurring and/or human activities or physical and natural environments. Dr. Huang's talk addressed approaches for smarter and more automated approaches for collecting such data, and described a scalable analytic framework for integrated air quality analysis during wildfires. A short paper by Dr. Huang, included in Wildfires-Appendix in the Poceedings, provides more details on the referenced framework and methods; and additional, supplementary information can be found in the presentation slides and videotaped presentation, posted in www.1dd das.org.

Prof. Kamran Mohseni, Univ of Florida, in his talk on *Dynamic Data Driven Applications for Atmospheric Monitoring and Tracking*, discussed methods for multi-UAVs coordination for monitoring atmospheric events such those related with wirdfires and smoke plumes. The work presented discussed how actual UAVs were deployed to test the efficacy of the DDDAS-based methods. Figure 8, shows remarkable agreement between the simulations and experimentally observed conditions of the actual plume concentrations, under background wind velocity and tubulence patterns. The Proceedings' Wildfires-Appendix includes the abstract of the talk and a representative set references; and additional, supplementary information can be found in the presentation slides and videotaped presentation, posted in www.1dddas.org.

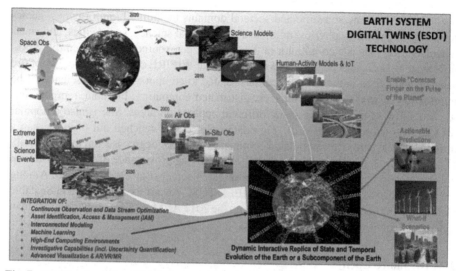

Fig. 7. NASA schematic of Earth System Digital Twin for continuous observation and environmental assessment (Source: https://esto.nasa.gov/aist/)

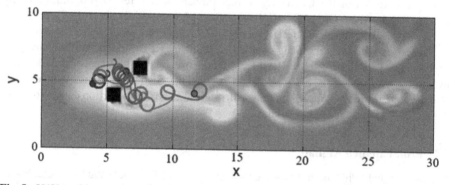

Fig. 8. UAV tracking as plume is advected by wind (Figure Courtessy of Pf. Kamran Mohseni).

3 Panelists' Bio-overviews

3.1 Prof. Ilkay Altintas

Ilkay Altintas, PhD, is the Chief Data Science Officer at the San Diego Supercomputer Center (SDSC), UC San Diego, where she is also the Founder and Director for the Workflows for Data Science Center of Excellence as well as the WIFIRE Lab, and a Fellow of the Halicioglu Data Science Institute (HDSI). In her various roles and projects, she leads collaborative multi-disciplinary teams with a research objective to deliver impactful results through making computational data science work more reusable, programmable, scalable and reproducible. Since joining SDSC in 2001, she has been a principal investigator and a technical leader in a wide range of cross-disciplinary projects. Her work has been applied to many scientific and societal domains including bioinformatics, geoinformatics, high-energy physics, multi-scale biomedical science, smart cities, and smart manufacturing. She is a co-initiator of the popular open-source Kepler Scientific Workflow System, the leader behind the now operational WIFIRE cyberinfrastructure for fire science, and the co-author of publications related to computational data science at the intersection of workflows, provenance, distributed computing, big data, reproducibility, and software modeling in many different application areas. She is also a popular MOOC instructor in the field of "big" data science, and reached out to more than a million learners across any populated continent. Her Ph.D. degree is from the University of Amsterdam in the Netherlands with an emphasis on provenance of workflow-driven collaborative science. She is an associate research scientist at UC San Diego. Among the awards she has received are the 2015 IEEE TCSC Award for Excellence in Scalable Computing for Early Career Researchers and the 2017 ACM SIGHPC Emerging Woman Leader in Technical Computing Award.

3.2 Dr. Janice Coen

Janice Coen, PhD, holds positions of Project Scientist in the Mesoscale and Microscale Meteorology Laboratory at the National Center for Atmospheric Research and Senior Research Scientist at the University of San Francisco. She received a B.S. in Engineering Physics from Grove City College and an M.S. and Ph.D. from the Department of Geophysical Sciences at the University of Chicago. She studies fire behavior and its interaction with weather using coupled weather-fire CFD models and flow analysis of high speed IR fire imagery. Her recent work investigated the mechanisms leading to extreme wildfire events, fine-scale wind extrema that lead to ignitions by the electric grid, and integration of coupled models with satellite active fire detection data to forecast the growth of landscape-scale wildfires.

3.3 Prof. Fatemeh Afghah

Fatemeh Afghah, PhD, is an Associate Professor with the Electrical and Computer Engineering Department at Clemson University and the director of the Intelligent Systems and Wireless Networking (IS-WiN) Laboratory. Prior to joining Clemson University, she was an Associate Professor with the School of Informatics, Computing and Cyber Systems, at Northern Arizona University. Her research interests include wireless communication networks, decision-making in multi-agent systems, UAV networks, security, and artificial intelligence in healthcare. Her recent project involves autonomous decision-making in uncertain environments, using autonomous vehicles for disaster management and IoT security. She is the recipient of several awards including the Air Force Office of Scientific Research Young Investigator Award in 2019, the NSF CAREER Award in 2020, NAU's Most Promising New Scholar Award in 2020, and the NSF CISE Research Initiation Initiative (CRII) Award in 2017. She is the author/co-author of over 130 peer-reviewed publications and served as the associate editor for several journals including Elsevier Journal of Network and Computer Applications, Ad hoc networks, Computer Networks, ACM Transactions on Computing for Healthcare (HEALTH), Springer Neural Processing Letters and the organizer and TPC chair for several international IEEE workshops in the field of UAV communications and AI, including IEEE INFOCOM Workshop on Wireless Sensor, Robot, and UAV Networks (WiSRAN'19), IEEE WOW-MOM Workshop on Wireless Networking, Planning, and Computing for UAV Swarms (SwarmNet'20&21&22), 2021 NSF Smart Health PI workshop on "Smart Health in the AI and COVID Era", 2022 NSF CPS Workshop on "AI in Healthcare".

3.4 Prof. Mrinal Kumar

Mrinal Kumar, PhD, is an Associate Professor and Elizabeth Martin Tinkham Endowed Professor of Aeronautical and Astronautical Engineering at The Ohio State University. At OSU, he serves as Director of the *Laboratory for Autonomy in Data-Driven and Complex Systems* (LADDCS), where his group conducts research on uncertainty quantification in complex engineering systems. Current federally funded projects at LADDCS include integration of autonomous unmanned aerial systems with prescribed wildland burns, trustworthy model building and predictive uncertainty quantification for space domain resiliency, and construction of a digital twin for prognostics of smart manufacturing systems. LADDCS is driven by one post-doctoral fellow, four Ph.D. and two Masters students, and 14 undergraduate researchers. To date, Dr. Kumar has guided the completion of 9 Ph.D. dissertations. He has published over 100 archival articles and peer reviewed proceedings. He is an Associate Fellow of the American Institute of Aeronautics and Astronautics. He is a recipient of the *NSF CAREER Award, AFOSR Young Investigator Award*, and the *Gerald M. Gregorek Excellence in Teaching Award* given by his home department at OSU. Dr. Kumar received his Ph.D. from Texas A&M University and bachelor's degree from the Indian Institute of Technology, Kanpur, both in aerospace engineering.

3.5 Dr./Prof. Milton Halem

Milton Halem, PhD, is a research professor in the UMBC Computer Science and Electrical Engineering department. He also holds an Emeritus position as Chief Information Research Scientist to the Director of the Earth Sciences Directorate at the NASA Goddard Space Flight Center. Prior to retiring in 2002, Dr. Halem served in the joint capacity as Assistant Director for Information Sciences and Chief Information Officer for the NASA Goddard Space Flight Center. Dr. Halem provided the strategic information science and technology focus and oversight for the entire mission critical programs and projects at the Center. He is most noted for his ground breaking research in simulation studies of space observing systems and for development of four dimensional data assimilation for weather and climate prediction. Over the years, his achievements have earned him numerous awards including the NASA Medal for Exceptional Scientific Achievement, the NASA Medal for Outstanding Leadership, and NASA's highest award; the NASA Distinguished Service Medal in 1996. Dr. Halem is also a noted screen printmaker of Art from Space.

3.6 Dr. Thomas Huang

Thomas Huang, PhD, is a Group Supervisor at the JPL's Instrument Software and Science Data Systems section, and the Strategic Lead for Interactive Analytics for the

National Space Technology Applications Program Office at JPL. Thomas is the NASA Principal Investigator for several big data analytic projects, and the System Architect for the NASA's Sea Level Change Portal. As an expert in large-scale, distributed intelligent data systems, Thomas led both planetary and Earth data system projects. As an advocate for free and open-source software, Thomas led the open sourcing of many NASA-funded technologies. He is the founder and creator of the Apache Science Data Analytics Platform (SDAP) technology as a community-driven, Cloud-based analytics framework. Thomas is a frequent invited speaker and panelist at various US and international events. He is the lead editor of a newly released book, titled *Big Data Analytics in Earth, Atmospheric, and Ocean Sciences*. It is part of the AGU Special Publication Series. Thomas is a member of the NOAA's Data Archive and Access Requirements Working Group of the NOAA's Science Advisory Board (SAB). Thomas is also a Computer Science lecturer at the California State Polytechnic University, Pomona, and member of its Industry Advisory Board.

3.7 Prof. Kamran Mohseni

Kamran Mohseni, PhD, is the W.P. Bushnell Endowed Chaired Professor in the Department of Electrical and Computer Engineering and the Department of Mechanical and Aerospace Engineering; and is the Director of the Institute for Networked Autonomous Systems. He received his B.S. degree from the University of Science and Technology, Tehran, Iran, his M.S. degree in Aeronautics and Applied Mathematics from the Imperial College of Science, Technology and Medicine, London, U.K., and his Ph.D. degree from the California Institute of Technology (Caltech), Pasadena, CA, USA, in 2000. He was a Postdoctoral Fellow in Control and Dynamical Systems at Caltech for almost a year. In 2001, he joined the Department of Aerospace Engineering Sciences, University of Colorado at Boulder, and in 2011, he joined the University of Florida, Gainesville, FL, USA as an endowed-chair professor.

3.8 Dr. Sreeja Nag (Co-Chair, Wildfires Panel)

Sreeja Nag, PhD, is a Senior Research Scientist at NASA Ames Research Center, contracted by BAER Institute, where she serves as the PI on the D-SHIELD project. She also leads Software Systems Engineering at Nuro, a Silicon Valley start- up that is building and deploying safe, self-driving robotic fleets for public roads. Sreeja completed her PhD from the Department of Aeronautics and Astronautics at Massachusetts Institute of Technology, Cambridge, USA. Her research interests include distributed space systems, space robotics for Earth observation, space traffic management, and vehicular robotics validation.

References

1. The National Interagency Fire Center (NIFC) statistics,Current Wildfires Burning in the U.S. | Fire, Weather & Avalanche Center (fireweatheravalanche.org). https://www.fireweatheraval anche.org/fire/current-list-of-us-wildfires:
2. Summer 2022: Exceptional Wildfire Season in Europe | EUMETSAT (July - August 2022)https://www.eumetsat.int/summer-2022-exceptional-wildfire-season-europe
3. NASA: NASA: Heatwaves and Fires Scorch Europe, Africa, and Asia. https://www.washingto npost.com/dc-md-va/2023/03/27/dc-smell-north-carolina-wildfires
4. Washington Post, March 27, 2023, https://www.washingtonpost.com/dc-md-va/2023/03/27/ dc-smell-north-carolina-wildfires

Using Dynamic Data Driven Cyberinfrastructure for Next Generation Disaster Intelligence

Ilkay Altintas[✉]

University of California, San Diego, La Jolla, CA 92093, USA
ialtintas@ucsd.edu

Abstract. Wildfires and related disasters are increasing globally, making highly destructive megafires a part of our lives more frequently. A common observation across these large events is that fire behavior is changing, making applied datadriven fire research more important and time critical. Significant improvements towards modeling wildland fires and the dynamics of fire related environmental hazards and socio-economic impacts can be made through intelligent integration of modern data and computing technologies with techniques for data management, machine learning and artificial intelligence. However, there are many challenges and opportunities in integration of the scientific discoveries and datadriven methods for hazards with the advances in technology and computing in a way that provides and enables different modalities of sensing and computing. The WIFIRE cyberinfrastructure took the first steps to tackle this problem with a goal to create an integrated infrastructure, data and visualization services, and workflows for wildfire mitigation, monitoring, simulation, and response. Today, WIFIRE provides an end-to-end management infrastructure from the data sensing and collection to artificial intelligence and dynamic data-driven modeling efforts using a continuum of computing methods that integrate edge, cloud, and high-performance computing. Through this cyberinfrastructure, the WIFIRE project provides data driven knowledge for a wide range of public and private sector users, enabling scientific, municipal, and educational use. This paper summarizes the talk reviewing our recent work on building this dynamic data driven cyberinfrastructure and impactful application solution architectures that showcase integration of a variety of existing technologies and collaborative expertise.

Keywords: Wildland Fire Science · Data Cyberinfrastructure · AI in Science

1 Introduction

Changing climate and increasing human activity at the wildland-urban interface, combined with accumulations of vegetation due to decades of fire suppression, has created a perfect storm of conditions resulting in frequent megafires across the globe. These extreme wildfires damage ecosystems and risk human life and property. A common observation across these large events is that fire behavior is changing, making applied data-driven wildland fire research more important and time critical 2.

© The Author(s), under exclusive license to Springer Nature Switzerland AG 2024
E. Blasch et al. (Eds.): DDDAS 2022, LNCS 13984, pp. 382–385, 2024.
https://doi.org/10.1007/978-3-031-52670-1_37

Wildland fire modeling is an important tool for understanding fire behavior, managing the risks of wildfires, and planning for fire resilient communities and ecosystems. While there are many models with varying strengths, complexities and uses [1], these computer-based models typically use scientific principles on how the fire interacts with landscape and atmosphere and data including weather (e.g., forecast, temperature, wind speed, and humidity), topographic data (e.g., slope, aspect, and elevation), vegetation and fuel data (e.g., fuel type, fuel moisture, density, and fuel load), and ignition data (e.g., location and ignition time). Across many purposes for their use (e.g., wildfire management, prescribed fire planning, risk analysis), fire models depend on varying data types and spatiotemporal resolutions. For examples, while a dynamic data-driven wildfire behavior model [9] depends on real-time data for understanding two-dimensional the fire rate of spread and direction, a prescribed fire model depends on threedimensional fuel characterization to allow more accurate modeling of fire behavior and outcomes, including spread, intensity, fuel consumption and smoke.

The data used by wildland fire models come from a variety of sources including weather stations and forecast models, ground-based sensors and cameras, satellites, and overhead observations (e.g., aircraft and drones), models of landscape and fuel, and other human observations related to the fire environment. Such a variety of data comes with typical big data challenges [2] ranging from data availability, complexity and dynamic computational scalability of data processing, quality, uncertainty and accuracy of data and related models, and interoperability of data with other data sources. Cataloging, curating, sharing and discovering data, and optimizing the integration of data sets for application-optimized modeling tools are needed across the continuum of models for further progress in data-driven wildland fire science. A Moore Foundation Community Workshop in April 2019 [3] identified "a shared, integrated platform for diverse sources of data, intelligence and information" as the top requirement for a "Fire Immediate Response System". The NSF-funded WIFIRE project [4, 5] took the first steps to tackle this problem, successfully creating an integrated system for wildfire monitoring, simulation, and response, creating a WIFIRE Commons [6] infrastructure for community-driven continuous capability enhancement, data curation and access for wildland fire data and models, and enablement of novel AI method development that integrate data and modeling with an evolving suite of tools in a flexible multi-fidelity application frameworks.

2   Data to Value in WIFIRE Commons

WIFIRE Commons catalogs, curates and integrates data and models for AI-driven fire science; maintains open programmatic access to data in a cloud-compatible form that can be integrated into the AI process through a gateway interface; and ensures provenance of data and models over time. Through this AI-enabled smart data andmodel integration, the long-term aim is to transform the agility of science-based wildland fire decision making, allowing for new kinds of models and data to be assimilated rapidly and allowing users to understand levels of uncertainty whether originating 3 from models or data. Figure 1 illustrates the data to solutions chain in the WIFIRE Cyberinfrastructure enabled by WIFIRE Commons.

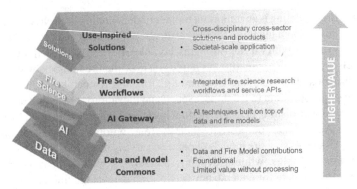

Fig. 1. WIFIRE Commons provides a foundation data, modeling and computing infrastructure for advanced fire science workflows and use-inspired decision support platforms.

The WIFIRE Commons is composed of three main components: (*i*) a *Data Commons* for organization FAIR [7] data and knowledge management services for disparate big data including from sensors, archives of many organizations and model products; (*ii*)} a *Model Commons* for FAIR software services for ingestion, set up, calibration and execution of physics (e.g., fire, fuel and weather), empirical (e.g., for uncertainty quantification), and machine learning/AI (e.g., inference from images) models; and (*iii*) an *AI Gateway* for community-driven development and application of AI models. A *provenance catalog* stored key executions of the AI Gateway and Model Commons for reproducibility and inspectability purposes.

Data Commons (https://wifire-data.sdsc.edu/) organizes datasets and services from government agencies including fire protection services, academic research networks and repositories, utility and other industry data, as well as outputs of model workflows, and enables efficient discovery, annotation, analysis, and configuration into model inputs through: (*i*) a data catalog of datasets (currently 2938 datasets) using CKAN [8]; (*ii*) WIFIRE Modeling Ontology (WFMO) (https://github.com/WIFIRE-Lab/WIFIREcom mons-ontology), and (*iii*) services for access and visualization of data.

Model Commons (https://wildfire.models.mint.isi.edu/) catalogs models and provides a registry of transformation services and other workflows for configuring, calibrating, and executing physics (e.g., fire, fuel and weather), empirical (e.g., for uncertainty quantification), and machine learning/AI (e.g., object identification and inferencing) models. Fire model configurations can be specified or Model Commons can choose the best configuration based on runtime parameters. AI models can be uploaded to the platform with a Jupyter Notebook to perform inference on new data within the Data Commons. WFMO was used as a semantic bridge linking datasets and models.

AI Gateway for community-driven development and application of AI models, is a portal to enable data exploration in the Data Commons, model building, and sharing in support of critical wildfire decision support. It serves as a tool to streamline and 4 simplify AI workflows for wildfire science in a JupyterHub environment on top of existing NSF resources.

Data and Model Commons catalogs are populated with relevant datasets, services and models that are used to discover items needed to create and execute modeling workflows. Access to data and modeling workflows in accordance with FAIR principles, i.e., ensuring that datasets and models can be discovered, accessed using commons browsing and visualization clients, made interoperable following standard protocols, and can be re-used across the different converging domains based on explicitly encoded definitions of variables and modeling procedures. These features enable use of data and models within fire modeling workflows and use-inspired decision support platform development (e.g., BurnPro3D - https://burnpro3d.sdsc.edu/index.html; and Firemap - https://firemap.sdsc.edu/).

3 Summary and Acknowledgements

The author would like to acknowledge the large group of individuals (https://wifire. ucsd.edu/commons-team) who contributed to the parts of the work highlighted in this talk summary. WIFIRE was funded by NSF 1331615 under CI, Information Technology Research and SEES Hazards programs, NSF 2040676 and 2134904 Convergence Accelerator program, DHS #311098–00001, SDG&E, CalOES, Intervalien Foundation, and various industry and fire agency partners.

References

1. Silva, J., et al.: A systematic review and bibliometric analysis of wildland fire behavior modeling. Fluids **7**, 374 (2022). https://doi.org/10.3390/fluids7120374
2. Lee, J.G., Kang, M.: Geospatial big data: challenges and opportunities. Big Data Res. **2**(2), 74–81 (2015). https://doi.org/10.1016/j.bdr.2015.01.003
3. Fire Immediate Response System Workshop Report, Moore Foundation (2019). https://www.moore.org/docs/default-source/default-document-library/2019-firs-workshopreport.pdf
4. Altintas, I., Bet al.: Towards an integrated cyberinfrastructure for scalable data-driven monitoring, dynamic prediction and resilience of wildfires. In: Proceedings of International Conference on Computational Science, ICCS 2015, pp. 1633–1642 (2015)
5. WIFIRE Homepage. https://wifire.ucsd.edu/, (Accessed April 2023)
6. WIFIRE Commons Homepage. https://wifire.ucsd.edu/commons, (Accessed April 2023)
7. Wilkinson, M., Dumontier, M., Aalbersberg, I., et al.: The FAIR guiding principles for scientific data management and stewardship. Sci Data **3**, 160018 (2016). https://doi.org/10.1038/sdata.2016.18
8. CKAN Homepage. https://ckan.org/, (Accessed April 2023)
9. Altintas, I.: Using dynamic data driven cyberinfrastructure for next generation disaster intelligence. In: Darema, F., Blasch, E., Ravela, S., Aved, A. (eds.) Dynamic Data Driven Applications Systems: Third International Conference, DDDAS 2020, pp. 18–21. Springer International Publishing, Cham (2020). https://doi.org/10.1007/978-3-030-61725-7_4

Autonomous Unmanned Aerial Vehicle Systems in Wildfire Detection and Management-Challenges and Opportunities

Fatemeh Afghah[✉]

Holcombe Department of Electrical and Computer Engineering, Clemson University, Clemson, SC 29631, USA
fafghah@clemson.edu

Abstract. Wildfires are one of the costliest and deadliest natural disasters in the United States, particularly in the Western USA. Wildfire frequency and severity are increasing due to climate change and urban sprawl. Detecting forest fires at early stages enhances the chance of in-time intervention and efficient fire management and evacuation strategies. Forest fires are commonly detected by sensor systems that are not widely available across the nation or using satellite images that offer global coverage but suffer from low temporal and spatial resolution, resulting in missing the forest fires in the early stages. Unmanned aerial systems (UAS) have been recently utilized for this purpose, noting their accessibility and low deployment cost. In this paper, we present some recent advances in drone-based fire detection and discuss current challenges toward a wide-scale deployment.

1 Introduction

The severity and frequency of wildfires have considerably increased over the last few decades and are expected to exacerbate due to more extreme heat and severe drought conditions. The United States has faced an annual average of 70,072 wildfires since 2000 that burned an annual average of 7 million acres of land across the nation [17]. This number is more than double the annual average of burned areas in the 1990s. While wildfires have a beneficial impact on ecological resources, massive wildfires, in particular the ones close to communities, severely endanger people's lives and wildlife, destroy agricultural lands, and negatively impact watersheds. The US Congressional Budget Office reported an average annual federal spending of $2.5 billion on fire suppression between 2016 and 2020 [22], not including the

This material is based upon work supported by the Air Force Office of Scientific Research under award number FA9550-20-1-0090 and the National Science Foundation under Grant Numbers CNS-2232048 and CNS-2204445.

E. Blasch et al. (Eds.): DDDAS 2022, LNCS 13984, pp. 386–394, 2024.
https://doi.org/10.1007/978-3-031-52670-1_38

funds spent on disaster assistance. The US National Institute of Standards and Technology (NIST) in a report published in 2017 estimated that the annual cost of wildfires will range from \$7.6 billion to \$62.8 billion, although the overall annual losses due to the wildfires were predicted to be in an order of magnitude higher, ranging from \$63.5 billion to \$285.0 billion [30].

Early detection of wildfires can result in developing in-time and optimal fire management strategies, reducing the cost and damage of wildfires to people, forests, and infrastructure. Existing fire monitoring technologies rely on satellite remote sensing and aircraft, or ground sensors and cameras deployed in forests. Satellite images do not offer the required time and spatial diversity to find wildfires at early stages, in particular in cloudy sky [5,10,11]. Sensors such as smoke detectors can operate under different weather and environmental conditions, however, are limited in deployment scale and have a limited sensing and communication range. Therefore, the current technologies leave uncertainties about a wildfire's location, condition, and behavior, especially during the early stages. The fast-increasing risk of wildfires calls for new technologies to allow early detection of wildfires and enable firefighters to have continuous observations of the fire growth to develop the best fire management strategies.

Offering unique features from 3D mobility, high speed, low-cost and easy deployment, patrolling drones can detect fire incidents at early stages, rapidly map and update imaging of the fire in near real-time, collect targeted observations to inform management solutions, and guide the firefighting tanker aircraft to drop their loads more accurately. Drones can also collect high spatial and temporal resolution measurements, hence, are capable of identifying the spot fires and clarifying ambiguities left in satellite observations when they are blocked by clouds or smoke. Besides wildfire monitoring, unmanned aerial systems (UAS) have been recently utilized in various natural and man-made disaster relief and post-disaster management missions to provide an agile mapping of the areas impacted by disasters, real-time video streaming of the impacted areas for first-responders, operating search and rescue in high-risk, hard-to-reach regions, delivering emergency kits and communication devices to the survivors, or serving as a temporary aerial based station when the infrastructure of cellular networks is damaged [20,21,23,31,33].

Although UAS provide peculiar features for wildfire management, there are still several challenges restraining the wide deployment of this technology. The restrictions on UAV flights including the requirement to operate under the visual line of sight of a human controller is one of the key factors limiting the operation of UAVs for early wildfire detection in remote forest areas, as it required the human operator to be present. Other factors include the limited flight time, the limited capability to operate in harsh weather conditions such as heavy wind, safety and privacy concerns, and the insufficiency of standards, and guidelines for the operation of UAS in disaster management missions.

In this paper, we discuss some of our recent work related to drone-based fire detection and management. In Sect. 2, we present recent models and use cases utilizing UAVs for forest fire management. In Sect. 3, we discuss the potential of using multiple autonomous or semi-autonomous drones in coordination with one another to overcome the limitations of missions based on using a single drone. Section 4 provides concluding comments.

2   Drone-Based Forest Fire Detection

Promptness is a key requirement in forest wildfire management when massive wildfires rapidly destroy the forest and residential areas, and may reach the communities. In these cases, early detection and real-time forest monitoring can result in in-time intervention and damage management solutions to reduce the damage. Satellite images have a low resolution or are not available in weather conditions such as cloudy sky, or in presence of heavy smoke and also suffer from a long delay which can result in missing forest fires in the early stages. On the other hand, wireless sensors cannot provide a wide observation noting the limited sensing distance range. UAV systems with low flight altitude (below the clouds) make them a valuable tool for fast and high-resolution mapping of wide forest areas as well as early detection of forest fires.

Real-time video streaming of surveyed forest areas to control stations for further processing requires a reliable long-range large-bandwidth communication between the UAVs and the ground station. The lack of availability of such payload communication for drones operating in remote areas calls for low-computation fire detection models that can be deployed onboard the drone, allowing to only transmit the fire map and location to fire management stations. Recent advances in embedded graphical processing units (GPUs) and tensor processing units (TPUs) facilitate onboard detection and object tracking at drones [12,24].

Several recent studies have focused on developing deep learning (DL)-based fire and smoke detection algorithms that can be utilized at commercial drones [16,35,36]. However, given the restrictions on using UAVs during wildfires, there are only few drone-collected fire image datasets available to be used for the development and evaluation of data-driven fire detection algorithms. Hence, most current DL-based algorithms are trained and tested on fire images which are collected from other sources such as terrestrial cameras systems (e.g., closed-circuit television (CCTV) cameras), aircraft or helicopters [6,13,29,34]. These images considerably differ from drone-collected images in terms of the types of cameras, as well as their angles and points of view, thereby, they are not appropriate to train or evaluate the drone-based fire monitoring and detection models.

Since the performance of supervised learning and particularly DL-based fire detection methods highly relies on availability of large and diverse training datasets, there is a significant need for the collection of drone-based fire image datasets. *Aerial Image Dataset for Emergency Response application (AIDER)* dataset published in 2020 includes image samples collected from world-wide-web search and a UAV platform for four different man-made and natural disaster events including Fire/Smoke, Flood, Collapsed Building, and Traffic Accidents, as well as normal images. In this dataset, the number of fire/smoke images that were collected by drones is limited.

Fire Luminosity Airborne-Based Machine learning Evaluation (FLAME) dataset, published in 2020 in [27] includes RGB as well as raw videos and thermal footage in Fusion, WhiteHot, and GreenHot modes collected by two drones (Phantom 3 Professional, and Matrice 200) during a prescribed burn in Northern Arizona. This dataset includes 47,992 RGB image frames that are labeled by two experts to "Fire" or "Non-Fire". This dataset also includes 2003 frames containing fire masks to be used as fire segmentation. This dataset is accompanied by i) a deep learning based fire classification algorithm using a modified Xception network (39,375 labeled images used for training and the remaining 8,617 frames are utilized for the test phase); ii) a fire segmentation algorithm to determine the fire location using a customized U-Net neural network [28]. This dataset has been used in several studies as a benchmark for DL-based fire detection and segmentation [7, 8, 19, 32, 37–39].

Since FLAME dataset was collected during a prescribed pile burn, the collected images do not reflect image samples similar to wildfires in terms of size, and intensity. A recent drone-collected image dataset named *Fire detection and ModeLing: Aerial Multi-spectral imagE (FLAME2)* dataset taken by a UAV platform during a prescribed fire in Northern Arizona [9]. One key feature offered by this dataset is presenting infrared and visible spectrum video pairs. Infrared images enable more accurate fire detection in cloudy or smoky sky or during nights and are more appropriate for detection of spot fires. A total of 53,451 image frames in this dataset were labeled by two human experts, where two sets of labeled images are available. In the first set of "Fire/No Fire", a frame was labeled as "Fire" if any indications of fire was observed in either RGB or IR frame. In the second set of "Smoke/No Smoke", an 'smoke' label was indicated when at least 50% of the RGB frame is filled with smoke. FLAME 2 includes a supplementary dataset on weather information, and georeferenced pre-burn point cloud data points, therefore, it can be used in data-driven fire modeling studies. Several deep-learning based fire detection and fire segmentation algorithms were studied in [4] utilizing fusion techniques taking advantage of side-by-side RGB and infrared images. While these studies take good steps toward developing data-driven fire detection and segmentation, there is still a great need for drone-collected wildfire images to be collected from different forest types and during various weather conditions.

3 Fleet of Firefighting Drones

Drones have significantly advanced several civilian and commercial applications such as aerial photography, package delivery, traffic monitoring, precision agriculture, and remote sensing in recent years. However, the current utilization of this technology is mostly limited to remotely radio-controlled or self-steering UAVs that are controlled by a ground station or a pilot located in manned aircraft relatively close to the danger-zone. As mentioned in Sect. 2, wildfire management missions such as fire detection, fire spread modeling and search and rescue of survivors are amongst the applications that significantly benefit from UAS. Nonetheless, concerns related to possible collisions of UAS with manned aircraft flying over the fire regions and the current regulations on the operation of drones under visual line of sight (VLoS) have been limiting factors toward wide-scale operations of drones in wildfire management. Small drones that fly at lower altitudes can alleviate concerns on conflict with manned aircraft, however, the limited flight time and payload prevent them from providing the expected services such as surveying large forest areas or carrying multiple cameras and sensors.

The full potential capabilities of UAV technology cannot be utilized unless the drones are used in an autonomous mode. Therefore, the future of drone-based disaster relief operations is expected to be dominated by smart, small and low-cost autonomous drones that cooperate and coordinate to perform complex missions such as wildfire management with no or minimal human interventions [25, 26]. A fleet of small and low-cost autonomous drones that cooperate and coordinate with another can recompense the sensing, communication, computation and short flight time constraints of small single drones to perform compound missions with no or minimal human interventions [3]. Other advantages of a cooperative drone-fleet is robustness against the UAV failure and extending the mission lifetime. Utilizing a multi-hop communication among the UAVs, where they operate as a relay for other teammates can be a solution for long-range communication link in remote areas [2].

In recent implementations of a fleet of multiple UAVs, the agents are often pre-programmed to perform a set of predefined tasks independently, where the initial tasks are defined and updated by a central controller who constantly observes the operation field [14, 15]. However, during a wildfire, the environment is subject to rapid and unexpected changes and is not fully observable in a real-time manner where only partial knowledge about the environment is available from prior images of the region (e.g., satellite images). Furthermore, the long range communication among the drones and the fire management station may not be feasible. Therefore, the UAV fleet need to comprehend the environment, identify and prioritize the required tasks and divide these tasks among themselves in the absence of a centralized unit.

A hierarchical leader-follower platform of multiple drones can be a practical solution to provide a long-term coverage of the fire with the required spatial and temporal resolutions. Compared to a fully distributed system, where each drone operates independently from the other ones, hence, being under by a heavy computation and sensing load, in a leader-follower approach, the drones coordinate with each other under the high-level control of a leader drone. In this model, high-altitude fixed-wing drones or high altitude drones (HAPs) can serve as leader, managing a fleet of low-altitude small UAS. This heterogeneous platform provides diverse features including HAP's long flight time, ability to fly in inclement weather, and small UAS' high-resolution imaging [1]. The leaders are responsible for managing the location and tasks and relaying small UAS' observations to the fire management center, and the follower UAVs are tasked with collecting high-resolution images while flying at low-altitudes. Developing an optimal leader-follower UAV system involves a complex problem to enable full coverage of the fire field in a timely manner with the minimum number of UAVs, taking into account their limited communication range and sensing capabilities. A quantum-inspired genetic algorithm for distributed coalition formation in UAV networks that finds the optimum coalitions, accounting for reliability and probability of UAVs' failure was developed in [18]. The idea behind this algorithm is to take advantage of both genetic algorithms and quantum computing mechanisms to find the most optimal solution in large-scale networks.

4 Conclusions

Exiting forest monitoring technologies including event-triggered sensors and satellite imaging do not offer a low-cost, and agile monitoring solution with the expected high spatial and spectral resolution. Drone-based fire detection, and monitoring offer fast and high-resolution mapping of wide forest areas considering unique features of UAVs in terms 3D mobility, flexibility, low flight altitude, and fast and easy deployment. In this paper, we presented several recent studies utilizing a single or multiple drone(s) in wildfire management. This paper also discusses the main challenges of using UAVs of fire detection and monitoring in forest areas such as the conservative flight regulations, the lack of availability of long-range high-transmission rate communication in remote forest areas, and the lack of sufficient large and diverse datasets to develop and evaluate data-driven fire detection and monitoring models.

References

1. Afghah, F., Razi, A., Chakareski, J., Ashdown, J.: Wildfire monitoring in remote areas using autonomous unmanned aerial vehicles. In: IEEE INFOCOM 2019 - IEEE Conference on Computer Communications Workshops (INFOCOM WKSHPS), pp. 835–840 (2019). https://doi.org/10.1109/INFCOMW.2019. 8845309
2. Afghah, F., Zaeri-Amirani, M., Razi, A., Chakareski, J., Bentley, E.: A coalition formation approach to coordinated task allocation in heterogeneous uav networks. In: 2018 Annual American Control Conference (ACC), pp. 5968–5975 (2018). https://doi.org/10.23919/ACC.2018.8431278
3. Bailon-Ruiz, R., Bit-Monnot, A., Lacroix, S.: Real-time wildfire monitoring with a fleet of uavs. Robot. Autonom. Syst. **152**, 104071 (2022). https://doi.org/10.1016/j.robot.2022.104071
4. Chen, X., et al.: Wildland fire detection and monitoring using a drone-collected RGB/IR image dataset. IEEE Access **10**, 121301–121317 (2022). https://doi.org/10.1109/ACCESS.2022.3222805
5. Coen, J.L., Schroeder, W., Rudlosky, S.D.: Transforming wildfire detection and prediction using new and underused sensor and data sources integrated with modeling. In: Blasch, E., Ravela, S., Aved, A. (eds.) Handbook of Dynamic Data Driven Applications Systems, pp. 215–231. Springer, Cham (2018). https://doi.org/10.1007/978-3-319-95504-9_11

6. Foggia, P., Saggese, A., Vento, M.: Real-time fire detection for video-surveillance applications using a combination of experts based on color, shape, and motion. IEEE Trans. Circuits Syst. Video Technol. **25**(9), 1545–1556 (2015)

7. Ghali, R., Akhloufi, M.A., Mseddi, W.S.: Deep learning and transformer approaches for UAV-based wildfire detection and segmentation. Sensors **22**(5) (2022). https://www.mdpi.com/1424-8220/22/5/1977

8. Guan, Z., Miao, X., Mu, Y., Sun, Q., Ye, Q., Gao, D.: Forest fire segmentation from aerial imagery data using an improved instance segmentation model. Remote Sens. **14**(13) (2022). https://doi.org/10.3390/rs14133159

9. Hopkins, B., et al.: Flame 2: fire detection and modeling: aerial multi-spectral image dataset (2022). https://doi.org/10.21227/swyw-6j78

10. Huang, Q., Razi, A., Afghah, F., Fule, P.: Wildfire spread modeling with aerial image processing. In: 2020 IEEE 21st International Symposium on "A World of Wireless, Mobile and Multimedia Networks" (WoWMoM), pp. 335–340. IEEE (2020)

11. Islam, S., Huang, Q., Afghah, F., Fule, P., Razi, A.: Fire frontline monitoring by enabling UAV-based virtual reality with adaptive imaging rate. In: 2019 53rd Asilomar Conference on Signals, Systems, and Computers, pp. 368–372 (2019). https://doi.org/10.1109/IEEECONF44664.2019.9049048

12. Jurado, J.M., Padrn, E.J., Jimnez, J.R., Ortega, L.: An out-of-core method for GPU image mapping on large 3d scenarios of the real world. Future Gen. Comput. Syst. **134**, 66–77 (2022). https://doi.org/10.1016/j.future.2022.03.022

13. Kyrkou, C., Theocharides, T.: Emergencynet: efficient aerial image classification for drone-based emergency monitoring using atrous convolutional feature fusion. IEEE J. Select. Topics Appl. Earth Observ. Remote Sens. **13**, 1687–1699 (2020)

14. Lacroix, S., Gancet, J.: Comets Project (2003). http://www.comets-uavs.org. Accessed 27 May 2021

15. Lacroix, S., Gancet, J.: Real-time coordination and control of multiple heterogeneous uavs: The comets project. In: 2006 IEEE/RSJ International Conference on Intelligent Robots and Systems, p. 9. IEEE (2006)

16. Lee, W., Kim, S., Lee, Y.T., Lee, H.W., Choi, M.: Deep neural networks for wild fire detection with unmanned aerial vehicle. In: 2017 IEEE International Conference on Consumer Electronics (ICCE), pp. 252–253. IEEE (2017)

17. Moore, A.: Climate Change is Making Wildfires Worse (2022)

18. Mousavi, S., Afghah, F., Ashdown, J., Truck, K.: Leader-follower based coalition formation in large-scale UAV networks, a quantum evolutionary approach. In: IEEE INFOCOM, Workshop on Wireless Sensor, Robot, and UAV Networks, Selected as Best Paper (2018)

19. Muksimova, S., Mardieva, S., Cho, Y.I.: Deep encoder–decoder network-based wildfire segmentation using drone images in real-time. Remote Sens. **14**(24) (2022). https://doi.org/10.3390/rs14246302

20. Namvar, N., Afghah, F.: Joint 3d placement and interference management for drone small cells. In: IEEE Asilomar Conference on Signals, Systems, and Computers (ASILOMAR) (2021)

21. Nelson, K.N., et al.: A multipollutant smoke emissions sensing and sampling instrument package for unmanned aircraft systems: development and testing. Fire **2**(2) (2019). https://doi.org/10.3390/fire2020032

22. Office, C.B.: WildFires (2022)

23. PÇka, M., Ptak, S., Kuziora, Å.: The use of UAV's for search and rescue operations. Procedia Eng. **192**, 748–752 (2017). https://doi.org/10.1016/j.proeng.2017. 06.129. 12th International Scientific Conference of Young Scientists on Sustainable, Modern and Safe Transport

24. Rad, P.A., Hofmann, D., Pertuz Mendez, S.A., Goehringer, D.: Optimized deep learning object recognition for drones using embedded GPU. In: 2021 26th IEEE International Conference on Emerging Technologies and Factory Automation (ETFA), pp. 1–7 (2021). https://doi.org/10.1109/ETFA45728.2021.9613590

25. Shamsoshoara, A., Afghah, F., Blasch, E., Ashdown, J., Bennis, M.: UAV-assisted communication in remote disaster areas using imitation learning. IEEE Open J. Commun. Soc. (2021)

26. Shamsoshoara, A., Afghah, F., Razi, A., Mousavi, S., Ashdown, J., Turk, K.: An autonomous spectrum management scheme for unmanned aerial vehicle networks in disaster relief operations. IEEE Access **8**, 58064–58079 (2020)

27. Shamsoshoara, A., Afghah, F., Razi, A., Zheng, L., Fulé, P., Blasch, E.: The FLAME Dataset: Aerial Imagery Pile Burn Detection Using Drones (UAVs) (2020). https://doi.org/10.21227/qad6-r683

28. Shamsoshoara, A., Afghah, F., Razi, A., Zheng, L., Fulé, P.Z., Blasch, E.: Aerial imagery pile burn detection using deep learning: the flame dataset. Comput. Netw. **193**, 108001 (2021)

29. Sudhakar, S., Vijayakumar, V., Kumar, C.S., Priya, V., Ravi, L., Subramaniyaswamy, V.: Unmanned aerial vehicle (UAV) based forest fire detection and monitoring for reducing false alarms in forest-fires. Comput. Commun. **149**, 1–16 (2020)

30. Thomas, D., Butry, D., Gilbert, S., Webb, D., Fung, J.: The costs and losses of wildfires. NIST Spec. Publ. **1215**(11) (2017)

31. Twidwell, D., Allen, C., Detweiler, J., Higgins, C.L.S.E.: Smokey comes of age: unmanned aerial systems for fire management. Front. Ecol. Environ. (2016)

32. Wang, J., Fan, X., Yang, X., Tjahjadi, T., Wang, Y.: Semi-supervised learning for forest fire segmentation using UAV imagery. Forests **13**(10) (2022). https://doi. org/10.3390/f13101573

33. Watts, A.C., Ambrosia, V.G., Hinkley, E.A.: Unmanned aircraft systems in remote sensing and scientific research: classification and considerations of use. Remote Sens. **4**(6), 1671–1692 (2012). http://libproxy.clemson.edu/login? url=https://www.proquest.com/scholarly-journals/unmanned-aircraft-systems-remote-sensing/docview/1537378479/se-2

34. Wu, H., Li, H., Shamsoshoara, A., Razi, A., Afghah, F.: Transfer learning for wildfire identification in UAV imagery. In: 2020 54th Annual Conference on Information Sciences and Systems (CISS), pp. 1–6. IEEE (2020)

35. Yuan, C., Liu, Z., Zhang, Y.: Aerial images-based forest fire detection for firefighting using optical remote sensing techniques and unmanned aerial vehicles. J. Intell. Robot. Syst. **88**(2–4), 635–654 (2017)

36. Yuan, C., Liu, Z., Zhang, Y.: Learning-based smoke detection for unmanned aerial vehicles applied to forest fire surveillance. J. Intell. Robot. Syst. **93**(1), 337–349 (2019)

37. Zhan, J., Hu, Y., Cai, W., Zhou, G., Li, L.: Pdam–stpnnet: a small target detection approach for wildland fire smoke through remote sensing images. Symmetry **13**(12) (2021). https://doi.org/10.3390/sym13122260

38. Zhan, J., Hu, Y., Zhou, G., Wang, Y., Cai, W., Li, L.: A high-precision forest fire smoke detection approach based on argnet. Comput. Electron. Agricult. **196**, 106874 (2022). https://doi.org/10.1016/j.compag.2022.106874
39. Zhang, L., Wang, M., Fu, Y., Ding, Y.: A forest fire recognition method using UAV images based on transfer learning. Forests **13**(7) (2022). https://www.mdpi.com/1999-4907/13/7/975

Role of Autonomous Unmanned Aerial Systems in Prescribed Burn Projects

Mrinal Kumar[1]([✉])(iD), Roger Williams[1](iD), and Amit Sanyal[2](iD)

[1] The Ohio State University, Columbus, OH 43210, USA
{kumar.672,williams.1577}@osu.edu
[2] Syracuse University, Syracuse, NY 13244, USA
aksanyal@syr.edu
https://mae.osu.edu/laddcs

Abstract. This short paper describes the potential for use of autonomous multi-UAS teams during all stages of a prescribed burn. The current state of practice of prescribed burning is labor intensive and based on numerous simplifying assumptions. UAS teams promise to increase efficiency and effectiveness, while creating the opportunity to develop new science related to fire behavior. Ingrained in the proposed UAS mission profiles is bi-directional feedback between sensing and computational components at multiple timescales, which is a hallmark of the Dynamic Data Driven Applications Systems (DDDAS) framework.

Keywords: Autonomous UAS · Prescribed Burns · Dynamic Data Driven Applications Systems (DDDAS)

1 Background

Fire is an integral part of the forest ecosystem. It is a natural disturbance and a key driver of forest ecosystem dynamics that assists in the recycling of nutrients, keeps pathogens at endemic levels, helps maintain healthy forest densities, provides growing space for tree and plant regeneration and makes forests less susceptible to devastating crown fires [12,21]. In 1910, a series of catastrophic fires occurred in Montana, Idaho, and Washington (the "Big Blowup") that burnt over 3 million acres in just two days and caused thousands of civilian fatalities. Unfortunately, this event convinced officials at the newly formed U.S. Forest Service (1905) that the only way forward was to institute a policy of total fire suppression, sometimes known as the wildfire exclusion paradigm [5]. Over the decades, this policy contributed to an imbalance between live and dead vegetation, detriment to forests' natural ability to regenerate and maintain biodiversity, and ironically, wildfires of increasing intensity.

Supported by National Science Foundation, Computer and Information Science and Engineering Directorate, under Grant Numbers 2132798 (MK, RW) and 2132799 (AS) and the Battelle Memorial Institute under the Battelle Engineering, Technology and Human Affairs (BETHA) Program (MK, RW).

E. Blasch et al. (Eds.): DDDAS 2022, LNCS 13984, pp. 395–402, 2024.
https://doi.org/10.1007/978-3-031-52670-1_39

A reevaluation of the wildfire exclusion paradigm became inescapable when the wildland-urban interface (WUI) was threatened and the "WUI fire problem" was nationally recognized in 1985 [5]. Today, WUI is the fastest growing land use in the nation, and there is an urgent need to restore balance between our wildlands and wildfire. *Prescribed burning* is a crucial part of the restoration plan and refers to the controlled application of fire by a team of fire experts under specified weather conditions that helps restore health to forest ecosystems. In the 1950's and 1960's, the first prescribed burning programs were (re)established in the coastal plains and lower Piedmont pine and grassland habitats, but the practice gained acceptance in the mountains only in the 1980's [20]. Given the relatively recent re-introduction of controlled burns into the forest management playbook, as well as distinct requirements for safety of operations and risk mitigation, the practice is almost exclusively driven by manual work that involves a very large number of human hours. There exist no guidelines for integration of robotics (ground or aerial) into prescribed burning despite recent advances.

On the other hand, while there have been major advancements in the field of autonomous UAS, their deployment in hazardous environments with unstructured and dynamic uncertainty remains limited. There are still many hurdles on the way to achieve assured autonomy of such systems, generally understood as safe, trustworthy and certifiable use, especially in the described environment and in beyond visual line of sight (BVLOS) [13,19]. In this short paper, we develop a unique use-case for fully autonomous UAS teams, one which creates opportunity to take humans out of danger, to assist in high-pressure decision making and risk mitigation, and help in creation of new science on wildland fire behavior. A well-designed prescribed burn is a multistage event that requires planning, survey, careful execution and elaborate post-burn activities. Multimodal dynamic sensor data related to processes (fire, forest regeneration) and the environment (local weather, heat flux contours) is generated across multiple timescales (fast: fire, smoke, heat flux, faster: wind, slow: regeneration) and must be incorporated into decision feedback loops for effective planning and execution of these UAS missions and subsequent fire and forest related modeling and analysis. These decision loops embody the principles of dynamic data-driven application systems (DDDAS) across all constituent timescales. For instance, multi-UAS missions during the prescribed burn establish a bi-directional feedback between weather (wind, smoke, temperature) and environmental situational awareness data (heat flux resource constraints) to determine task allocation and optimal path-planning solutions; which in turn drive situational awareness updates and indicate potential sensing conflicts. On a slower timescale, DDDAS decision loops drive large-scale survey of forest lands to track regeneration and wildland ecosystem health. Below, we describe integrated efforts to advance the state of practice in each stage of a prescribed burn, and to advance assured autonomy of UAS in such an environment [14].

2   UAS Missions in Various Stages of a Prescribed Burn

2.1   Pre-Burn

A controlled burn begins with a survey of the forest lands. The firing pattern and window of burn are decided by the nature and concentration of fuel across the burn site, in addition to topographic and weather conditions. Presently, this survey is done by human labor by walking across the burn site in a grid pattern. Fuel samples are collected and their concentration and chemical properties are determined. Unsurprisingly, the survey is never comprehensive, especially in large and/or difficult terrains. Assumptions are made that are extended to the entire burn site, based on feeding the manually collected data into fire propagation models like BehavePlus™.

UAS Missions: Cooperative multi-UAS path planning can achieve scalable survey of forest lands to identify optimal candidate sites for controlled burning classified in terms of fuel load, disease, invasive species, etc. Forests have a three dimensional profile comprised of layers of understory grass, shrub and forb vegetation [15] underneath the dominant canopy of overstory trees. UAS equipped with advanced multispectral and hyperspectral imaging [8,10,11] can help with identification of plant and tree species and mapping individual species and overall tree density. Once the optimal burn site is identified, hyperspectral imaging can help map and temporally track foliage with immaculate detail, starting as early as the spring season. Appropriate classification algorithms can help create detailed maps of foliage species, density and chemical composition. This data is essential for determining relationships between fuel distribution and fire temperature and rate of spread in different geographical locations of the United States. In addition to cooperative, distributed path-planning for surveying large domains [2,6,9], UAS must carry certified advanced detect and avoid algorithms that enable beyond visual line of sight (BVLOS) flight amidst a highly dense network of tree branches, including under the forest canopy. The UAS mission exploit the DDDAS paradigm on a relatively slow timescale: analysis of sensor data helps identify candidate sites for the burn, followed by a more in-depth survey of species and density distribution mapping of the chosen burn site.

2.2   During the Burn

The current practice of controlled burns is almost entirely devoid of robotic participation. Burn events are onerous and require nimble decision-making skills and coordination, orchestrated by the burn boss and her crew leaders. The burn boss' priorities are keeping the event on schedule, containing the fire within the burn area perimeter, managing smoke outflow, and monitoring the fire intensity and weather conditions to call into effect appropriate emergency plans when needed. Controlled burns are highly dynamic events that stand to gain from improved situational awareness created by multimodal fire related information.

UAS Missions: Multi-UAS missions during the burn embody the DDDAS paradigm at a relatively fast timescale, and exert a significant burden on the mission's computational elements to keep up with the fire process and the environment. Physical sensing is achieved mainly by temperature and multispectral vision and IR sensing. The DDDAS sensor reconfiguration loop is driven by the computational ground station, which maintains up to date process and environmental situational awareness through the data assimilation loop (see Fig. 1).

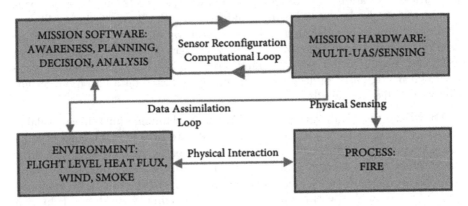

Fig. 1. Schematic of the DDDAS Framework in Multi-UAS Missions During a Prescribed Burn.

UAS operate out of an unrivaled vantage point in tracking the progression of the burn. Collaborative multi-agent UAS missions [1] will provide uninterrupted situational awareness of fire, weather, and smoke conditions, thus assisting the burn boss with decision-making, mitigating smoke-related disruptions and the incidence of fire escape conditions [16–18]. If uncontrolled fire spread does occur, UAS generated situational awareness will assist in optimal resource management for containment. Following the determination of optimal target pose for a given UAS (derived from situational awareness modules), a resource-constrained path planning problem must be posed that involves a path-dependent integral loading constraint given in Eq. (1) below

$$\underbrace{\text{``}\delta(t) = 1\text{''}}_{\text{``damage''}} \equiv \underbrace{\int_0^t \underbrace{\mathcal{F}_{\mathcal{L}}(\tau, \mathbf{s}(\tau), u(\tau))}_{\text{rate of damage}} \, d\tau}_{\text{path load}} \leq \underbrace{\mathcal{L}^\star}_{\text{loading limit}} \tag{1}$$

In the above equation, the the rate of damage to the platform is given by the function $\mathcal{F}_{\mathcal{L}}(\cdot, \cdot, \cdot)$ and is context dependent. When considering the example of flight-over-fire, it represents a normalized heat flux as follows

$$\mathcal{F}_{\mathcal{L}}(t, \mathbf{s}(t), \mathbf{u}(t)) = \frac{A}{mc_p}\phi(\mathbf{s}(t)) \tag{2}$$

the variables above have the usual meaning (m = vehicle mass, A = incident area, c_p = heat capacity, $\phi(\cdot)$ = heat flux). Heat flux *can be* allowed to be negative (cooling effect), allowing the planner to perform load-shedding by interspacing flight through hot regions with flight over cooler areas, thus keeping the total rise in temperature under loading limits:

$$\dot{\phi}(x(t), y(t)) = \begin{cases} \dot{\phi}_{\text{rad}}(x(t), y(t)) = \dot{q}'', & \text{if } \dot{\phi}_{\text{rad}}(x(t), y(t)) > 0 \\ \dot{\phi}_{\text{cool}}(x(t), y(t)) = -h(T - T_{\text{min}}), & \text{otherwise} \end{cases}$$

(3)

One may consider only non-negative loading constraints such that loading relief is not possible. In other words, $\dot{\phi}_{\text{cool}}(x(t), y(t)) = 0$. While convection and radiation are the primary modes of heat transfer from the wildfire, radiation becomes the dominant form of heat transfer for surfaces exceeding $400\,^{\circ}\text{C}$ [7]. As the flame temperatures of a wildfire can vary between $800\,^{\circ}\text{C}$ and $1000\,^{\circ}\text{C}$, the fire's radiative heat that extends up into the atmosphere is the primary concern as it affects the safety of the UAS. Figure 2 shows results from a recently developed *backtracking hybrid* $A^{\star}$ graph search technique (Fig. 2(b)) compared against a traditional hybrid $A^{\star}$ search (Fig. 2(a)) to account for heat loading resource constraints shown using colored contours [9], resulting in scalable, close to optimal solutions for this NP-hard problem.

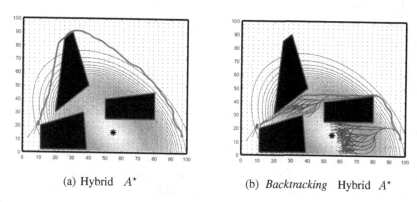

(a) Hybrid $A^{\star}$ (b) *Backtracking* Hybrid $A^{\star}$

Fig. 2. Hybrid $A^{\star}$ vs Backtracking Hybrid $A^{\star}$ for Resource Constrained Path Planning.

In terms of disturbance rejection and path following, UAS possess the capability to handle unsteady and turbulent airflow, higher temperatures and variable density of air in such regions. These effects lead to uncertainties in the flight dynamics, affecting both the translational and rotational motion of UAVs, and brings adverse effects in the control performance of these systems [3]. Therefore, it becomes critical in these applications to ensure nonlinearly stable and robust flight, with guaranteed stability margins. Moreover, the high temperatures can also cause updrafts which can lead to flight instability. Recent work by authors of this paper has provided an asymptotically stable tracking control scheme using a finite-time stable disturbance observer in the feedback loop, for UAS modeled as a rigid body. The dynamics of the system is discretized using a Lie group variational integrator in the form of a "grey box" dynamics model that also accounts for unknown additive disturbance force and torque. These disturbance terms are estimated using the finite-time stable disturbance observer in real-time and then compensated by the control scheme. Further details can be found in Ref. [3]. Additional details of UAS missions in a prescribed burn, including situational awareness and real-time mission realignment are skipped due to space constraints. In short, a significant amount of foundational and integrative work is needed to achieve certifiable UAS missions over a live prescribed burn.

Over the next few years, it is expected that UAS missions will generate unprecedented datasets of fire spread, intensity and smoke that can be employed for analysis after completion of the burn. Autonomous missions with RGB and IR imaging and chemical sensing [4] will profile fire intensity, fire perimeter, smoke density and chemical composition. These datasets must be correlated with pre-burn survey missions to discover new scientific relationships between fire and smoke behavior and topography, weather, and fuel. There is tremendous opportunity to perform novel fire science that will increase the safety and efficiency of future controlled burns and help mitigate destructiveness of wildfires.

2.3 Post-Burn

A significant number of human hours are spent post-burn to verify burn objectives, assess unintended damage and mop-up of the burn perimeters. The team must validate that there is no existing danger of fire escape or further smoke generation. There is no record of use of certified autonomous ground or aerial robots assisting in this work.

UAS Missions: In the hours to weeks following the controlled burn, UAS equipped with high resolution imaging at appropriate wavelengths can evaluate the quality of the controlled burn against a range of metrics.Machine learning tools can help map fuel consumption, top-kill in the understory, and discoloration of foliage in the overstory to determine the percentage of crown scorch. Similarly, learning tools applied to UAS data can help map the extent of soil and root damage [20]. UAS mounted smoke sensors can help determine air quality across altitudes for an extended period of time [4]. Finally, autonomous UAS surveillance will help track the regeneration process to determine the long-term benefits of controlled burning in a given region.

3 Summary

In the near future, we expect that autonomous UAS teams will leverage the DDDAS paradigm to empower prescribed burn teams and drive a paradigm shift in forest land management through UAS assisted burning and mitigation of destructive wildfires. Research and development work must occur in collaboration with stakeholders, i.e., states' Departments of Natural Resources to advance assured autonomy of collaborative UAS in unstructured hazards and dynamic uncertainty of a wildland burn.

References

1. Aggarwal, R., Soderlund, A., Kumar, M.: Multi-UAV path planning in a spreading wildfire. In: Guidance, Navigation, and Control Conference at SciTech. Virtual, 11–20 January 2021
2. Aggarwal, R., Soderlund, A., Kumar, M., Grymin, D.J.: Risk-aware path planning for unmanned aerial systems in a spreading wildfire. AIAA J. Guid. Control Dyn. **45**(9), 1692–1708 (2022)
3. Bhale, P., Kumar, M., Sanyal, A.: Finite-time stable disturbance observer using grey-box dynamics model. In: American Control Conference, Atlanta GA, 8–10 June 2022
4. Burgues, J., Marco, S.: Environmental chemical sensing using small drones: a review. Sci. Total Environ. **748**, 141172 (2020)
5. Cohen, J.: The WUI fire problem. In: Forest History Today (2008)
6. Cortez, A., Ford, B., Nayak, I., Narayanan, S., Kumar, M.: Path planning for a dubins agent with resource constraints and dynamic obstacles. In: Guidance, Navigation, and Control Conference at AIAA Scitech. National Harbor DC, 23–27 January 2023
7. Drysdale, D.: An Introduction to Fire Dynamics. Wiley, Hoboken (2011)
8. Esposito, F., et al.: An integrated electrooptical payload system for forest fires monitoring from airborne platform. In: IEEE Aerospace Conference, pp. 1–13 (2007)
9. Ford, B., Aggarwal, R., Kumar, M., Manyam, S.G., Casbeer, D., Grymin, D.J.: Backtracking hybrid a^* for resource constrained path planning. In: Guidance, Navigation & Control Conference at AIAA Scitech Forum, San Diego CA, 3–7 January 2022
10. Grenzdörffer, G., Engel, A., Teichert, B.: The photogrammetric potential of low-cost UAVs in forestry and agriculture. Int. Arch. Photogramm. Remote. Sens. Spat. Inf. Sci. **31**(B3), 1207–1214 (2008)
11. Klingbeil, A.E., et al.: Two-wavelength mid-IR absorption diagnostic for simultaneous measurement of temperature and hydrocarbon fuel concentration. In: Proceedings of the Combustion Institute. Elsevier (2009)
12. Knapp, E.K., Keeley, J.E., Ballenger, E.A., Brennan, T.J.: Fuel reduction and coarse woody debris dynamics with early season & late season prescribed fire in a Sierra Nevada mixed conifer forest. Forest Ecol. Mgmt. **208**, 383–397 (2005)
13. Medlin, R., et al.: Trustworthy autonomy: a roadmap to assurance. Part I: system effectiveness. Technical report, Institute for Defense Analyses, May 2020
14. Mueller, J.M.: The ABCS of assured autonomy. In: Cunningham, M., Cunningham, P. (eds.) International Symposium on Technology in Society Proceedings. IEEE (2019)
15. Smith, E.: Ecological relationships B/W overstory & understory vegetation in ponderosa pine forests of the southwest. Technical report, The Nature Conservancy (2011)
16. Soderlund, A., Kumar, M.: Estimating the spread of wildland fires via evidence-based information fusion. IEEE Trans. Control Syst. Technol. (2022). https://doi.org/10.1109/TCST.2022.3183645
17. Soderlund, A., Kumar, M., Yang, C.: Autonomous wildfire monitoring using airborne and temperature sensors in an evidential reasoning framework. In: Intelligent Systems Conference @AIAA SciTech, San Diego, CA, 7–11 January 2019
18. Soderlund, A.A.: Characterization of wildland fires through evidence-based sensor fusion and planning. Ph.D. thesis, The Ohio State University (2020)

19. Topcu, U., Bliss, N.: Assured autonomy: path toward living with autonomous systems we can trust. Technical report, Computing Community Consortium (2020)
20. Waldrop, T.A., l. Goodrick, S.: Introduction to prescribed fires in southern ecosystems. science update SRS-054. Technical report, U.S. Department of Agriculture Forest Service, Southern Research Station, Asheville, NC (2012)
21. Williams, R.A.: Effects of fire exclusion and the need to reintroduce fire (2021). Lecture Notes: ENR 3335.01 The Ohio State Univesity

Towards a Dynamic Data Driven Wildfire Digital Twin (WDT): Impacts on Deforestation, Air Quality and Cardiopulmonary Disease

M. Halem[1]([✉]), A. K. Kochanski[2], J. Mandel[3], J. Sleeman[4], B. Demoz[1], A. Bargteil[1], S. Patil[1], S.Shivadekar[1], A. Iorga[1], J. Dorband[1], J. Mackinnon[6], S. Chiao[5], Z. Yang[5], Ya. Yesha[1], J. Sorkin[7], E. Kalnay[8], S. Safa[8], and C. Da[8]

[1] University of Maryland, Baltimore County, USA
halem@umbc.edu
[2] San Jose State University, San Jose, USA
[3] University of Colorado Denver, Denver, USA
[4] JHU/Applied Physics Lab, Laurel, USA
[5] Howard University, Washington, USA
[6] NASA/Goddard Space Flight Center, Greenbelt, USA
[7] University Maryland Medical School, Greenbelt, USA
[8] University Maryland, College Park, USA

Abstract. Recent persistent droughts and extreme heatwave events over the Western states of the US and Canada are creating highly favorable conditions for mega wildfires. The International Program of Climate Change AR6 report suggests that such extreme events will continue occurring with increasing frequency and intensity over forested regions, globally. While humangenerated fires for farming in the Amazon are at a potential tipping point, wildfires in the Northern Hemisphere are comparably generating broad regions of deforestation. The smoke from recent mega wildfires in California, driven by atmospheric and fuel conditions controlling their intensity, has been observed to penetrate the planetary boundary layer, stay in the atmosphere for a long time, and travel long distances. The wildfire smoke from such events has the potential to reach distant cities and towns over the Eastern US, significantly reducing the air quality of these distant communities, and to adversely impact human health by increasing Covid-19 morbidity as well as the number of respiratory and smoke-related heart diseases.

In this paper, we will apply the concepts of a dynamical data-driven wildfire system to implement a real-time Wildfire Digital Twin (WDT) simulation at sub-km resolution to enable the study of mega wildfire smoke impact scenarios at various distant locations from the occurring wildfires over western N. America. WDT provides a valuable planning tool to implement parameter impact scenarios by season, location, intensity, and atmospheric state. We augment the NASA Unified WRF (NUWRF) model with a dynamic fire spread parameterization (SFIRE) coupled to GOCART, CHEM, and HRRR5 physics. We implement a data-driven, near-time continuous assimilation scheme for ingesting and assimilating observations from the NOAA satellite instruments, VIIRS, and ABI and from a streaming sensor web of radars, ceilometers, and satellite lidar observational systems into the nested regional NUWRF model. We accelerate the high-resolution nested

E. Blasch et al. (Eds.): DDDAS 2022, LNCS 13984, pp. 403–410, 2024.
https://doi.org/10.1007/978-3-031-52670-1_40

NUWRF model performance to make it suitable for forecasting applications by emulating the WRF microphysics and GOCART parameterizations with a deep dense transform machine learning neural net architecture, FourCastNet, that can maintain a simulated hourly atmospheric forecast in seconds. The WDT can also model the development of data-driven smoke from plumes and track the smoke across the US as it penetrates the planetary boundary layer, subsequently increasing the surface PM2.5. The SFIRE model spread and plume interaction with the atmosphere is a unique contribution by the WDT, fully enabling the interaction of smoke aerosols with observed clouds, the microphysics precipitation, convection, and the GOCART Chem, currently unavailable in other fire and smoke forecasting models.

Keywords: Wildfire · Digital Twin · SFIRE · Deforestation · Smoke Impacts

1   Introduction

Currently, record breaking annual mega wildfires are occurring globally. Climate change is responsible for prolonged droughts and extreme heating that create favorable environments for breakouts of such mega wildfires according to the IPCC AR6, IPCC Special Report 2019 [1, 2]. Deforestation reduces carbon cycle effect of removing CO2 from the atmosphere over the deforested areas and further reduces precipitation. If wildfires continue to occur more frequently, in response to such climate change conditions, scientists are raising concerns of consequential climate change and societal problems accompanying such mega wildfires. Based on simulation studies by Noble, C. et al. 2016 [3], the Amazon has 2 tipping points, a 4^0 C temperature change and a 40% deforestation area. Deforestation could lead to wide areas of savannahs leading to changes in the hydrological cycle and reductions in biodiversity and the acquaculture. Lovejoy, T. and Nobre, C. 2021 [4] alerted the community that we may already be at a tipping point resulting from deforestation caused by human and natural wildfires not just in the Amazon but in extratropical latitudes. In Fig. 1, Butler, R. 2021 [5] shows tree cover loss, data obtained from Hansen, M.C.2021 [6], indicating deforestation loss in 2020 greater in the extra tropics than in the equatorial forests.

Smoke from mega wildfires is being transported great distances carrying chemical aerosols from the fires fuel that can act to enhance cases of air quality pollution. Recent research by Dominici, F. 2020 [7], Wu, C. 2019 [8], and O'Dell, K. 2021 [9] have observed significant increases in COVID-19 cases as well as increases in morbidity from cardio-pulmonary disease.

1.1   A Wildfire Digital Twin

Recent advances in computer technologies have led to the development of real-time digital twin simulations of manufacturing systems. More recently, digital twin technology has been applied to the study of weather and climate processes governing the planet Earth [10], and the Antarctic processes [11]. "A digital twin is a virtual representation of an object, process or system that spans its lifecycle, is updated from real-time data, and uses simulation, machine learning and reasoning to help decision-making" (fromWikipedia).

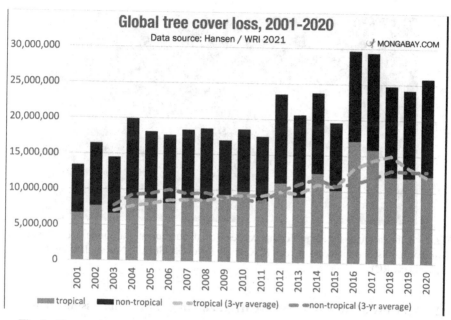

Fig. 1. Hectares loss due to wildfires for the tropics and extra tropics in the 21st century.

We present here the initial strategy for developing, implementing, and evaluating a regional Wildfire Digital Twin (WDT) model for the simulation of fire, its spread and impacts of smoke on air quality, and cardiopulmonary disease on local and distant communities. We build this approach on the integration of the exceptional works developed by the authors from the different universities. WDT will embed within the NASA Unified Weather Research Forecasting (NUWRF) model [12], a data driven, chemical based, interactive atmospheric fire spread resolving model WRF-SFIRE [13] and test the performance and long-term stability of deep, dense, physics aware machine learning emulations of the HRRR model [14] compatible with a rapid refresh hourly prediction.

To test the model, we have chosen as a case study, the August Complex wildfire from August 10 to Oct. 10, 2020 [15], the largest complex wildfire recorded in California's modern history of record keeping with nearly 10,000 fires that burned over 4.2 million acres of the state's 100 million acres of land. September 2020 was also a record-breaking heat wave and the August Complex fire became the first "gigafire" with a burned area exceeding 1 million acres. More than 367 known fires were reported as of August 17. Another reason for choosing this case study was the availability of plume data height distributions from the Multi-angle Imaging Spectroradiometer (MISR) on NASA's Terra satellite [16]. MISR has nine different cameras pointing toward Earth at different angles. As Terra passed over the August Complex Fire on Aug. 31, MISR collected stereo snapshots of the smoke plume from different angles. The highest parts of the plume from the August Complex Fire reached approximately 2.5 miles (4 km) into the atmosphere—putting it above the boundary layer. Fire traveled more than 500 km to the west and over 750 km east of the source (Fig. 2).

Fig. 2. In the image, is the August complex fires west of Sacrement, California, actively burning to the north, west and southeast. (Credit R. Kahn, K.J. Noyes GSFC, A. Nastan, JPL and J.Tackett, J.P. Vernier LRC).

1.2   WDT Use of NUWRF-SFIRE-CHEM and Innovative Data Assimilation

The NASA-Unified WRF (NUWRF) modeling system [17] is an observation-driven integrated modeling system representing aerosol transport, cloud, precipitation, and land surface processes at satellite-resolved scales. NU-WRF is intended as a superset of the standard NCAR Advanced Research WRF (WRF-ARW) [18]. The SFIRE is a dynamical fire spread model and fuel moisture model that has been coupled with WRF and WRF-CHEM by Mandel et al. 2011, 2014 [19] and Kochanski et al. 2015 [33]. SFIRE resolves weather driven fire progression, computes fire emission and heat fluxes that are fed into the WRF or WRF-CHEM model, which explicitly resolve plume rise and smoke dispersion. The WRF-CHEM-SFIRE model has been validated against a summer 2015 megafire in N. California by Kochinski et al., 2019 [20] and a Wildfire Smoke-Plume Rise RxCADRE aircraft campaign [21] employing Large Eddy Simulations (LES) [22] at scales of 100 m/layer extending to 3–5 km. by Moiseeva et al., 2021 [23]. The Coupled NUWRF-CHEM–SFire model will be used to conduct plume rise studies that represent the largest uncertainty in modeling of smoke transport systems.

NUWRF-SFIRE is a 4-level nested AWR model that couples a 2-D fire spread with a 3-D nested mesoscale weather model (Figs. 3a and 3b).

The NUWRF-SFIRE-Chem model treats meteorology within the dynamical core of the Weather Research and Forecast model (WRF); Skamarock et al. 2008 [24] and fire spread is based on a high resolution 2D-Grid at 30 m based on a lagged ODE representation of the Fuel Moisture Content by Vejmelka et al. [25] as shown in Figs. 4a and 4b.

This newer version of Sfire provides (i) better fuel moisture contents forecast, (ii) an automatic regional calibration, (iii) the potential to add other inputs for a more accurate forecast, such as canopy, sunlight, and wind, and estimate their added influence from data, without developing and coding entire new models for a more accurate forecast and (iv) a better more completely automated fire and smoke simulations.

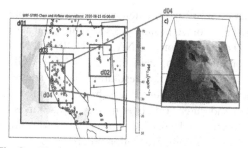

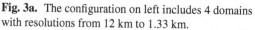

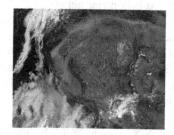

Fig. 3a. The configuration on left includes 4 domains with resolutions from 12 km to 1.33 km.

Fig. 3b. Domain created for the August Complex

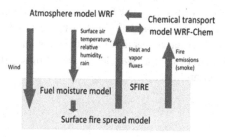

Fig. 4a. The overall scheme of WRF-SFIRE

$$\frac{dm}{dt} = \frac{E_w + \Delta E - m(t)}{T} \text{ if } m(t_k) < E_w + \Delta E,$$

$$\frac{dm}{dt} = \frac{E_d + \Delta E - m(t)}{T} \text{ if } m(t_k) > E_d + \Delta E,$$

$$\frac{dm}{dt} = 0 \text{ if } E_w + \Delta E \leq m(t_k) \leq E_d + \Delta E, \quad \frac{d\Delta E}{dt} = 0.$$

Fig. 4b. Fuel moisture content m(t) in wood represented as an ODE at each node of surface grid.

We plan to explore the feasibility of performing the strongly coupled data assimilation scheme recently developed by Kalnay et al. [26] to successfully improve the coupled Ocean and Atmospheric predictive models with the 2-D SFIRE surface grid model coupled to WRF atmospheric 3-D grid model. The strongly coupled ensemble data assimilation framework was able to use ocean observations to improve the atmospheric analysis and the use of atmospheric observations improved the ocean analysis and forecast. This conclusion was confirmed using a hierarchy of coupled models, ranging from the simple coupled Lorenz model to the state-of-the-art coupled general circulation model GFSv2 at NOAA [27]. We will test the strongly coupling of the NOAA VIIRS and GEOS ABI instrument data set and the Radar and Lidar data sets for the WRF model to improve the SFire spread model forecasts and the SFire burn data sets to improve the WRF model smoke forecasts.

1.3 Making WDT a Relevant Technology

The NUWRF based Wildfire Digital Twin model will be the first to fully interact with the HRRR model, cloud radiative processes, the NOAH Land surface model, surface fuel chemistry transports and the planetary boundary layer physics. We are employing a near time continuous EnKF data assimilation scheme for the NOAA ground radar, a physics aware machine learning aerosol backscatter algorithm (K. Patel et al., [28]) and streaming the Lidar backscatter concentration for derived pblh from a NASA supported unified ceilometer network (J. Sleeman et al., [29], an extended aerosol algorithm to augment the coverage of inferred pblh from ceilometers and NASA satellite Lidar aerosol backscatter instruments, including AOD data sets from the NASA MODIS and NOAA GOES, VIIRS operational satellites for inference of PM2.5, and the NOAA GEOS ABI instrument with near continuous IR coverage at 1km resolution. Nevertheless, the performance time for WDT predictions when implemented at sub-km resolution with the inner most nested interval of 300 m and outer grid covering the US at 3km is not fast enough to provide timely forecasts to support the decision makers with probabilistic estimates to homeowners for life and property saving evacuation orders.

To meet this need, we are initiating tests of a machine learning Nvidia based global FourCastNet model (Phatak et al. [30]) to emulate the WRF-SFIRE WDT model that will be trained with the 8-year HRRR4 hourly data operational cycle, not a reanalysis. Thus far, we have implemented the inference global 10-day weather forecast tests [31] and were able to show running times on a laptop conduct accurate 1–2 h forecasts in seconds. As a result, our effort is to develop a regional AI based version of WRF-SFire using the FourCastNet model or related Neural Operators such as wavelet based if improved model performance is demonstrated for another architecture.

1.4 Wildfire Smoke Health Impact

One health benefit outside the normal scope of modeling is tracking the dispersal of wild-fire related particulate matter and the combustion derived gases that enables epidemio-logical evaluations of their effect on cardiopulmonary disease. It will allow comparisons of health metrics in areas previously been exposed to fire-related atmospheric conditions with areas that have not been exposed. Health related metrics will be derived by examining ICD-10 codes obtained from Centers for Medicare & and Medicaid Services (CMS) claims data. ICD-10 codes [32], which are used for medical billing, describe all diagnoses and services provided to patients, both in-hospital and in a doctor's office.

Acknowledgements. The work presented in this paper was mainly supported in part by the NASA ESTO FIRET-QRS-22-0001 Program Office and partly by the NSF funded Center for Accelerated Real Time Analytics at UMBC. We wish to thank the AIST program office manager Dr. J. LeMoigne for initially awarding the AIST and M. Seablom for selecting this proposal as the initial Fire Tech grant. The views and conclusions contained herein are those of the authors and should not be interpreted as necessarily representing the official policies or endorsements, either expressed or implied, of the NASA Fire-Tech Program or the U.S. Government, or any other funding entities.

References

1. IPCC. Climate change 2021: the physical science basis. In: Masson-Delmotte, V., (eds.) Contribution of Working Group I to the Sixth Assessment Report of the Intergovernmental Panel on Climate Change. Cambridge University Press (2021)
2. IPCC Special Report on Climate Change and Land. Chapter 2. Exec. Summary (2019)

3. Nobre, C.A., Sampaio, G., Borma, L.S., Castilla-Rubio, J.C., Silva, J.S., Cardoso, M.: Land-use and climate change risks in the Amazon and the need of a novel sustainable development paradigm. Proc. Natl. Acad. Sci. USA **113**(39), 10759–10768 (2016). https://doi.org/10.1073/pnas.1605516113

4. Lovejoy, T.E., Nobre, C.: Amaxon tipping point: last chance for action. Sci. Adv. (2021) https://www.science.org/journal/sciadv

5. Butler, R.A.: Global Forest Loss Increase in 2020, Mongobay Series: Global Forests, Planetary Boundaries (2021)

6. Hansen, M.C., et al.: Global maps of the twenty first century forest carbon flux. Nat. Clim. Chang. **11**, 234–240 (2021)

7. Domenici, F., et al.: Air pollution and COVID-19 mortality in the United States: strengths and limitations of an ecological regression analysis. Sci. Adv. Cornovirus **4** (2020)

8. Wu, C., et al.: Evaluation of Different Machine Learning Approaches to Forecasting PM 2.5 Mass Concentrations (2019). https://aaqr.org/articles/aaqr-18-12-oa-0450

9. O'Dell, K., et al.: Estimated mortality and morbidity attributable to smoke Plkumes in the United States: not just a western problem. Geo Health **5**(9) (2021). https://doi.org/10.1029/2021GH000410

10. ECMWF: Destination Earth Digital Twin. https://www.ecmwf.int/en/about/what-we-do/environmental-services-and-future-vision/destination-earth

11. Antarctica Digital Twin. https://www.ed.ac.uk/informatics/news-events/stories/2020/scientists-to-create-a-digital-twin-of-antarctica

12. NASA Unified Weather Research Forecast Model (NUWRF). https://earth.gsfc.nasa.gov/index.php/meso/models/nu-wrf

13. WRF-SFIRE. https://wiki.openwfm.org/wiki/WRF-SFIRE

14. Hi-Resolution Rapid Refresh (HRRR) Model. https://rapidrefresh.noaa.gov/hrrr/

15. August Complex Wildfire (2020). https://en.wikipedia.org/wiki/August_Complex_fire

16. https://misr.jpl.nasa.gov/mission/misr-instrument

17. A Description of the Advanced Research WRF Model (WRF-ARW). https://opensky.ucar.edu/islandora/object/opensky:2898

18. Mandel, J., Beezley, J.D., Kochanski, A.K.: Coupled atmosphere-wildland fire modeling with WRF 3.3 and SFIRE 2011. Geosci. Model Develop. **4**, 591–610 (2011). https://doi.org/10.5194/gmd-4-591-2011

19. Mandel, J., et al.: Recent advances and applications of WRF-SFIRE. Nat. Hazards Earth Syst. Sci. **14**, 2829–2845 (2014). https://doi.org/10.5194/nhess-14-2829-2014

20. Kochanski, A.K., Mallia, D.V., Fearon, M.G., Mandel, J., Souri, A.H., Brown, T.: Modeling wildfire smoke feedback mechanisms using a coupled fire-atmosphere model with a radiatively active aerosol scheme. J. Geophys. Res. Atmos. **124**, 9099–9116 (2019). https://doi.org/10.1029/2019JD030558

21. Kochanski, A.K., Jenkins, M.A., Mandel, J., Beezley, J.D., Krueger, S.K.: Evaluation of WRF-Sfire performance with field observations from the FireFlux experiment. Geosci. Mod. Develop. **6**(1109–1126), 2013 (2013). https://doi.org/10.5194/gmd-6-1109-2013

22. Wells, G.: Commission, Capturing Fire: RxCADRE Takes Fire Measurements to Whole New Level (2013)

23. Moiseeva, N., Stull, R.: Wildfire smoke-plume rise: a simple energy balance parameterization. Atmos. Chem. Phys. **21**(1407–1425), 2021 (2021). https://doi.org/10.5194/acp-21-1407-2021

24. Skamarock, W.C., et al.: A Description of the Advanced Research WRF Version 3. NCAR Technical Note 475 (2008)

25. Vejsmelka, M., Kochanski, A., Mendel, J.: Data assimilation of fuel moisture in WRF-SFIRE. In: Proceedings of 4th Fire Behavior and Fuels Conference, 18–22 February 2013, Raleigh, North Carolina (2013)

26. Kalnay, E., Sluka, T., Yoshida, T., Da, C., Mote, S.: Towards strongly coupled data assimilation with additional improvements from machine learning. Nonl. Process. Geophys. (2023). https://npg.copernicus.org/preprints/npg-2023

27. Da, C., Kalnay, E., Chen, T.-C.: Multi-layer observation localization for hyperspectral infrared observations in the LETKF: ideal experiments with a multi-layer quasi-geostrophic model. In: 103rd Annual AMS Meeting, Denver (2023)

28. Patel, K., Sleeman, J., Halem, M.: Physics-aware deep edge detection network. Remote Sens. Clouds Atmos. XXVI **11859**, 32–38 (2021)

29. Sleeman, J., Caicedo, V., Zaiei, D., Halem, M., Demoz, B., Delgado, R.: Using Machine Learning to Identify Planetary Boundary Layer Heights for Ceilometer-Based Aerosol Backscatter Retrievals AGU Fall Meeting (2020)

30. Pathak, J., et al.: Fourcastnet: a global data-driven high-resolution weather model using adaptive Fourier neural operators. arXiv preprint arXiv:2202.11214v1 [physics.ao-ph] (2022)

31. Hamer, S., Halem, M., Saha, S.: A data driven machine learning approach to OSSEs. In: 103rd AMS Conference (2023)

32. Center for Medicare & Medicaid Services ICD-10 Code. https://www.cms.gov/Medicare/Coding/ICD10

33. Kochanski, A.K., Jenkins, M.A., Yedinak, K., Mandel, J., Beezley, J., Lamb, B.: Toward an integrated system for fire, smoke and air quality simulations. Int. J. Wildland Fire (2015). https://doi.org/10.1071/WF14074

An Earth System Digital Twin for Air Quality Analysis

Thomas Huang(✉) , Sina Hasheminassab, Olga Kalashnikova, Kyo Lee, and Joe Roberts

NASA Jet Propulsion Laboratory, California Institute of Technology, Pasadena, CA, USA
thomas.huang@jpl.nasa.gov

Abstract. Severe drought and wildfires have become grim indicators of the severity of the weather and climate extremes our planet is increasingly facing. The latest Intergovernmental Panel on Climate Change (IPCC) report describes the unprecedented rate at which the global climate has warmed in the last 200 years (UN News 2021), resulting in increasing ocean temperature, rising sea levels, intensifying rains and floods, new records for heatwaves and droughts, and ever-growing stress on freshwater availability. The growing collections of multi-discipline, high resolution spatiotemporal data requires us to be smarter and more automated, about scalable analytic framework and what data to incorporate into an analysis. This paper presents an open-source Earth System Digital Twin (ESDT) architecture being developed at the NASA's Jet Propulsion Laboratory (JPL) for integrated air quality analysis during wildfires.

Keywords: air quality · analytic collaborative framework · earth system digital twins · new observing strategies · open-source · wildfires

1 Introduction

To understand this complex Earth system, an extensible, integrated, scalable big data architecture is necessary to enable rapid analysis and machine learning (ML)-driven actionable predictions. AI-based predictions and resource management can help identify what data to ingest, what analysis will be needed, what assets can be dispatched, and what actions we can be taken to mitigate possible disastrous events. As defined by the NASA's Advanced Information Systems Technology (AIST) program (https://esto.nasa.gov/aist/), an Earth System Digital Twin (ESDT) is a dynamic, interactive, digital replica of the state and spatiotemporal evolution of Earth's climate system. It integrates multiple numerical models along with observational data, and connects them with analysis, artificial intelligence (AI), and visualization tools. Together, these enable

Supported by NASA Earth Science Technology Office (ESTO) Advanced Information Systems Technology (AIST) program.

E. Blasch et al. (Eds.): DDDAS 2022, LNCS 13984, pp. 411–414, 2024.
https://doi.org/10.1007/978-3-031-52670-1_41

users to explore the current state of Earth's climate system, predict future conditions, and run hypothetical scenarios to better understand how the system would evolve under various assumptions. Recognizing the need, the NASA Air Quality Analytic Collaborative Framework (AQACF) [2] and the Integrated Digital Earth Analysis System (IDEAS) projects are two technology investments to establish sustainable, extensible, and reusable open-source framework for air quality data analysis with advanced numerical simulations and ML-based predictions for dynamic data acquisition and value-added product generation.

2   Components of Air Quality Digital Twin

Components of the AQDT include

Data and Services Assets: An extract-transform-load (ETL) workflow component for metadata harvesting, error detection and correction, regridding/reprojecting, Analysis Ready Data (ARD) transformation, and generation of pre-computed statistics before the data can be put into an Analysis Optimized Storage (AOS) to be used by all service components of the digital twin architecture (see Fig. 1).

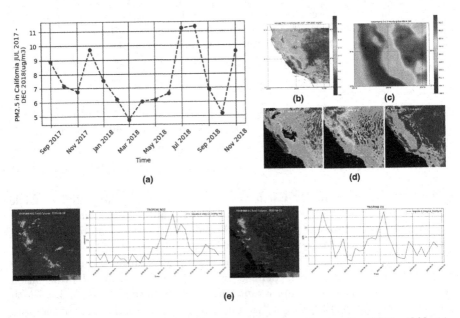

Fig. 1. Example air quality analysis and visualizations during the 2018 wildfire in California. (a) PM2.5 time series; (b) Area Averaged Map of PM2.5; (c) Modern-Era Retrospective analysis for Research and Applications, Version 2 (MERRA-2; Gelaro et al., 2017) Planetary Boundary Layer Height (PBLH) Anomaly; (d) Dynamic generated animation of PM2.5; (e) The ETL automates re-gridding of TROPOMI products (NO2 and CO shown above), visualization generation, and generation of ARDs for cloud-based, interactive analysis.

Integrated Multi-physics, Multiscale, Probabilistic Models: One of the advanced atmospheric compositions for local and global scale models supported by our effort is the GEOS-Chem (https://geos-chem.seas.harvard.edu/) or its cloud variant, the GEOS-Chem with the high-performance option (GCHP) (see Fig. 2).

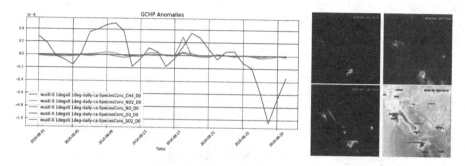

Fig. 2. Integration with GCHP for dynamic job submission and onboard output of different chemical species (CH4, NO2, NO, O3, and SO2), and dynamic generated animation.

AI/ML and Advanced Analytic: The application of AI/ML enables long-term prediction, data classification, process orchestration and management, etc. Currently, the project can predict PM2.5 in Los Angeles in near-real-time with high accuracy (see Fig. 3).

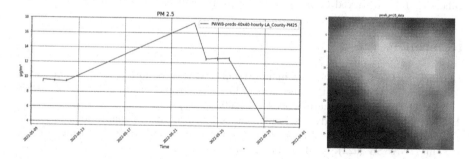

Fig. 3. Streamline training and deployment of the PM2.5 prediction model using AWS SageMaker. The model offers near-real-time PM2.5 prediction for the City of Los Angeles.

New Observation and Analysis: Using an advanced numerical model and ML-based method, this project is currently developing a dynamic data acquisition and integration capability. As part of the validation effort, this project established an in-situ air quality station at the JPL. The station collects various air quality species that are ready for immediate matchup with other satellite and model data. As part of the AIST New Observing Strategies (NOS) effort, this capability will be expanded in 2023 to simulate re-tasking the multiangle camera that is onboard of NASA's upcoming Multi-Angle Imager for Aerosols instrument [1] (see Fig. 4).

	Instrument	Measured Parameter(s)	Measurement Type	Sampling Interval	Data Latency
①	Airphoton Nephelometer MMP100	Polarized Aerosol Scattering	Continuous	2 minutes	Real-time
②	Aeroqual AQY	NO2, O3, PM2.5	Continuous	1 hour	Real-time
③	Davis Vantage Pro2 Weather Station	Temperature, Relative Humidity, Wind Speed, Wind Direction, etc.	Continuous	15 minutes	Real-time
④	AethLabs microAeth MA350	Black Carbon	Continuous	1 minute	Real-time
⑤	PurpleAir PA-II	PM1, PM2.5, PM10 Mass	Continuous	2 minutes	Real-time
⑥	GRIMM EDM 164	Size-resolved PM Mass and Number (0.25 - 32 μm)	Continuous	15 minutes	Real-time
⑦	Aerosol Mass and Optical Depth (AMOD)	PM2.5, Aerosol Optical Depth	Continuous	2 minutes	Real-time
		PM2.5 Chemical Components (e.g., Sulfate, Nitrate, Elemental Carbon, Organic Carbon, Dust, etc.)	24-hr Integrated	Every 3rd Day	3-6 Months
⑧	AQMesh	CO, NO2, O3, SO2, PM1, PM2.5, PM10	Continuous	15 minutes	Real-time
⑨	QuantAQ Modulair-PM	Size-resolved PM Mass and Number (0.35 - 40 μm)	Continuous	1 minute	Real-time
⑩	AirPhoton S5Si (as part of SPARTAN network)	PM2.5 Chemical Components (e.g., Sulfate, Nitrate, Elemental Carbon, Organic Carbon, Dust, etc.)	24-hr Integrated	Every 3rd Day	3-6 Months

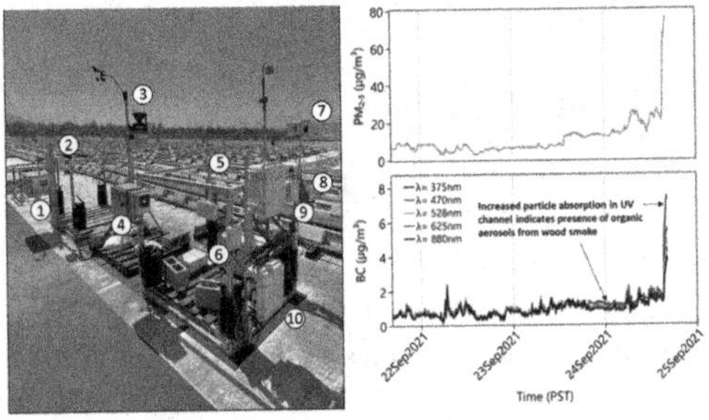

Fig. 4. In-Situ air quality station at the NASA JPL detected smoke plume from wildfire.

Acknowledgement. The research was carried out at the Jet Propulsion Laboratory, California Institute of Technology under a contract with the National Aeronautics and Space Administration (80NM0018D0004) through the NASA Earth Science Technology Office (ESTO) Advanced Information Systems Technology (AIST) program.

References

1. Diner, D.J., et al.: Advances in multiangle satellite remote sensing of speciated airborne particulate matter and association with adverse health effects: from MISR to MAIA. J. Appl. Remote Sens. **12**, 042603 (2018)
2. Huang, T., et al.: An advanced open-source platform for air quality analysis, visualization, and prediction. In: International Geoscience and Remote Sensing Symposium (IGARSS 2022), Kuala Lumpur, Malaysia (2022)

Workshop on Climate, Life, Earth, Planets

Dynamic Data-Driven Downscaling to Quantify Extreme Rainfall and Flood Loss Risk

Anamitra Saha[1]($\boxtimes$), Joaquin Salas[2], and Sai Ravela[1]

[1] Massachusetts Institute of Technology, Cambridge, MA, USA
anamitra@mit.edu
[2] Instituto Politécnico Nacional, Santiago de Querétaro, Mexico

Abstract. The adverse socio-economic effects of natural hazards will likely worsen under climate change. Modeling their risk is essential to developing effective adaptation and mitigation strategies. However, climate models typically do not resolve the detailed information that risk quantification demands. Here, we propose a *dynamic data-driven* approach to estimate extreme rainfall-induced flood-loss risk. In this approach, coarse-resolution climate model outputs ($0.25° \times 0.25°$) are downscaled to high-resolution ($0.01° \times 0.01°$) rainfall. After that, rainfall, historical insurance loss provided by The Federal Emergency Management Agency (FEMA), and other geographic data train a flood-loss model. Our approach shows promise for quantifying flood-loss risk, showing a weighted average value of $R^2 = 0.917$ for Cook County, Illinois, USA.

Keywords: Flood loss · Extreme rainfall · Downscaling

1 Introduction

Floods are devastating natural hazards with rising frequency and severity in the changing climate. While flood forecasting is essential for disaster response, flood-risk projections are critical for climate change adaptation and mitigation. However, the emergence of future flooding is a non-stationary nonlinear stochastic process. Even if predictability were not an issue, nature has not yet revealed the deep tails of its extremes. Instead of relying on historical statistics alone, climate model projections of specific scenarios can help determine risk. However, risk assessments demand fine-scale local information that climate models typically lack. High-resolution ensemble simulations accounting for model uncertainties and imperfections remain prohibitively expensive. Thus, estimating future flood-loss risk is challenging.

Consequently, interest is emerging in climate model downscaling (super-resolution) methods. Here, we consider a workflow that downscales coarse

The authors acknowledge support from Liberty Mutual (029024-00020), ONR (N00014-19-1-2273), The MIT Weather Extreme and CREWSNET Climate Grand Challenge projects, and the generosity of Eric and Wendy Schmidt by recommendation of Schmidt Futures as part of its Virtual Earth System Research Institute (VESRI).

E. Blasch et al. (Eds.): DDDAS 2022, LNCS 13984, pp. 417–424, 2024.
https://doi.org/10.1007/978-3-031-52670-1_42

climate-model input rainfall fields to high-resolution output rainfall fields and, in conjunction with other variables (*e.g.*, topography, insurance claims) over a region, learn to estimate event-wise losses. A probability distribution over an area for a window of time aggregates event- and point-wise losses into flood-loss risk, and time-varying distributions over sliding time windows drive decision support. Many workflow variations are possible, including using physical rainfall-driven inundation modeling, incorporating cyclone-induced rain, pluvial and fluvial interactions, integrating with coastal storm surges, waves, and tides, and modeling the effect of compound events. These are beyond this paper's scope, focusing on a dynamic data-driven approach to rainfall downscaling applied directly to flood-loss risk modeling.

A small ensemble simulation of high- and low-resolution climate model pairs covering various climate scenarios can train downscaling functions in the long-range setting. To estimate flood-loss risk, one can retrieve fine-scale details of rainfall fields and their uncertainties from a large, rapidly simulated coarse rainfall field ensemble. One must be cautious in such a framework, especially as there is no ground truth, and we accept that all models, coarse and fine, are in error. In contrast, the short-range setting (*e.g.*, inter-annual time scale), the focus of this paper, allows present-condition coarse resolution models and high-resolution observations to train downscaling, which is both verifiable and immediately functional. The insurance industry, for example, requires such downscaling for underwriting insurance a year or two ahead. We posit that pairing high-resolution observational data with coarse climate model variables also provides, via transfer learning, a methodological pathway to the long-range climate setting.

Estimating a downscaling function even for the "short-horizon flood-loss risk" problem is non-trivial. On the one hand, purely machine learning (and statistical) techniques that regress across resolutions often fail to incorporate basic physical principles. On the other hand, the training data needs of learning machines are often prohibitive for modeling rare extremes! To alleviate these problems, we propose a novel dynamic data-driven approach. Our system (see Fig. 1) consists of four steps. A simplified orographic lift model is the first downscaling step, which is applicable in any climate and is unconstrained by training data resolution limits. A dynamic data-driven approach that actively seeks nearby coarse model fields (from ECMWF Reanalysis v5, ERA5 [1]) and relevant fine-resolution (Daymet [2]) fields to construct a downscaling function is the second step. This step exemplifies DDDAS: The ERA5 input instantiates a conditional ensemble-based Gaussian Process downscaling model [3], which, applied in reverse, *i.e.*, upscaling, improves the conditional-Gaussian Process for the next iteration - in this way, dynamically producing a kernel for downscaling with few well-chosen data.

The estimates from these two steps prime a generative adversarial learning machine in the third step, which blends the physics and statistics and, consequently, requires little training data to produce high-resolution rainfall data from coarse climate model inputs. In the fourth step, using a stochastic process to quantify downscaled rainfall uncertainties, optimal estimation for bias correction is performed. We show that the risk of extreme annual rainfall captured by our downscaled rainfall maps closely matches the observed while quantifying uncertainty, ameliorating data needs, and incorporating known physics.

Historical data from the Federal Emergency Management Agency (FEMA) and high-resolution gridded rainfall maps of extreme rainfall events (Daymet), then train an event-wise flood loss prediction model. Extreme Gradient Boosting [4] is our choice of regressor for loss modeling since it outperforms other approaches in our experiments. Following an additional bias correction procedure, the predicted flood-loss risk distribution compares well to the observed distribution in testing.

The rest of the paper is as follows. Section 2 presents methodology and Sect. 3 presents results. Section 4 concludes the paper with a brief discussion of the significance of this work.

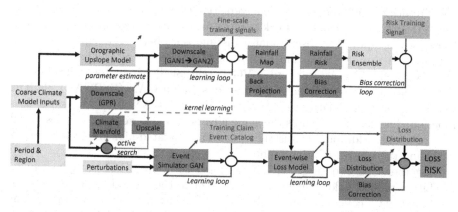

Fig. 1. Our approach uses orographic lifting, active conditional Gaussian Processes and Generative Adversarial Networks for downscaling. Rainfall return-period distributions computed from downscaled rainfall maps train bias correction functions. Bias corrections are back-projected in runtime for higher fidelity spatial rainfall estimates. High-resolution rainfall maps and insurance claim datasets train a synthetic claim simulator, an event-wise flood-loss model and a corresponding bias correction process to estimate flood-loss risk. During runtime, synthetic claims and high-resolution rainfall maps simulate flood-loss distributions for a period of interest. In the illustration, the light green boxes denote inputs, dark green boxes indicate potential outputs of the system. Yellow boxes are required only during training, blue boxes denote learning components, gray boxes denote static components, orange boxes represent dynamic data-driven components, black lines denote forward paths, red lines denote feedback paths, and orange lines denote active search paths. The pathways depicted by dashed lines are ongoing or planned future work. (Color figure online)

2   Methods

Our approach consists of several parts (see Fig. 1), including a dynamic data-driven conditional Gaussian Process regressor (CGP) for data-driven downscaling, a physics module to estimate the orographic component of rainfall, and

Generative Adversarial Networks (GAN) that combine the information gained from physics and statistics to produce high-quality fine-scale rainfall fields.

The CGP step queries training data indexed on a manifold [5] to retrieve nearest neighbors and estimate a downscaling function. After that, in reverse, the upscaled prediction errors target new data for downscaling in the next iteration and, thus, actively, upon convergence, produce a "first-guess" downscaled rainfall field. The physics model uses the upslope method [6] to estimate orographic rainfall, assuming that precipitation is proportional to the total condensation rate induced by the vertical wind in a saturated atmospheric column. See Sect. 4 for additional discussion.

The upslope and dynamic data-driven CGP estimates are combined to train a two-stage generative adversarial network (GAN), yielding a "second-guess" rainfall super-resolution field. In either stage, two deep networks, a generator, and a discriminator compete in the adversarial learning approach. The generator uses a Deep Convolutional Network to reconstruct a super-resolution rainfall field from a given input. The discriminator distinguishes between a super-resolution output from the generator and high-resolution ground truth. This study uses a ResNet-style [7] generator and a VGG-style [8] discriminator, following the network architecture of Enhanced Super-Resolution GAN (ESRGAN) [9]. The discriminator uses Relativistic Average GAN's (RaGAN) [10] adversarial loss function. The generator reuses the same adversarial loss, adding an ℓ_1 loss term that promotes additional consistency between simulated high-resolution output fields and their low-resolution inputs.

Together, the two stages transform low-resolution ERA5 data into high-resolution rainfall fields using Daymet rainfall as the ground truth. Because of model and data collection biases, ERA5 and Daymet precipitation maps do not correspond daily, even though one is a reanalysis and the other is an observation. Training without correspondence is challenging, which the two-stage approach overcomes. In the first stage (GAN-1), rainfall is downscaled from 0.25° to 0.1° resolution using the ERA5 field as the predictor and the corresponding ERA5-Land [11] field as the ground truth. In the second step (GAN-2), upscaled Daymet rainfall fields become predictors to the matched high-resolution Daymet fields for training downscaling from 0.1° to 0.01° resolution. In sequence, GAN-1 and GAN-2 provide the pathway from ERA5 predictors to Daymet-resolution downscaled rainfall. The transformation from ERA5-Land to Daymet may contain bias, which our approach automatically corrects in its last bias-correction step.

Due to the lack of one-to-one correspondence between ERA5 and Daymet, the final downscaled rainfall and ground truth are not event-wise comparable. However, one can compare rainfall risk. A two-parameter Generalized Pareto distribution [12] is fit to the annual return periods estimated from the empirical cumulative distribution function (ECDF) of rainfall from each grid point of each spatial grid and time window of interest. The averaged annual rainfall-return period curves over the test period represent the overall risk of extreme rainfall (see Fig. 2 (j)). However, ERA5 generally underestimates rainfall risk due to its tendency to underestimate the intensity of severe storms, leading to bias. Sampling observed rainfall distribution at the extremes (higher than 99.9th percentile), the statistical excess beyond the underestimated projections, stochastically injects perturbations to the deterministic downscaling. Evaluations show that the expected risk under this injected excess reduces bias.

The injected stochastic perturbations yield higher-order moments. In addition to the mean, the return-period curve's covariance is also available. Since the observed (from Daymet) extreme rainfall return period curve also has a covariance, an optimal estimate fuses the two. The three curves: the deterministically downscaled return period curve from GAN, the mean of the stochastic injections (the prior), and the optimal estimate (posterior) in Fig. 2(j) show that both the bias (colored curve) and the uncertainty (shaded region) reduce. The optimal estimated return period curve provides a final bias correction that is = back-projected onto the spatial rainfall fields to improve high-resolution fidelity and quantify uncertainties.

The flood loss risk model estimates event-wise flood losses and aggregate flood-loss risk, using public insurance data from the National Flood Insurance Program (NFIP) [13] and a high-resolution rainfall field. The NFIP database contains 2,547,311 records of insurance claims filed with FEMA from 1970–2021. We pre-processed them by filling in the missing values using the expectation-maximization method, adjusting for inflation, normalizing continuous values, and transforming categorical variables to their one-hot representation. Our purpose is to predict the variable describing the amount paid on building flooding claims using the rest of the variables as predictors.

We use extreme gradient boosting (XGB) [4] as the regressor for our prediction model. In XGB, we try to approximate the regression function using decision trees, and the difference between the approximation and the reference values is the target of subsequent decision trees. The process continues iteratively, adjusting the weights for the most challenging samples to incorporate into the current model. In addition to XGB, we experimented with a few other potential regressors, such as conditional GAN (CGAN) and Gaussian Processes (GP). GP models a stochastic process assuming that the interaction among the random variables follows a multivariate normal distribution. Consequently, second-order statistics become crucial.

CGAN tightly couples the approximation capabilities of regular neural network-based regression with the guided distribution modeling capabilities at the core of the generative adversarial learning framework. However, XGB outperforms the alternatives. We experimented with two methods for choosing the amount of data to build the regressor. In the shifting-window method, the regressor trains on a fixed ten-year lagged period, making predictions for the eleventh year. The training period grows as new data becomes available in the expanding-window method. We also compare the performance of our model with and without the high-resolution rainfall information incorporated into the predictors.

Evaluating the difference between the reference distribution of values of the response variable and the inferred distribution allows bias assessment. In particular, we model the response variable as a Burr Pareto distribution [14] and use the inverse CDF method to perform the bias correction.

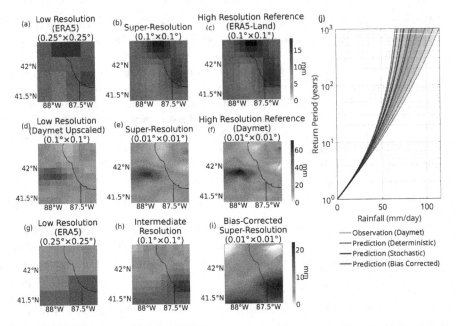

Fig. 2. Evaluation of multi-step downscaling for extreme events. **Top Row:** Downscaling model GAN-1 transforms rainfall from (**a**) ERA5 (0.25°) to (**b**) 0.1° resolution, compared to (**c**) ERA5-Land (reference, 0.1°). **Middle Row:** Downscaling model GAN-2 transforms rainfall from (**d**) upscaled Daymet (0.1°) to (**e**) 0.01° resolution, compared to (**f**) Daymet (observed, 0.01°). **Bottom Row:** Combined downscaling model transforms rainfall from (**g**) ERA5 (0.25°) to (**h**) intermediate resolution (0.1°) and (**i**) bias-corrected fine resolution (0.01°). (**j**) Annual mean rainfall-return period curves compared for observed rainfall risk (green) with deterministic (red), stochastic (blue), and bias-corrected (magenta) prediction. Generalized Pareto distributions fit to observations or predictions to estimate return period curves. Solid lines denote the mean, and the shaded areas indicate standard error. (Color figure online)

3   Results

We test our methods in Cook County (Chicago), Illinois, a flood prone area. For downscaling, meteorological data within a 1° × 1° bounding box between 41.4°N-42.4°N and 88.25°W-87.25°W are extracted. The data is split into training (1981–1999), validation and bias correction (2000–2009), and testing (2010-2019) sets. Figure 2 (a–i) illustrates a qualitative downscaling performance assessment of low-resolution ERA5 input for a particular extreme event. Notice that the downscaled rainfall fields capture the spatial pattern of observed rainfall well. In Fig. 2 (j), shows the Pareto-distribution simulated mean annual extreme rainfall return period curve for observations, deterministic downscaling, stochastic downscaling, and the optimal estimate (see Methods). Bias correction closes the risk gap that deterministic prediction substantially underestimates. For the Chicago area, the final annual risk projections show around 6.8% bias and less than 10% standard error even at an extreme 1000-year return period between Pareto-fit Daymet and model downscaling, respectively. Please note that the data does not represent a 1000-year rainfall event.

Figure 3 compares the reference and model-predicted aggregate annual flood-loss distributions before and after bias correction for the testing period. A Kolmogorov-Smirnov test [15] between the reference and bias-corrected distributions resulted in a p value of 0.725. Thus, we accept that the bias-corrected predicted claim distribution is also Burr Pareto at a 0.05 significance level. Figure 3 shows the R^2 performance of the fitted and bias-corrected Burr Pareto flood-loss distributions. For the shifting-window method, XGB could not construct a sound regressor for 2010 and 2017 (NFIP + Daymet) and 2017 (NFIP). Excluding years without a regressor, the testing population-weighted average R^2 is 0.544 (NFIP) and 0.917 (NFIP+Daymet). For the expanding-window method, with the inclusion of Daymet, XGB could not construct a valid regressor for 2010 and 2015 (NFIP+Daymet) and 2015 (NFIP). For the cases where it was possible to build a regressor, the mean performance was 0.432 (NFIP) and 0.674 (NFIP+Daymet).

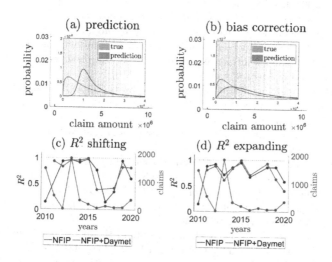

Fig. 3. Reference and predicted loss distribution for the testing dataset split **(a)** before and **(b)** after bias correction and distribution-wise flood loss model performance (R^2) evaluated for **(c)** shifting and **(d)** expanding windows, with (NFIP+Daymet) and without (NFIP) rainfall.

4   Discussion and Conclusion

This study uses Daymet, ERA5, orographic upslope dynamics, and machine learning (ML) in a dynamic data-driven manner to estimate flood-loss risk. High-resolution rainfall, rainfall risk, event-wise flood losses, and flood loss risk are estimated. Priming GANs with active CGP mitigates the paucity of historical extreme rainfall data. Orographic priming additionally enables the physical consistency of the result. While GAN generates fine-scale information, however, additional work adapting the loss function to resolve finer scales is needed. Downscaled rainfall, flood-loss and their return periods compare well with the validation or observational fields and their risk, also when compared to recent work [16]. Although this work applies to short-term projections, training with few high-resolution climate models makes it applicable to long-term climate scenarios. Detailed versions of this article are found in arXiv [17, 18].

References

1. Hersbach, H., et al.: The ERA5 global reanalysis. Q. J. R. Meteorolog. Soc. **146**(730), 1999–2049 (2020)
2. Thornton, P., et al.: Gridded daily weather data for North America with comprehensive uncertainty quantification. Sci. Data **8**(1), 1–17 (2021)
3. Trautner, M., Margolis, G., Ravela, S.: Informative neural ensemble Kalman learning. arXiv:2008.09915 (2020)
4. Chen, T., Guestrin, C.: XGBoost: a scalable tree boosting system. In: International Conference on Knowledge Discovery and Data Mining (2016)
5. Ravela, S.: Dynamically deformable resampled random manifolds for high-dimensional, nonlinear inference in geoscience in the presence of uncertainty. In: AGU Fall Meeting Abstracts, p. IN13C-1670 (2016)
6. Roe, G.: Orographic precipitation. Annu. Rev. Earth Planet. Sci. **33**, 645–671 (2005)
7. He, K., Zhang, X., Ren, S., Sun, J.: Deep residual learning for image recognition. In: IEEE CVPR, pp. 770–778 (2016)
8. Simonyan, K., Zisserman, A.: Very deep convolutional networks for large-scale image recognition. arXiv:1409.1556 (2014)
9. Wang, X., et al.: ESRGAN: enhanced super-resolution generative adversarial networks. In: Leal-Taixé, L., Roth, S. (eds.) ECCV 2018. LNCS, vol. 11133, pp. 63–79. Springer, Cham (2019). https://doi.org/10.1007/978-3-030-11021-5_5
10. Jolicoeur-Martineau, A.: The relativistic discriminator: a key element missing from standard GAN. arXiv:1807.00734 (2018)
11. Muñoz-Sabater, J., et al.: ERA5-Land: a state-of-the-art global reanalysis dataset for land applications. Earth Syst. Sci. Data **13**(9), 4349–4383 (2021)
12. Hosking, J., Wallis, J.: Parameter and quantile estimation for the generalized pareto distribution. Technometrics **29**(3), 339–349 (1987)
13. Dombrowski, T., Ratnadiwakara, D., Slawson, C.: The FIMA NFIP's redacted policies and redacted claims datasets. J. Real Estate Lit. **28**(2), 190–212 (2021)
14. Burr, I.: Cumulative frequency functions. Ann. Math. Stat. **13**(2), 215–232 (1942)
15. Smirnov, N.: Table for estimating the goodness of fit of empirical distributions. Ann. Math. Stat. **19**(2), 279–281 (1948)
16. Lin, C., Cha, E.: Hurricane freshwater flood risk assessment model for residential buildings in Southeast US coastal states considering climate change. Nat. Hazard Rev. **22**(2), 04020024 (2021)
17. Saha, A., Ravela, S.: Downscaling extreme rainfall using physical-statistical generative adversarial learning. arXiv:2212.01446 (2022)
18. Salas, J., Saha, A., Ravela, S.: Learning inter-annual flood loss risk models from historical flood insurance claims and extreme rainfall data. arXiv:2212.08660 (2022)

Keynotes-Appendix

Computing for Emerging Aerospace Autonomous Vehicles

Sertac Karaman(✉)

Massachusetts Institute of Technology, Cambridge, USA
sertac@mit.edu

Abstract. Aerospace autonomous vehicles are going through a renaissance. Consumer drones, enabled by autonomous visual navigation capabilities, are serving a rapidly-growing number of applications, ranging from entertainment to inspection. The fast-decreasing cost of access to space is enabling miniature autonomous satellites towards better communication, Earth observation, or even launching in-orbit manufacturing and maintenance. This talk presents a set of bleeding-edge technologies enabled by new advances in algorithms, computer architecture, integrated circuits and sensor design [1]. These technologies will enable a spectrum of new vehicles, ranging from the most miniature and long-endurance to the fastest and the most agile, with exciting new applications in aerospace autonomous vehicles [2, 3]. Some of the applications and recent progress towards enabling them are outlined in the presentation.

Keywords: Dynamic Data Driven Applications Systems · XXXX (from Keynotes Scope)

References

1. Ryou, G., Tal, E., Karaman, S.: Multi-fidelity black-box optimization for time-optimal quadrotor maneuvers. Int. J. Robot. Res. **40**(12–14), 1352–1369 (2021)
2. Tal, E., Karaman, S.: Accurate tracking of aggressive quadrotor trajectories using incremental nonlinear dynamic inversion and differential flatness. IEEE Trans. Control Syst. Technol. **29**(3), 1203–1218 (2020)
3. Tal, E., Ryou, G., Karaman, S.: Aerobatic trajectory generation for a vtol fixed-wing aircraft using differential flatness. arXiv preprint arXiv:2207.03524 (2022)

From Genomics to Therapeutics: Single-Cell Dissection and Manipulation of Disease Circuitry

Manolis Kellis[✉]

Massachusetts Institute of Technology, Cambridge, USA
manoli@mit.edu

Abstract. Disease-associated variants lie primarily in non-coding regions, increasing the urgency of understanding how gene-regulatory circuitry impacts human disease. To address this challenge, we generate comparative genomics, epigenomic, and transcriptional maps, spanning 823 human tissues, 1500 individuals, and 20 million single cells. We link variants to target genes, upstream regulators, cell types of action, and perturbed pathways, and predict causal genes and regions to provide unbiased views of disease mechanisms, sometimes reshaping our understanding. We find that Alzheimer's variants act primarily through immune processes, rather than neuronal processes, and the strongest genetic association with obesity acts via energy storage/dissipation rather than appetite/exercise decisions. We combine single-cell profiles, tissue-level variation, and genetic variation across healthy and diseased individuals to map genetic effects into epigenomic, transcriptional, and function changes at single-cell resolution, to recognize cell-type-specific disease-associated somatic mutations indicative of mosaicism, and to recognize multi-tissue single-cell effects of exercise and obesity. We expand these methods to electronic health records to recognize multi-phenotype effects of genetics, environment, and disease, combining clinical notes, lab tests, and diverse data modalities despite missing data. We integrate large cohorts to factorize phenotype-genotype correlations to reveal distinct biological contributors of complex diseases and traits, to partition disease complexity, and to stratify patients for pathway-matched treatments. Lastly, we develop massively-parallel, programmable and modular technologies for manipulating these pathways by high-throughput reporter assays, genome editing, and gene targeting in human cells and mice, to propose new therapeutic hypotheses in Alzheimer's, obesity, and cancer. These results provide a roadmap for translating genetic findings into mechanistic insights and ultimately new therapeutic avenues for complex disease and cancer.

Keywords: Dynamic Data Driven Applications Systems · XXXX (from Keynotes Scope)

References

1. Mohammadi, S., Davila-Velderrain, J.: Kellis, M.: A multiresolution framework to characterize single-cell state landscapes. Nat. Commun. **11**(1), 5399 (2020). https://doi.org/10.1038/s41467-020-18416-6

E. Blasch et al. (Eds.): DDDAS 2022, LNCS 13984, pp. 428–429, 2024.
https://doi.org/10.1007/978-3-031-52670-1

2. Park, Y., et al.: Single-cell deconvolution of 3,000 post-mortem brain samples for eQTL and GWAS dissection in mental disorders. bioRxiv 426000 (2021). https://doi.org/10.1101/2021.01.21.426000

3. Boix, C.A., James, B.T., Park, Y.P., Meuleman, W., Kellis, M.: Regulatory genomic circuitry of human disease loci by integrative epigenomics. Nature **590**, 300–307 (2021). https://doi.org/10.1038/s41586-020-03145-z. PMID 33536621

4. Lopes, N., et al.: Distinct metabolic programs established in the thymus control effector functions of gamma-delta T cell subsets in tumor microenvironments. Nat. Immunol. **22**(2), 179–192 (2021). https://doi.org/10.1038/s41590-020-00848-3

Data Augmentation to Improve Adversarial Robustness of AI-Based Network Security Monitoring

Nathaniel D. Bastian[✉]

United States Military Academy (USMA), West Point, NY, USA
nathaniel.bastian@westpoint.edu

Abstract. Cyber security is an international challenge that is increasingly important as the inter-connectedness of the world grows. The reliance of systems on computational assets makes them vulnerable to attack. Traditionally, networks and systems are monitored by cyber security operators who rely on intrusion detection systems to provide indicators of compromise via alerts. With the growing number and frequency of alerts and the increasing sophistication of attacks, human operators are incapable of keeping pace. Data-driven and statistical tools, such as algorithms from artificial intelligence (AI), have the potential to assist in this area. These AI technologies enable the collection of synchronized, real-time capabilities to discover, define, analyze, and mitigate cyber threats and vulnerabilities with limited human intervention. Network intrusion detection systems (NIDS) are a primary component of the broader practices in network security monitoring for cyber security operations at enterprise, operational and tactical environments. Today, integrating AI components into NIDS can help identify patterns associated with known threats or detect abnormal behavior. These AI-based NIDS, however, are susceptible to adversarial AI evasion attacks. This talk will address how data augmentation techniques can be used to improve the adversarial robustness of AI-based NIDS for network security monitoring [1]. Particularly, the talk will address how meta-heuristics can be used to generate adversarial examples to then be combined as part of a meta-learning adversarial training framework [2, 3].

Keywords: Dynamic Data Driven Applications Systems · XXXX (from Keynotes Scope)

References

1. Alhajjar, E., Maxwell, P., Bastian, N.: Adversarial machine learning in network intrusion detection systems. Expert Syst. Appl. **186**, 115782 (2021)
2. Farrukh, Y.A., Khan, I., Wali, S., Bierbrauer, D., Pavlik, J.A., Bastian, N.D.: Payload-byte: a tool for extracting and labeling packet capture files of modern network intrusion detection datasets. TechRxiv. Preprint (2022). https://doi.org/10.36227/techrxiv.20714221.v2
3. Chalé, M., Bastian, N.D.: Generating realistic cyber data for training and evaluating machine learning classifiers for network intrusion detection systems. Expert Syst. Appl. **207**, 117936 (2022)

E. Blasch et al. (Eds.): DDDAS 2022, LNCS 13984, p. 430, 2024.
https://doi.org/10.1007/978-3-031-52670-1

Improving Predictive Models
for Environmental Monitoring Using
Distributed Spacecraft Autonomy

Sreeja Nag[✉]

NASA Ames Research Center, Mountain View, USA
sreejanag@alum.mit.edu

Abstract. Inter-connected, heterogeneous constellations of spacecraft can be exceptionally good at environmental monitoring applications when equipped with decision autonomy and platform agility. Autonomous spacecraft operations can improve our predictive models of natural phenomena by optimizing future observations in quick response to past observations and forecasts. Better predictive models and smartly targeted data improve situational awareness of evolving phenomena and help actionate timely response. The value of a smart sense-predict-plan-act will be demonstrated via the D-SHIELD (Distributed Spacecraft with Heuristic Intelligence to Enable Logistical Decisions) project. D-SHIELD's scheduler is optimized such that the collection of observational data and their downlink, constrained by the constellation constraints (orbital mechanics), resources (power, data) and subsystems (instruments, attitude control), results in maximum science value for a selected scenario. Constellation topology, spacecraft and ground network characteristics serve as inputs to operations design. The scheduler can run autonomously onboard the spacecraft, or at the ground station with resultant schedules uplinked to the spacecraft for execution. D-SHIELD includes a science simulator to inform the scheduler of the predictive value of observations or operational decisions, which is custom built as a function of the use case, using novel data assimilation methods and predictive algorithms. We have applied the D-SHIELD simulation framework to several use cases spanning urban floods, cyclones, and most recently, a global soil moisture monitoring scenario using a 6 satellite constellation carrying P and L band radars, and L band radiometers. Platform agility allows the satellites to slew in any direction by changing their roll and look angle. Decision autonomy schedules the satellites for what to look at, at what time, and with what instruments and parameters. Results show potential for reduction in predictive uncertainty of global soil moisture compared to flagship missions. The framework is currently being extended to monitor wildfire spread for responsive control.

Keywords: Dynamic Data Driven Applications Systems · XXXX (from Keynotes Scope)

E. Blasch et al. (Eds.): DDDAS 2022, LNCS 13984, pp. 431–432, 2024.
https://doi.org/10.1007/978-3-031-52670-1

References

1. Levinson, R., Niemoeller, S., Nag, S., Ravindra, V.: Planning satellite swarm measurements for earth science models: comparing constraint processing and MILP methods. In: Proceedings of the International Conference on Automated Planning and Scheduling, vol. 32, pp. 471–479 (2022)
2. Lammers, R., Li, A.S., Ravindra, V., Nag, S.: Prediction models for urban flood evolution for satellite remote sensing. J. Hydrol. **603**, 127175 (2021)
3. Nag, S., Murakami, D., Marker, N., Lifson, M., Kopardekar, P.: Prototyping operational autonomy for space traffic management. Acta Astronaut. **180**, 489–506 (2021)
4. Ravindra, V., Nag, S., Li, A.S.: Ensemble guided tropical cyclone track forecasting for optimal satellite remote sensing. IEEE Trans. Geosci. Remote Sens. (TGRS). **59**, 3607–3622 (2020). https://doi.org/10.1109/TGRS.2020.3010821
5. Nag, S., Li, A.S., Merrick, J.H.: Scheduling algorithms for rapid imaging using agile cubesat constellations. COSPAR Adv. Space Res. Astrodynamics **61**(3), 891–913 (2018)

Wildfires-Appendix

Simulating Large Wildland & WUI Fires with a Physics-Based Weather-Fire Behavior Model: Understanding, Prediction, and Data-Shaped Products

Janice Coen[1,2(✉)]

[1] University of San Francisco, Sa Francisco, CA 94117, USA
mohseni@ufl.edu

[2] National Center for Atmospheric Resaarch (NCAR), Boulder, CO 80305, USA

Abstract. Accurate prediction of onset and propagation of large-scale wildires and their after effects for example impact in surrounding communities (the Wildfire-Urban Interface – WUI), requires comprehensive, multiphysics, multimod; Coupled weather-fire behavior modeling is a physically-based fire modeling approach that uses prognostic computational fluid dynamics (CFD) models of the atmospheric environment coupled at each time step with a module of algorithms that parameterize subgrid-scale wildland fire processes including a fire's rate of spread and fuel consumption rates based on laboratory and field experiments. This coupling allows winds, as they vary in time and space, to shape fire growth, and the heat fluxes from the fire to shape atmospheric motions, notably creating intense "fire-induced winds". Coupled weather-fire behavior model have been used to simulate large wildland fires that may reach thousands to hundreds of thousands of hectares in size. We present results from the CAWFE (Coupled Atmosphere-Wildland Fire Environment), a coupled weather-fire model, which has been used to model dozens of large, high impact wildland fire events and has been adapted to run faster than real time on a single processor. By choosing NWP numerical approaches that accurately simulating fine-scale circulations grid spacing of a few hundred meters (an order of magnitude finer than standard weather forecasts) and include the heat flux feedbacks from the fire, we have found that not only the expansion of the fire, but changes in direction and intensity, and the formation of fire phenomena such as large fire whirls can be reproduced. Importantly, these dynamic models can reproduce extreme wind event and plume-driven fires - conditions where traditional kinematic tools are weakest. Recent work investigated the impact of community conflagrations on fire behavior. As fire growth is a foundation for other applications such as estimating fire emissions for smoke transport, we discuss how this coupled modeling, integrated with data assimilation of remote sensing products, is being used to improve community fire data products [Refs 1–5].

Keywords: Coupled Fire-Weather modeling · CAWFE · NWP · Prognostic Computational Fluid Dynamics (CFD) models · Fire-induced winds · Fine-scale circulation · Heat flux feedbacks · Plume-driven fires

© The Author(s), under exclusive license to Springer Nature Switzerland AG 2024
E. Blasch et al. (Eds.): DDDAS 2022, LNCS 13984, pp. 435–436, 2024.
https://doi.org/10.1007/978-3-031-52670-1

References

1. Prein, A.F., Coen, J., Jaye, A.: The character and changing frequency of extreme California fire weather. J. Geophys. Res. Atmos. **127**, e2021JD035350 (2022). https://nam12.safelinks. protection.outlook.com/?url=https%3A%2F%2Fdoi.org%2F10.1029%2F2021JD035350& data=05%7C01%7C%7Cfe44cd81a879429436eb08dac13b75a5%7C84df9e7fe9f640afb435aa aaaaaaaaaa%7C1%7C0%7C638034758360623661%7CUnknown%7CTWFpbGZsb3d8eyJW IjoiMC4wLjAwMDAiLCJQIjoiV2luMzIiLCJBTiI6Ik1haWwiLCJXVCI6Mn0%3D%7C3000 %7C%7C%7C&sdata=ZdJjuP8LFpiZ9oxTlp1hWTjuweIYBZ5oHf74cQUeLng%3D& reserved=0
2. Coen, J., Cruz, M., Rosales-Giron, D., Speer, K.: Coupled fire-atmosphere model evaluation and challenges. In: Speer, K., Goodrick, S. (eds.) Wildland Fire Dynamics: Fire Effects and Behavior from a Fluid Dynamics Perspective, in press. Cambridge University Press, Cambridge (2022)
3. Coen, J.L., Schroeder, W., Conway, S., Tarnay, L.: Computational modeling of extreme wildland fire events: a synthesis of scientific understanding with applications to forecasting, land management, and firefighter safety. J. Comput. Sci. **45**, 101152 (2020). https://nam12. safelinks.protection.outlook.com/?url=https%3A%2F%2Fdoi.org%2F10.1016%2Fj.jocs.2020. 101152&data=05%7C01%7C%7Cfe44cd81a879429436eb08dac13b75a5%7C84df9e7fe 9f640afb435aaaaaaaaaaaa%7C1%7C0%7C638034758360623661%7CUnknown%7CTWFpb GZsb3d8eyJWIjoiMC4wLjAwMDAiLCJQIjoiV2luMzIiLCJBTiI6Ik1haWwiLCJXVCI6Mn 0%3D%7C3000%7C%7C%7C&sdata=BJQaXfDybCKlOfEp%2B%2FGJtCPFC1P8zv Ztc1QZwfzlC8Q%3D&reserved=0
4. Coen, J.L., Schroeder, W., Quayle, B.: The generation and forecast of extreme winds during the origin and progression of the 2017 Tubbs Fire. Atmosphere **9**, 462 (2018)
5. Blasch, E.P. Darema, F., Ravela, S., Aved, A.J. (eds.) Handbook of Dynamic Data Driven Applications Systems. 2nd edn, vol. 1. Springer, Cham (2021). https://doi.org/10.1007/978-3-031-27986-7

Dynamic Data Driven Applications for Atmospheric Monitoring and Tracking

Kamran Mohseni[✉]

University of Florida, Gainsville, FLA, USA
mohseni@ufl.edu

Abstract. This presentation addresses instrumentation capabilities through adaptive coordination of collections of small unmanned aerial vehicles (UAVs), for monitoring and tracking atmospheric features such as fires (wildfires and prescribed), plumes, etc. [Ref 1, 2] Such capabilities can create enhanced methods, cognizant of time-varying and real-time conditions for fire behaviors and effects, by for example, supporting DDDAS-based bi-directional feedback-control across modeling and instrumentation, and which span multiple modalities and timescales. In this context. [Ref 3–5] The following are of particular interest and are addressed in this presentation: a) the development of customized UAVs for such applications; and b) challenges with the curse of dimensionality in such problems are considered and addressed through the development of a structure preserving model reduction technique, namely the symplectic orthogonal decomposition. [Ref 6] These approaches are discussed in the context.

Keywords: unmanned aerial vehicles (UAVs) · adaptive coordination · monitoring · tracking · curse of dimensionality · structure preserving model reduction · orthogonal decomposition

References

1. Silic, M., Mohseni, K.: An experimental evaluation of radio models for localizing fixed-wing UAVs in rural environments. IEEE Trans. Veh. Technol. (2023)
2. Silic, M., Mohseni, K.: Field deployment of a plume monitoring UAV flock. IEEE Robot. Autom. Lett. 4(2), 769–775 (2019)
3. Peng, L., Silic, M., Mohseni, K.: A DDDAS plume monitoring system with reduced Kalman filter. Procedia Comput. Sci. 51, 2533–2542 (2015)
4. Lipinski, D., Mohseni, K.: Micro/miniature aerial vehicle guidance for hurricane research. IEEE Syst. J. 10(3), 1263–1270 (2015)
5. Mohseni, K.: Dynamic data-driven UAV network for plume characterization. University of Florida Gainesville, United States (2016)
6. Peng, L., Mohseni, K.: Symplectic model reduction of Hamiltonian systems. SIAM J. Sci. Comput. (Publisher: Society for Industrial and Applied Mathematics) 38(1), A1–A27 (2016)

E. Blasch et al. (Eds.): DDDAS 2022, LNCS 13984, p. 437, 2024.
https://doi.org/10.1007/978-3-031-52670-1

Author Index

E. Blasch et al. (Eds.): DDDAS 2022, LNCS 13984, pp. 439–441, 2024.
https://doi.org/10.1007/978-3-031-52670-1